Proceedings of the Probabilistic Safety Assessment and Management (PSAM) 12 Conference - Volume 1

Foreword

It is was our honor to welcome you to Honolulu, Hawaii, for the twelfth rendition of the Probabilistic Safety Assessment and Management (PSAM) Conference. The planning for PSAM Honolulu began back in 2007 (before PSAM 9 in Hong Kong), when we looked at several locations around the United States, included Arizona, California, Boston, and even considered locations in Oceania. Based upon the feedback both during and after the conference, PSAM 12 proved to be a great suc-cess.

We would like to thank all of the volunteers, those that served before, during, and after the Conference. Members of the Technical Program Committee, the Organizing Committee, the session chairs, and the presenters have our gratitude for making PSAM 12 the most memorable PSAM yet.

This publication represents the technical proceedings for the Conference. Due to the large number of published papers (a total of 391), we have subdivided the technical content (papers) into multiple volumes.

On behalf of the International Association for Probabilistic Safety Assessment and Management Board of Directors, we hope that this publication will provide a valuable technical resource in addition to a reminder of the memorable stay in the Hawaiian Islands.

Dr. Curtis Smith   Dr. Todd Paulos
Technical Program Chairs   General Chair

Number of Papers Presented at PSAM 12 (by country)

Sponsors

Sponsors

EPRI Assesses Seismic Resistance of Electronic Components

Together... Shaping the Future of Electricity

Technical Program Committee

Technical Program Chair: Curtis Smith, INL USA

Assistant Technical Program Chairs: Steve Epstein, Lloyd's Register Japan
                                        Vinh Dang, PSI Switzerland
                                        Ted Steinberg, QUT Australia

We would like to thank the members of the PSAM 12 Technical Program Committee. These individuals helped to make PSAM 12 a success by reviewing abstracts, technical papers, organizing sessions, and providing technical leadership for the conference.

Technical Committee Members:

Roland Akselsson

S. Massoud (Mike) Azizi

Tito Bonano

Ronald Boring

Roger Boyer

Mario Brito

Kaushik Chatterjee

Vinh Dang

Claver Diallo

Nsimah Ekanem

Steve Epstein

Fernando Ferrante

Federico Gabriele

Ray Gallucci

S. Tina Ghosh

David Grabaskas

Katrina Groth

Seth Guikema

Steve Hess

Christopher J. Jablonowski

Moosung Jae

Jeffrey Joe

Vyacheslav S. Kharchenko

James Knudsen

Zoltan Kovacs

Ping Li

Harry Liao

Francois van Loggerenberg

Jerome Lonchampt

Soliman A. Mahmoud

Diego Mandelli

Donoval Mathias

Zahra Mohaghegh

Thor Myklebust

Cen Nan

Mohammad Pourgolmohammad

Marina Roewekamp

Clayton Smith

Shawn St. Germain

Ted Steinberg

Kurt Vedros

Smain Yalaoui

Robert Youngblood

Enrico Zio

Organizing Committee

General Chair: Dr. Todd Paulos

General Vice Chair: Prof. Stephen Hora, USC

Technical Program Chair: Curtis Smith, INL USA

Webmaster, Registration,
Support for Papers/Abstracts
Submission and Review: Hanna Shapira, TICS

Table of Content

Table of Content

Table of Content

Page   Paper

Nuclear refugees after large early radioactive releases

Ludivine Pascucci-Cahen[a]

[a] Institut de Radioprotection et de Sûreté Nucléaire, Fontenay-aux-Roses, France

Abstract: However improbable, large early radioactive releases from a nuclear power plant would entail major consequences for the surrounding population. In Fukushima, 80,000 people had to evacuate the most contaminated areas around the NPP for a prolonged period of time. Had they remained where they lived, they would have received doses dangerous for their health in the long run. These people have been called "nuclear refugees".

The paper first argues that the number of nuclear refugees is a better measure of the severity of radiological consequences than the number of fatalities, although the latter is widely used to assess other catastrophic events such as earthquakes or tsunami. It is a valuable partial indicator in the context of comprehensive studies of overall consequences.

Section 2 makes a clear distinction between long-term relocation and emergency evacuation and proposes a method to estimate the number of refugees.

Section 3 examines the distribution of nuclear refugees with respect to weather and release site. The distribution is asymmetric and fat-tailed: unfavorable weather can lead to the contamination of large areas of land; large cities have in turn a higher chance of being contaminated. Variability with respect to site is quite intuitive; however, results show that simulations are far superior to an approach based on population living within 20 or 30 km around the site.

Keywords: PSA, accident consequences, nuclear refugees.

1. INTRODUCTION

1.1. Fatalities

In the literature on disasters and emergency situations, different disasters are often compared on the basis of prompt fatalities. Car accidents or plane crashes can thus be compared with earthquakes or tsunamis on this basis. Nuclear accidents are sometimes included in such comparisons although the number of prompt fatalities in nuclear accidents is quite low: none are attributed to the Fukushima accident whereas about 30 are registered for the Chernobyl accident.

Nuclear accidents entail a large number of other damaging consequences; their total cost may reach hundreds of billions of dollars [1]. It has therefore been pinpointed that prompt fatalities are a poor indicator of overall accident severity as far as the nuclear sector is concerned. Total cancer fatalities consecutive to exposition to ionizing radiations have therefore been proposed as a more comprehensive and apt indicator. It was argued that considering only prompt fatalities is a way to minimize the total number of casualties.

From an economic point of view, this argument is not entirely valid. When attempting to quantify the loss involved in a prompt fatality, economists consider that, from the point of view of the nation, it is the loss of a "statistical life" i.e. the loss of a number of years equal to half the average lifespan, say between 40 and 45 years. In comparison, cancer fatalities induced by ionizing radiation typically take place 20 years after the event and thus entail a loss of life closer to half a statistical life.

In addition, fatalities due to cancer are profoundly different from prompt fatalities: they cannot be counted. They are estimated on the basis of a complex calculation of doses followed by a simple application of the "ICRP coefficients". Relying in this way on the no threshold linear relationship is officially not recommended by ICRP. And in most cases, even 20 years after the disaster, observers will be unable to provide any evidence: the bulk of radiological cancers cannot be distinguished from other cancers and statistics will be inconclusive. Indeed the most severe accident scenarios should "only" cause tens of thousands of cancer fatalities spread out over some thirty years in the worst cases while total deaths attributed to cancers in a country like France are about 150,000 per year. Cancer fatalities due to a nuclear accident should therefore only represent of minor percentage of total cancer deaths each year. It will generally be impossible to detect them.

In brief, fatality statistics are not helpful to gauge nuclear accidents.

Is there a way to provide a better indicator than fatalities? An indicator that would be easily understandable by each and every one? Which could be readily observed and would describe the extent of human suffering involved?

1.2. Damage indicators in Katarisk

KATARISK is a Swiss tool aimed at understanding all possible sources of disaster [2]. It distinguishes five broad categories of events (CE) of increasing severity (disasters manageable at a local, regional, federal or European scale).

Figure 1: Classification of events in KATARISK

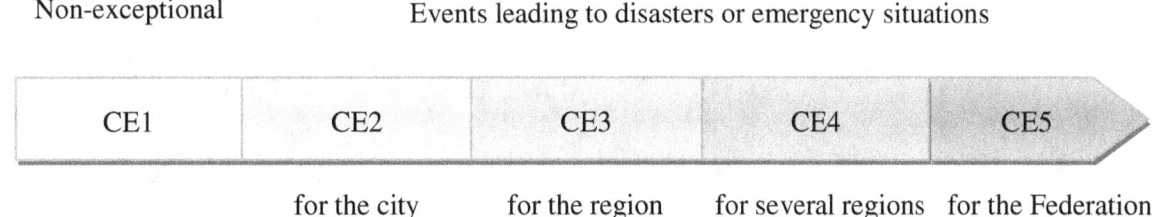

KATARISK uses a number of damage indicators applicable to a large spectrum of events, ranging from railway accidents to a nuclear accident, including earthquakes, droughts, floods, dam overflows, hurricanes, and epidemics. Indicators are not limited to fatalities:

Table 2: Example of damage indicators proposed by KATARISK

Fatalities, injuries during events causing serious damage over a wide area (number of people)

Fatalities, injuries, illnesses during epidemics (number of people)

Number of persons evacuated

Persons in need (refugees, homeless, people requiring care)

Impairment of vital resources: damage to agricultural land, water and forest (number of km²)

For each indicator, limit values are suggested for each event category irrespective of the nature of the emergency (see Table 3).

For nuclear accidents, fatalities are not tremendously descriptive as argued in the introduction. The number of persons evacuated is a short term emergency indicator which does not address long term effects. Thus the number of refugees and the "impairment of vital resources" both appear the most relevant in this list. There is a clear difference between these two indicators, however: the number of contaminated km² depends on the considered level of contamination — a map of contamination will display several colors — and the resulting figures are more complex and open to interpretation than a single figure. They are more difficult to assess whereas radiological refugees can readily be counted. Therefore, the number of refugees appears as the preferred candidate indicator.

In the nuclear industry, the number of nuclear refugees is a valuable indicator. Its meaning is easy to understand for everyone. The reality it describes is directly observed; it conspicuously exposes the suffering of victims and is necessarily the focus of the media. It is important, however, to keep in mind that these remarkable assets cannot transform it into a comprehensive measure of accident losses: it is a very useful indicator in the context of comprehensive studies of overall consequences.

Table 3: Values for damage indicators proposed by KATARISK for comparison of disasters

CE1	CE2	CE3	CE4	CE5	
Fatalities, injuries during events causing serious damage over a wide area (number of people)					
1	100	1,000	10,000	100,000	1,000,000
Fatalities, injuries, illnesses during epidemics (number of people)					
1	10,000	50,000	100,000	1,000,000	7,000,000
Number of persons evacuated					
1	1,000	10,000	100,000	1,000,000	10,000,000
Persons in need (refugees, homeless, people requiring care)					
1	10,000	100,000	1,000,000	10,000,000	
Impairment of vital resources: damage to agricultural land, water and forest (number of km²)					
0.1	5	50	500	5,000	50,000

2. NUCLEAR REFUGEES: DEFINITION AND ESTIMATION METHOD

2.1. Definition

After Chernobyl, the criterion for relocation was: ground activity concentration above 555 kBq/m² of Cs-137. After the Fukushima accident, Japanese authorities enforced a threshold of 500 kBq/m² of Cs-137, although the criterion was expressed in terms of doses. This was broadly consistent with the feedback from Chernobyl.

In general, refugees are those people who have to leave their home for many years due to excessive contamination. This refers to a threshold level declared by the authorities; it can be expressed in terms of doses but it is equivalent and simpler to refer to levels of ground contamination. The present study uses the figure of 555 kBq/m² of Cs-137. The precise level chosen does not affect the calculation method or the nature of results.

Emergency evacuations are generally performed to protect the population living in the immediate vicinity of the NPP from the radioactive plume. Evacuees may come back home fairly rapidly once the emergency is over provided contamination levels allow. Conversely, all refugees may not have been evacuated. Thus refugees and evacuees do not refer to identical populations.

Estimating the number of nuclear refugees involves combining deposits with population data. Since contamination heavily depends on climatic conditions, so does the number of refugees. It is probabilistic in nature.

2.2. Estimation of deposits

Deposits are estimated at each grid point of a predefined grid using atmospheric dispersion models. These require:

1) **A source term** which details the quantity, nature and discharges of radioactive elements into the environment. As far as accident costs and nuclear refugees are concerned, the most important element is Cesium-137 as it contaminates the environment for a prolonged period of time. For this study, this is the only required radioelement (see above definition).

2) **Meteorological data:** the wind direction and its possible changes during the course of the plume determine the areas affected by the fallout. Rain leaches the plume and causes greater deposition of radioactive particles in some places.

In this study, source terms are based on the IRSN level 2 PSA for the French 900 MWe reactors. Activity of released aerosols is the physical indicator used to assess the severity releases. It varies between less than 1E+15 Bq to more than 1E+19 Bq. Two categories of nuclear accidents with large off-site impacts have been distinguished:

1) Severe accidents, with controlled and filtered releases of radionuclides into the environment, for which the activity of released aerosols varies between 1E+15 Bq and 1E+16 Bq; and

2) Major accidents with massive radioactive releases to the environment, comparable in severity to Fukushima or Chernobyl. The activity of released aerosols ranges from 1E+16 Bq to 1E+19 Bq and above.

Figure 3: Broad categories of accidents and release of aerosols in the case of a 900 MWe French reactor

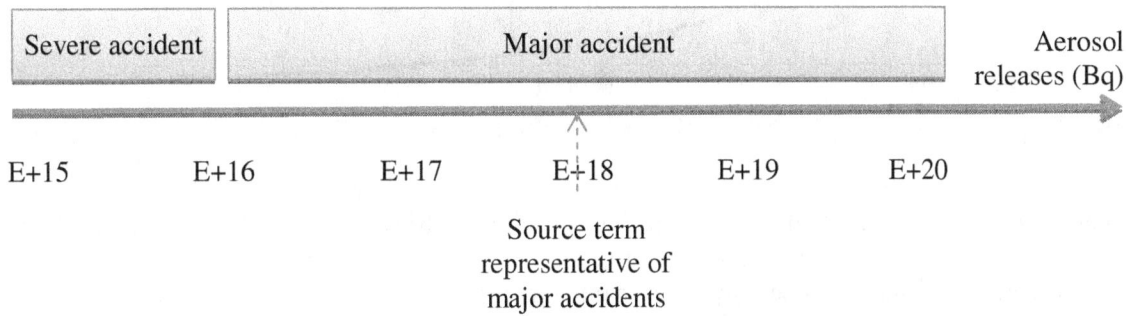

This paper addresses major accidents with large early releases. To give a realistic picture and avoid focusing on extreme cases, a median source term was considered which corresponds to the release of 1E+18 Bq of aerosols.

To address meteorological variability, 6,000 runs have been calculated spanning 10 years (2002 to 2011) of actual 3D weather data. Two consecutive runs are 12.5 hours apart and each run lasts up to several days.

2.3. Order of magnitude of radiological refugees for a major accident in France

With these definitions, a major accident in France would involve the relocation of about 100,000 nuclear refugees. This figure is subject to large variations due to weather conditions. It also varies significantly from site to site (there are 19 NPP sites in France).

3. THE NUMBER OF NUCLEAR REFUGEES HEAVILY DEPENDS ON WEATHER

3.1. Asymmetric and fat-tailed

The 6,000 runs calculated as described provide a distribution of radiological refugees with respect to weather. Very high numbers can then be observed; these extreme values are unlikely but they point to potentially disastrous situations. In other words, the distribution is extreme: asymmetric and fat-tailed. More and more unfavorable weather conditions lead to larger and larger areas of land being contaminated and higher and higher probabilities for a large city to be contaminated. Each time this happens, the number of nuclear refugees jumps up, the so-called "fall off the cliff effect".

The geographical distribution of population is extreme which implies an extreme distribution of refugees. Zajdenweber (2001) [3] argues that it is the reason behind a number of extreme distributions, for instance the distribution of wealth.
Results computed for a typical French nuclear site are shown Figure 4 below.

Figure 4: Exceedance probability curve

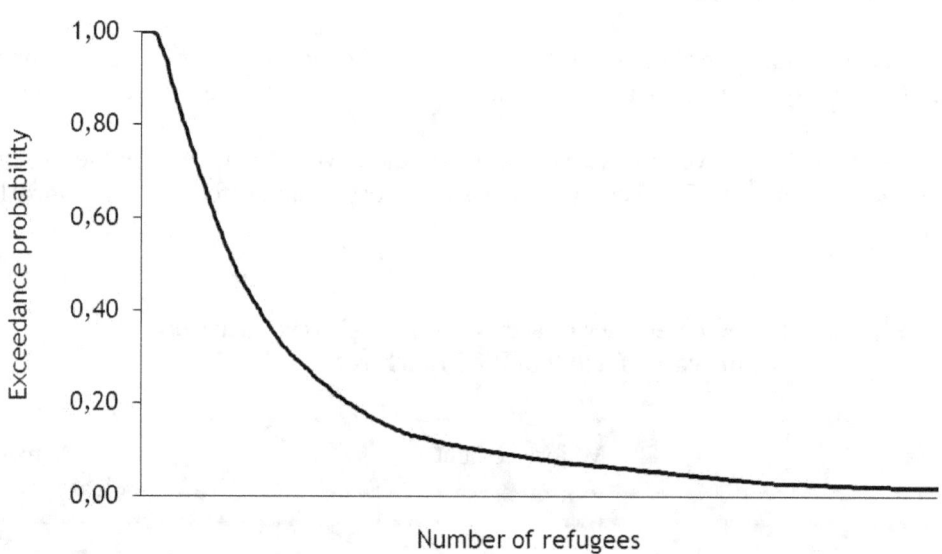

The median value can be multiplied several times when particularly adverse weather prevails at the time of the accident. For example, 400,000 refugees is by no means an inconceivable figure. Conversely, favorable weather considerably reduces the number of nuclear refugees.

3.2. A representative met-sample needs to be fairly large

Extreme distributions are only poorly identified with limited samples. This is the case with radiological refugees, as shown on Figure 5.

Figure 5 shows the effect of limiting the met-sample to one year of (actual) data i.e. 600 runs over 1 year of weather data. The worse year within the full 10 years of data is 2002 with significantly higher numbers of refugees. In contrast, the best year is 2011. Overall, up to ± 30 % deviations from the reference distribution are observed.

One year of meteorological data cannot correctly "represent" long-term weather trends. The met-sample needs to span several years of data; this is particularly true when focusing on extreme values of the probability distribution function, for example the 5% to 95% percentiles. How many years are required?

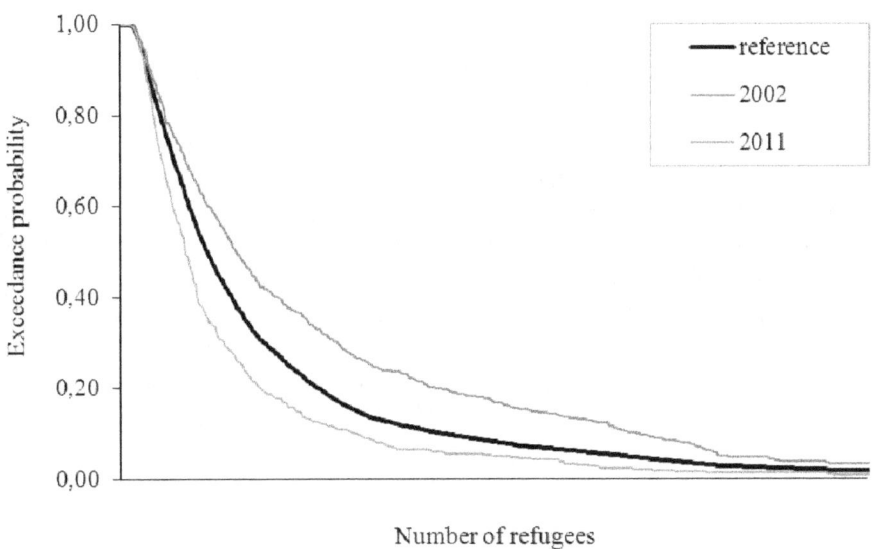

An optimal met-sample should be sufficiently large to be "representative" and sufficiently small to guarantee acceptable execution times. An experiment was carried out by running computations for increasing sizes of the met-sample. These ranged from 10 to 6,000 runs of weather data randomly drawn from the total 6,000 calculated at the outset. The distribution of nuclear refugees was computed for each of these draws; it tends toward the reference distribution which appears in Figures 4 and 5. For instance, Figure 6 represents median values for increasing sizes of the met-sample in this experiment.

Figure 6: Median value of the distribution of refugees for increasing sizes of met-sample

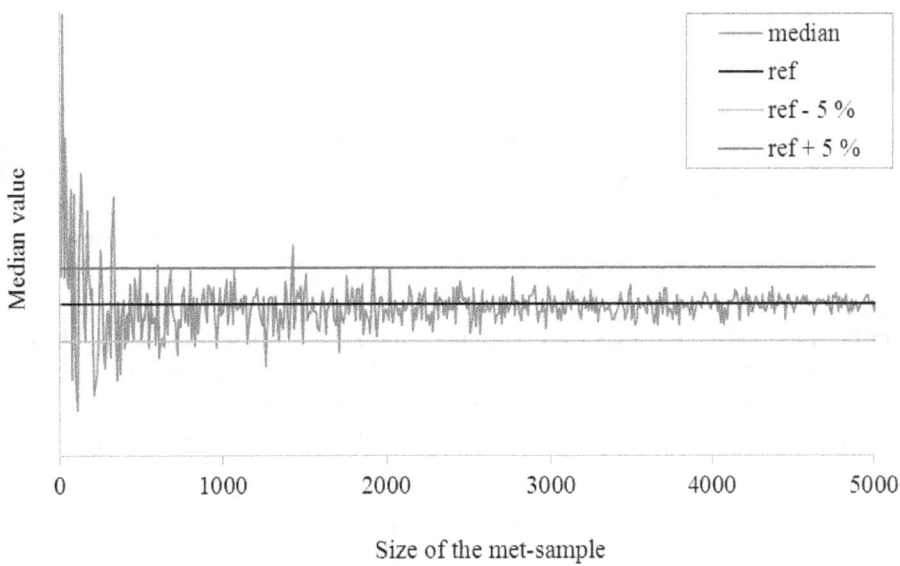

The Figure suggests that a 2,000 met-sample could be large enough, as the deviation from the reference value then falls within a ± 5% bracket. Similar calculations for the 5-percentile, mean and 95-percentile values suggest a 4,000 threshold above which all these figures remain within a ± 5% deviation from the reference value.

Such a result is source term-specific, site-specific and experiment-specific. It seems reasonable to recommend using a 10-year, 6000 runs sample and to check it is sufficiently representative by conducting the type of experiment mentioned above.

4. THE NUMBER OF NUCLEAR REFUGEES DEPENDS ON THE RELEASE SITE

The general result states that a major accident occurring on a French nuclear reactor leads to a median relocation in the order of 100,000 nuclear refugees. This median is subject to large variations with respect to the actual location of the accident because potentially affected populations greatly vary from one NPP to another. For example, accidents at seafront NPPs should imply less nuclear refugees, as winds can direct the plume to the sea — precisely what happened several times during the releases from the Fukushima Dai-ichi site. Figure 7 illustrates this variability in the case of France.

Figure 7: Nuclear refugees from various NPP sites (median values)

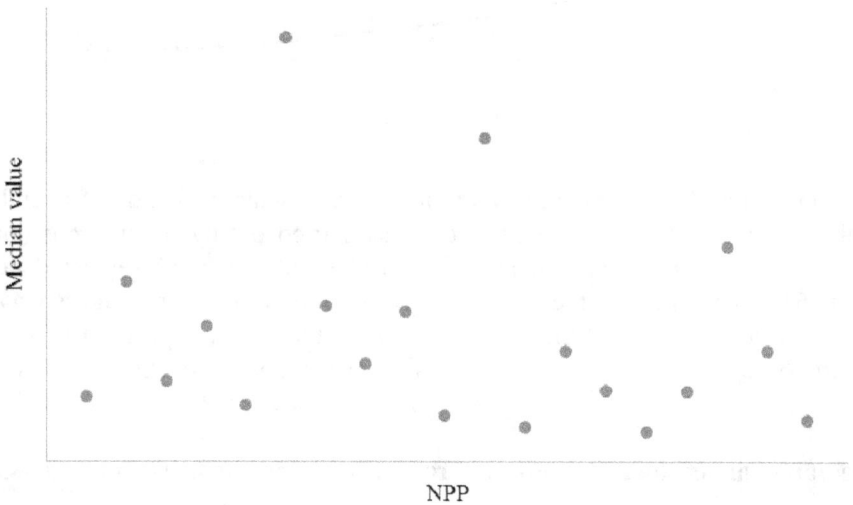

It has sometimes been suggested to measure the vulnerability of different sites with respect to population by drawing circles around NPP sites and counting the population within such circles [4]. This however, is unable to offer more than a rough preliminary idea. A good understanding of phenomena requires assessing the complex chain of phenomena which lead from the release to the contamination of land. Dominant wind directions and topography can significantly affect the dispersion of radionuclides in the atmosphere and their deposition onto the ground [5; 6].

A word of caution before concluding: one should avoid jumping to conclusions on the basis of such data as depicted in Figure 7. The risk attached to different NPPs cannot be compared on the sole basis of estimates of nuclear refugees. It depends on other factors such as the probabilities of radioactive releases; these in turn could depend on differences in reactor design; on safety enhancements carried out since they were commissioned; on the likelihood of external threats (earthquakes, floods…); etc.

4. CONCLUSION

The number of expected radiological refugees due to nuclear accidents is fairly simple to calculate. It is easy to understand and communicate. It could be quite useful to help perceive the consequences of major accidents. There are obvious limits, however, to the use of this indicator: it should not be as helpful for less severe accidents (with few refugees); it gives no indication as to image costs or possible modifications in the electricity production system resulting in higher prices for consumers.

For major nuclear accidents in France, the number of expected radiological refugees is high; highly depend on climatic conditions; and variable from site to site. Both size and variability depend on the presence of large cities within the relocation zone.

This raises the question of how best to manage the contamination of large cities. On the one hand relocating the entire population should reduce long-term health effects of exposition to residual ground shine. However, mass relocation can cause great suffering for displaced populations as exemplified by the Fukushima case. Since cities are easier to decontaminate than agricultural areas, they could perhaps benefit from a thorough decontamination effort with a view to allow city dwellers to return home as soon as possible and thus avoid the hardships of temporary sheltering and those of being transplanted.

References

[1] Pascucci, L. and Momal, P. *"Massive radiological releases profoundly differ from controlled releases"*, Eurosafe, (2012).

[2] Federal office of civil protection and disaster assistance, *"Method of risk analysis for civil protection"*, (2011).

[3] Zajdenweber, D. *"Économie des extremes"*, Flammarion, 2001.

[4] Butler, D. *"Reactors, residents and risk"*, Nature, (2011).

[5] Quelo D. et al., *"Validation of the Polyphemus platform on the ETEX, Chernobyl and Algeciras cases"*, Atmospheric Environment, Volume 41, Issue 26, pp. 5300–5315, (2007).

[6] BMTA et al., *"State of the model to simulate the Fukushima Daiichi nuclear power plant accident"*, Pollution atmosphérique, Volume 217, (2013).

Multidimensional risk evaluation: assigning priorities for actions on a natural gas pipeline

Mônica Frank Marsaro[a*], Marcelo Hazin Alencar[a], Adiel Teixeira de Almeida[a] and Cristiano Alexandre Virgínio Cavalcante[a]

[a] Universidade Federal de Pernambuco, UFPE, Recife, Brazil

Abstract: This paper presents a multicriteria decision model application to define actions with a view to mitigating the risks involved in this mode of transportation. Natural gas is a fossil fuel that is important for society and is transported through pipelines. It is used for different purposes in industrial and civil applications. Although pipelines are one of the safest transport systems, some accidents involving natural gas have occurred. The Multicriteria decision model described in this paper is put forward as a means to minimize such possibilities. It incorporates MAUT (Multi-attribute Utility Theory), which considers a decision maker's preferences and some aspects of the Decision Theory approach. Three dimensions of risk, namely the human, financial and environmental ones - are targeted in the context of probabilistic consequences. As an important result, the information obtained from the model is shown to be important in order to define how resources should best be allocated and to establish maintenance policies for managing and mitigating risk.

Keywords: Multicriteria, risk, natural gas pipeline.

1. INTRODUCTION

Due to the increasing in the global demand for energy (which may well triple in the first half of the 21st century) and extensive and prolonged blackouts having occurred around the world, research on the use of other energy sources has been undertaken to seek solutions to these problems [1]. Moreover, due to the need to preserve and conserve the environment, greater attention is being given to cleaner sources of energy, of which Natural Gas is one [2]. Thus, natural gas can be used in the chemical industry, in the process of producing electricity, as fuel in automobiles, homes, and in other applications.

The most common means of transporting Natural Gas is by pipeline. Pipeline extensions may range from running for a few meters to hundreds of kilometers. Long pipelines pass through different locations, each with specific characteristics e.g. agricultural, industrial or residential areas, preserved environmental sites, commercial districts, etc.). Many authors consider the transport of flammable substances in pipelines as the safest and most economic among existing modes of transportation, especially when compared to road and rail [3].

Pipelines are subject to different actions that may damage their structure. Corrosion, third party actions and human errors during operation and maintenance are examples of events that can cause holes or disruptions in pipeline, thus producing a gas leak. The release of gas can lead to accidents with catastrophic consequences that adversely affect human beings and property, causing financial loss and harming a natural gas company's image.

Most studies on risk in the natural gas pipeline environment have been conducted using different methods of evaluation which have included qualitative and quantitative approaches [4]. This paper, specifically, uses a multi-criteria decision model for risk assessment which has a risk value hierarchy and considers three dimensions of risk: human, environmental and financial. Multi-attribute Utility Theory (MAUT) is used in this model. The results are used as input in the risk management process to support the decision under conditions of resource constraint. Thus, resources can be used by the decision maker (DM) so that risk is mitigated and managed more befittingly with reality.

The paper has the following sections: Section 2 presents a review of the literature on Multicriteria risk analysis; Section 3 discusses some concepts of multicriteria decision making, including MAUT; Section 4 sets out the multicriteria decision model; Section 5 describes an application of the model; and Section 6 draws some conclusions.

2. MULTICRITERIA RISK ANALYSIS

Many risk definitions have been made over the years and especially take into account the relationship between the probability of an accident occurring and the consequences arising [5]. Therefore, [4] considers that risk can be quantified, measured and expressed as a mathematical relationship with the help of available data, with the results being applicable in widely different areas [6]. This analysis can be used to define the probability of potential accidents and their causes, and examining the measures necessary to mitigate risk.

However, there is a perceptible a tendency on assessing risk for only the number of people affected or the financial impact on the resulting scenario to be considered. However, if only one of these factors is considered, this is insufficient due to the complexity and seriousness of the issues involved in this type of analysis [7].

Since a pipeline normally extends over different areas, as mentioned, an accident in one area probably would not have the same consequences as a similar type of accident in another area. For example, an accident in an uninhabited area does not affect humans in the same way when compared to an accident in a residential or commercial area. So it is very important to use multiple criteria in the model presented in order to have a result that incorporates multiple dimensions and provides a risk analysis in accordance with the actual characteristics of the pipeline, since for each section analyzed, there will be different impacts on each of the dimensions considered. Moreover, it is of paramount importance to assess the structure of the DM's preferences and judgments [8]. In the next section a brief introduction will be given to multicriteria decision making.

3. MULTICRITERIA DECISION MAKING

A multicriteria decision problem consists of a decision problem in which more than one alternative is analyzed in a context of conflicting objectives [9]. This decision may be taken by one DM or a group of DMs, who accept responsibility for the decision and its possible consequences. According to [10], multiple criteria decision making can work with different aspects related to the DM's characteristics, and it takes into account conflicting criteria used in the analysis.

Therefore, the set of methods, techniques and concepts of multicriteria decision support (MCDA - MultiCriteria Decision Aid) seek to assist people and organizations in solving problems where multiple views can be evaluated [11], such that information provided by the DM can be synthesized and organized, so that decisions are taken more effectively. [12].

MAUT (Multi-attribute Utility Theory) is used to develop the model discussed in this paper, taking into account the probabilistic context in which pipeline risk analysis is inserted. MAUT presents a very well defined axiomatic structure, based on the axioms of utility theory. [13]. Furthermore, MAUT enables the utility function to be elicited by structured elicitation protocols.

4. MULTIPLE CRITERIA DECISION MODEL

The decision model considered in this paper uses a multiple risk dimension based evaluation for each section of a pipeline network, thereby providing an idea of the hierarchy of risk between each pipeline section by comparing these sections, while taking environmental, human and financial risks into consideration as is the DM's character as to his/her aversion to or propensity for risk for each of the dimensions. Figure 1 shows the process for applying the model and illustrates its steps which are discussed below. This model uses only one decision maker. It is important to note that the decision maker's value judgments are influenced by his/her own values and by incorporating, on his/her own, the values of the stakeholders involved. Figure 1 presents the final structures of the multicriteria decision model.

Figure 1: Model structure (adapted from [14])

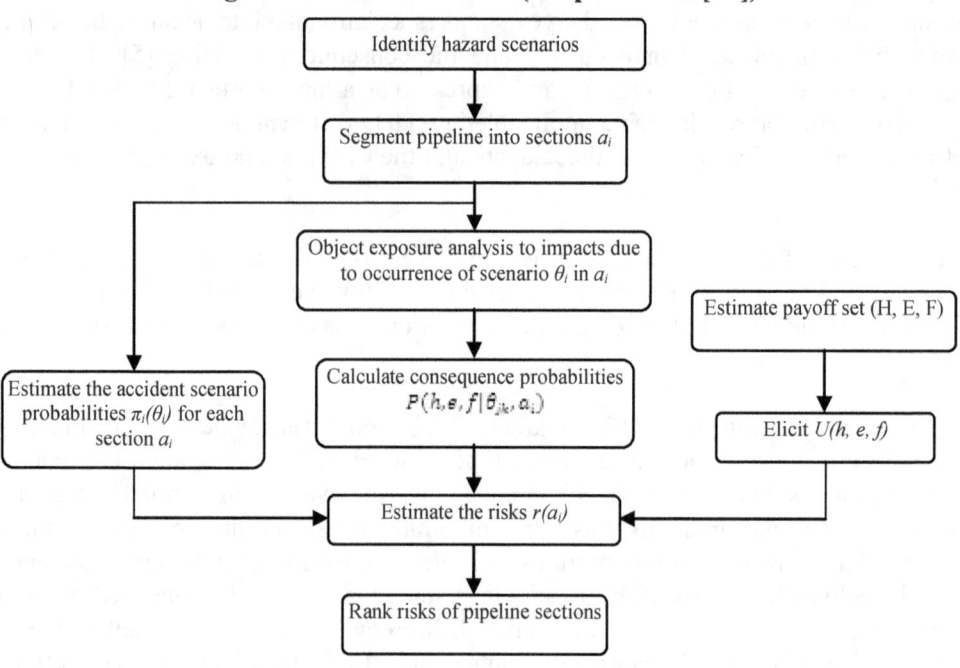

Additionally, hazard scenarios that can occur in natural gas pipelines are listed as shown in Figure 2.

Figure 1: Event tree for a natural gas pipeline (Adapted from [14])

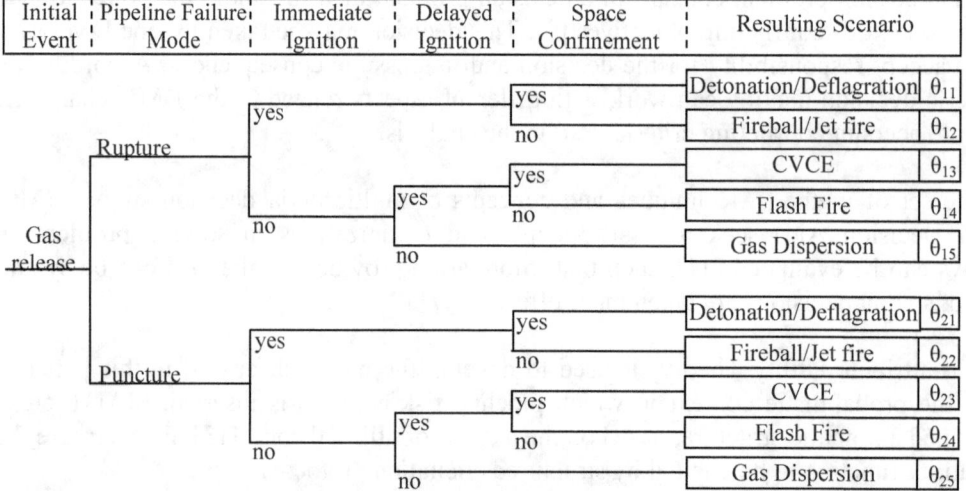

Pipelines can be divided into sections that have homogeneous characteristics such as soil type, age of piping, pipeline diameter, pressure and cathodic protection.

The probability of the occurrence of an accident for each hazard scenario ($\pi_i(\theta_{jk})$) is accounted for and evaluated in one of the steps of the model as is the probability of a normal scenario ($\pi_i(\theta_N)$), where no damage occurs. When no historical data are available for evaluation purposes, the model uses the DM´s knowledge as per *a priori* probability distributions.

The consequences due to an accident occurring are evaluated in a given set of payoffs for the deemed dimensions of risk e.g. possible losses due to the energy and heat released by overheated gas, flames or explosions. As to the human dimension, the maximum number of people present in the danger zone is considered. As for the environmental dimension, what is considered is the area of vegetation scorched by flames from the burning gas released through a hole or rupture of the pipeline, while for the financial dimension monetary losses associated with the accident should be considered.

When calculating the impact on the human and environmental dimensions, the average radiation flux due to deflagration is obtained. According to [15] η is the ratio of the irradiated heat over the total heat released; τ_a is the atmospheric transmissivity; H_c is the combustion heat of the natural gas; r is the radius of the critical danger radius; and Q_{eff} is the effective rate of the gas leak. Thus, the average flux of radiation flux is as per the following Equation (1):

$$I = \frac{\eta \times \tau_a \times Q_{eff} \times H_c}{4 \times \pi \times (r)^2} \tag{1}$$

To evaluate the consequences on the financial dimension the sum of losses due to three aspects is considered: suspension of invoicing ($F(t_Q)$, caused by revenue losses incurred due to the gas supply being interrupted; salvage and restoration works; ($W(t_Q)$); damage to property, fines and compensation payments for injuries ($W(t_Q)$) resulting from the accident in the pipeline, as per Equation (2):

$$P_f = F(t_Q) + W(t_Q) + M(t_Q) \tag{2}$$

The model also has a well-structured process for evaluating the scale constants and utility function, in which the DM's preferences as regards risk are taken into account, namely he/she be averse to, neutral about taking risks or may have a propensity towards them. An additive utility function is considered, where the additive Independence property of $U(h,e,f)$ implies the existence of preferential independence between the set of payoffs. A utility function is defined by the equation (3) [16].

$$U(h,e,f) = k_h U(h) + k_e U(e) + k_f U(f) \tag{3}$$

To calculate the risk, a consequence function is evaluated for each dimension of risk. It is considered there is no statistical correlation between human, environmental and financial damage, thereby enabling the consequences to be evaluated separately in each of the risk dimensions [14].

The estimation of risks is carried out considering the losses occasioned by a hazard scenario occurring in the section evaluated in the pipeline, where the utility function and probability density function are combined, as shown by Equation (4) [7].

$$L(\theta_{jk}, a_i) = -\int_h \int_e \int_f [P(h,e,f \mid \theta_{jk}, a_i) \times U(h,e,f)] \, dh \, de \, df \tag{4}$$

The risk is then calculated by Equation (5):

$$r(a_i) = \left\{ \sum_j \sum_k \left\{ - \begin{bmatrix} +\int_h f(h|\theta_{jk},a)k_h u(h)dh \\ +\int_e f(e|\theta_{jk},a)k_e u(e)de \\ +\int_f f(f|\theta_{jk},a)k_f u(f)df \end{bmatrix} \right\} \cdot \pi_i(\theta_{jk}) \right\} + (-1).\pi_i(\theta_N) \tag{5}$$

5. NUMERICAL APPLICATION

In this section, a numerical application is presented in order to illustrate the multiple criteria decision model developed for assessing risk in a natural gas pipeline.

The pipeline investigated in this application has ten sections, the total length of which is 25,300 km, divided according to the characteristic of land use in the surrounding area. Figure 3 shows one stretch of this pipeline.

Figure 3: Diagram of the stretch of the pipeline analyzed and the characteristics of land use of the surrounding areas

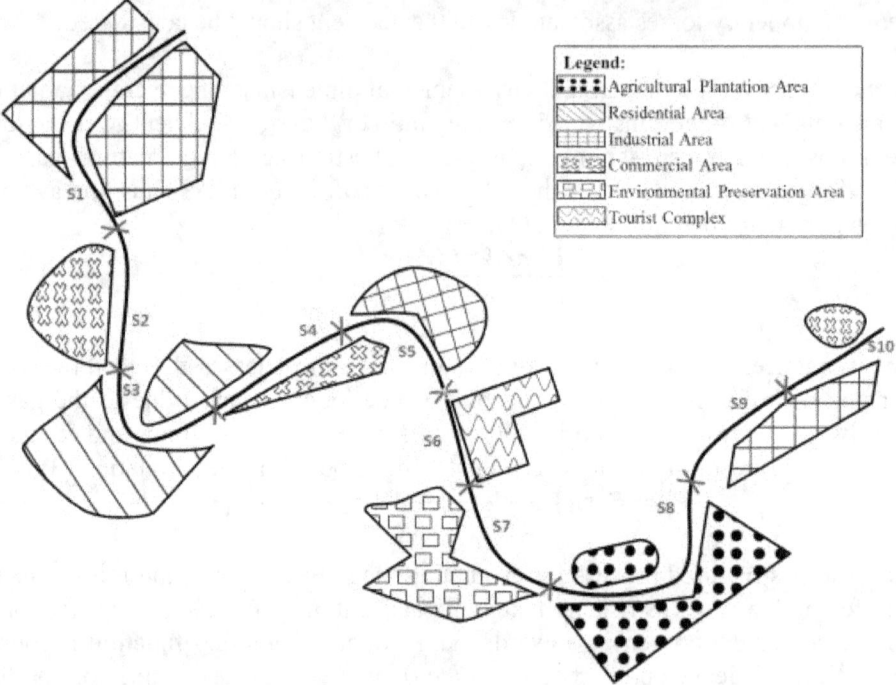

The DM´s attitude towards risks was evaluated and scale constants were elicited. Thus the utility function U(h, e, f) could be evaluated, where the last step in risk analysis is to check the value of risk by Equation (5). From this result some interpretations can be made.

One of these interpretations is that an absolute difference index (AD) for each section should be adopted and calculated from the ratio "r(x$_i$)-r(x$_{i+1}$)" where "x$_i$" and "x$_{i+1}$" represent the positions in the ranking of given sections of the pipeline. Another interesting index for the risk analysis is the difference ratio (DR), expressed by Equation (6).

$$DR = \frac{r_x(a_i) - r_{x+1}(a_j)}{r_{x+1}(a_i) - r_{x+2}(a_j)} \qquad (6)$$

These values are presented in Table 1.

Tabela 1: Absolute difference and difference ratio for the sections of the pipeline

Ranking position (x)	Section	AD	DR
1	S$_7$	11.14	9.32
2	S$_9$	1.19	2.33
3	S$_8$	0.51	0.13
4	S$_6$	3.69	1.37
5	S$_5$	2.69	2.41
6	S$_4$	1.11	0.09
7	S$_{10}$	11.71	3.89
8	S$_1$	3.014	2.5
9	S$_2$	1.2	-
10	S$_3$	-	-

From Table 1, it is possible to verify that Section seven (S$_7$) has a higher risk when all three risk dimensions are considered: human, financial and environmental. Higher losses in this section are expected because an accident when transporting gas occurred in this pipeline. On using the interval scale from the utility function, it can be inferred that the increment in risk for the risk from S_7 to S_9 is 9.32 times greater than that from section S_9 to S_8, in the same way that the increment on risk from S_{10} to S_1 is 3.89 times in relation of S_1 to S_2.

This analysis of the increment in risk enables the DM to evaluate how best to allocate resources, should these be scare. Moreover, it enables the DM to evaluate how to mitigate risk more efficiently, since a ranking for risk was generated and from this it is possible to identify which sections are most prone to the largest consequences, should an accident occur, especially in sections with homogenous characteristics.

6. CONCLUSION

This paper has demonstrated that it is possible to evaluate the risk from transporting Natural Gas by pipeline by means of a multicriteria model using Multi-attribute Utility Theory, which takes the DM's behavior into account in relation to risk in three risk dimensions: human, environmental and financial, besides considering his/her preferences when eliciting the scale constant. Thus, the interval scale of the utility theory for evaluating the increment on risk from one section to another was considered.

Acknowledgments

This study was supported by the Brazilian National Research Council (CNPq) and CAPES (a Brazilian Research Agency under the Ministry of Education of Brazil).

References

[1] Bentham, J. *The scenario approach to possible futures for oil and natural gas*, Energy Policy, pp. 87-92, (2014).
[2] Koku, O., Perry, S., & Kim, J. *Techno-economic evaluation for the heat integration of vaporisation cold energy in natural gas processing*, Applied Energy, pp. 250-261, (2014).
[3] Jamshidi, A., Yazdani-Chamzini, A., Yakhchali, S. H., & Khaleghi, S. *Developing a new fuzzy inference system for pipeline risk assessment*, Journal of Loss Prevention in the Process Industries, *26*, pp. 197-208, (2013).
[4] Marhavilas, P. K., Koulouriotis, D., & Gemeni, V. *Risk analysis and assessment methodologies in the work sites: On a review, classification and comparative study of the scientific literature of the period 2000-2009*, Journal of Loss Prevention in the Process Industries, *24*, pp. 477-523, (2011).
[5] Aven, T. *On how to define, understand and describe risk*, Reliability Engenireeng Safety System, pp. 623-631, (2010).
[6] Ma, L., Li, Y., Liang, L., Li, M., & Cheng, L. *A novel method of quantitative risk assessment based on grid difference of pipeline sections*, Safety Science, 59, pp. 219-226, (2013).
[7] Alencar, M. H., & Almeida, A. T. *Assigning priorities to actions in a pipeline transporting hydrogen based on a multicriteria decision model.* International Journal of Hydrogen Energy, 35, pp. 3610-3619, (2010).
[8] Garcez, T. V., & Almeida, A. T. *A risk measurement tool for an underground electricity distribution system considering the consequences and uncertainties of manhole events*, Reliability Engineering & System Safety, 124, pp. 68-80, (2014).
[9] Cailloux, O., Mayag, B., Meyer, O., & Mousseau, V. *Operational tools to build a multicriteria territorial risk scale with multiple stakeholders*, Reliability Engineering & System Safety, 120, pp. 88-97, (2013).
[10] Ferreira, R. J., Almeida, A. T., & Cavalcante, C. A. *A multi-criteria decision model to determine inspection intervals of condition monitoring based on delay time analysis*, Reliability Engineering & System Safety, 94(5), pp. 205-912, (2009).
[11] Vincke, P. *Mutlicriteria decision-aid*. John Wiley & Sons, (1992).

[12] Belton, V., & Stewart, T. J. *Multiple Criteria Decision Analysis: An integrated Approach.* Kluwer Academic Publishers, (2002).

[13] Keeney, R. L., & Raiffa, H. *Decision with Multiple Objectives: Preferences and value trade-offs.* John Wiley & Sons, (1976).

[14] Brito, A. J., Almeida, A. T., & Mota, C. M. (2009). A multicriteria model for risk sorting of natural gas pipelines based on ELECTRE TRI integrating Utility Theory. *European Journal of Operational Research, 200,* pp. 812-821.

[15] Jo, Y. D., & Ahn, B. J. *Analysis of hazard areas associated with high-pressure natural gas pipelines,* Journal of Loss Prevention in the Process Industries, 15, pp. 179-186, (2002).

[16] Lins, P. H., & Almeida, A. T. *Multidimensional risk analysis of hydrogen pipelines based on multiattribute utility function,* International Journal of Hydrogen Energy, 37, pp. I3545 - I3554, (2012).

[17] Saltelli, A., Ratto, M., Andres, M., Campolongo, F., Cariboni, J., Gatelli, D., et al. *Gloal sensitivity analysis: the primer.* John Wiley & Sons Ltd, (2008).

[18] Saltelli, A., Tarantola, S., Campolongo, F., & Ratto, M. *Sensitivity analysis in practice: A guide to Assessing Scientific Models,* England: John Wiley & Sons Ltd., (2004).

[19] Mechri, H. E., Capozzoli, A., & Corrado, V. *Use of the ANOVA approach for sensitive building energy design,* Applied Energy, 87(10), pp. 3073-3083, (2010).

[20] Schouten, M., Verwaart, T., & Heijman, W. *Comparing two sensitivity analysis approaches for two scenarios with a spatially explicit rural agent-based model,* Environmental Modelling & Software, 54, pp. 196-210, (2014).

[21] Tian, W. *A review of sensitivity analysis methods in building energy analysis,* Renewable and Sustainable Energy Reviews, 20, pp. 411-419, (2013).

Development of Accident Consequence Assessment Scheme using Accident Cost and Consideration of Decontamination Model

Kampanart Silva[a], Koji Okamoto[b], Yuki Ishiwatari[b,c], Shogo Takahara[d] and Jiraporn Promping[a]

[a] Thailand Institute of Nuclear Technology, Nakhon Nayok, Thailand
[b] The University of Tokyo, Tokyo, Japan
[c] Hitachi-GE Nuclear Energy, Ltd., Ibaraki, Japan
[d] Japan Atomic Energy Agency, Ibaraki, Japan

Abstract: Severe accident at nuclear power plants, including the Fukushima accident in March 2011, wreak various kinds of consequences, including health effects, economic, social and environmental impacts. The authors developed the scheme of the accident consequence assessment using "accident cost", aiming for it to be an index that is as comprehensive as possible. Normalized accident costs of all accident sequences along with their breakdowns, and the breakdown of the average accident cost are presented. The radiation effect cost, the decontamination cost and the relocation cost are the three major components that dominate the accident cost. The decontamination model was reconsidered since decontamination effects were taken into account by very simple assumptions and decontamination cost was estimated by a rough calculation scheme in the former model. 99 decontamination-related parameters were selected and the model is formed. A sensitivity analysis was performed to identify parameters with large influence on accident cost calculation and large extent of interactions with other parameters. Parameters with high importance tend to have large extent of interactions with other parameters. Parameters influential to accident cost, e.g., the dose of setting decontamination target area, a number of waste management-related parameters, are identified.

Keywords: Consequence Assessment, Accident Cost, Decontamination Model, Sensitivity Analysis

1. INTRODUCTION

Severe accident at nuclear power plants, including the Fukushima accident in March 2011, wreak various kinds of consequences, including health effects, economic, social and environmental impacts. Earlier studies [such as 1-3] on severe accident consequence assessment concentrated on mostly health effects as one of indices of consequences. This maybe because the probabilistic safety criteria or goals related to the consequence of severe accidents which were (and still are) commonly used by the regulatory bodies and utilities in several countries are the acute and the chronic doses [4], thus it was necessary to conduct a research that enable to evaluate the doses and propose measures to achieve the goals and fulfill the criteria.

As an index of consequences is assessed, countermeasures are usually proposed to minimize those particular consequences. Minimization of an index of consequences, however, does not necessarily minimize other consequences, in some cases it even increases other consequences. For example, decontamination which is a measure to reduce the dose received by the public may increase the economic impact of the accident as it may cost a great deal. A common index that can take into account various consequences is therefore needed to enable minimization of overall consequences. "Accident cost" (also called "cost per severe accident) has been used for this purpose as it can cover a large scope of consequences and it is simple to understand. In ExternE [5], Hirschberg et al. [6] and IAEA technical reports series no. 394 [7], many kinds of consequences are evaluated in terms of monetary value, referring to the consequences of the Chernobyl accident. As the objective of these studies was to perform a comparative accident consequence assessment among the electricity generation systems, the consequences selected are those can be commonly evaluated in all systems, and there is a possibility for consequences particular to nuclear severe accidents to be overlooked. Park [8] also estimated the total damage cost of severe accidents in particular conditions, but the assumption was way too conservative and the cost associated with decontamination is not included.

The authors have been developing the scheme of the accident consequence assessment using "accident cost", aiming for it to be an index that is as comprehensive as possible [9-11]. We have modified the accident cost calculation scheme based on the updates of the Fukushima accident in March 2011, and comments from experts who associate with recovery after the accident. The latest version of the accident cost calculation scheme and its results will be introduced in the Section 2.

Though our previous studies on estimation of accident cost [9-11] provide significant insights, which would help to comprehensively assess the consequence of severe accident and to optimize the radiation protection and severe accident management countermeasures, there is still room for improvement. Since the formulation of the assessment scheme was the primary objective, the values of the parameters are determined without adequate data collection and enough consideration, which may crucially affect the results. Although a sensitivity analysis was performed in order to check the validity of the values selected for parameters that are believed important, ceteris paribus technique, where a single parameter is varied at a time while all other parameters are fixed to a constant, was used to perform the sensitivity analysis. As all other parameters are constant when a parameter is examined, the sensitivity of the parameters cannot be systematically evaluated. This makes it difficult to consider the interaction among the parameters which is very important in a non-linear system like severe accident consequence assessment.

A global sensitivity analysis which can take into account the changes of many parameters at a time is therefore needed for this accident cost calculation model. Values of all important parameters must also be reconsidered based on information obtained from review of literatures related severe accidents in the past and updates from the Fukushima accident, in order to get a more realistic consequence assessment. However, if the scope of the sensitivity analysis and the number of parameters incorporated are too large, the uncertainty of the model maybe too large which may obstruct identification of important parameters from non-important ones, and the model itself maybe too complicate to comprehend.

The authors finally decided to focus on decontamination model in this study. This is because: (1) the decontamination cost is one of the three important cost components of the accident cost, i.e., radiation effect cost, relocation cost and decontamination cost, and, (2) decontamination is currently one of the most important the Fukushima accident-related issues, and the insights from this study may help identify the factors that need careful consideration during the decision making process.

The objectives of this study are: (1) to select parameters that are necessary for evaluation of decontamination cost, and formulate the decontamination model for accident cost calculation (Section 3), and, (2) to collect adequate data to set the distribution of each parameter, and perform a sensitivity analysis to identify parameters with large influence on accident cost calculation and large extent of interactions with other parameters which require careful attention, and parameters with negligible influence of which the value can by fixed to constant (Section 4).

2. OVERVIEW OF ACCIDENT COST AND OBJECTIVES OF STUDY ON DECONTAMINATION MODEL

2.1. Methodology

This section shows the overview of calculation of accident cost. Detail of the calculation methodology is provided in K. Silva et al. [9] First of all, the type of the nuclear reactor and its location are determined. Severe accident sequences are defined in a manner that can cover all conceivable severe accidents. Then only accident sequences that lead to release of radioactive materials from the containment vessel are selected. After that, the source term data of each sequence, including the release time, release duration and the amount of the released radionuclides, are taken from the level 2 PRA results. Also the radiation protection scenario is set. This includes the conditions of sheltering, evacuation, relocation and restriction of food intake. At this stage, containment failure frequencies (CFFs) of representative accident sequences are taken from the level 2 PRA results. The CFFs are used to weight the accident sequences in the calculation of the average accident cost in order to prioritize the accident sequences according to their probabilities of occurrence (see Equation (1)). The reason that the CFFs are used to represent the accident occurrence probabilities is that the CFFs are the probabilities that the containment fails to confine the radioactive materials. They have stronger relations than the core damage frequencies (CDFs) with the probability of release of radioactive material to the environment which could determine the extent of the consequences of the accidents.

In the next step, the consequence analysis is performed using the level 3 PRA code, OSCAAR (Off-Site Consequence Analysis of Atmospheric Releases of radionuclides) [12], which was developed by the Japan Atomic Energy Agency (JAEA). OSCAAR estimates the periods of the radiation protection countermeasures, i.e., sheltering, evacuation, relocation and restriction of food intake, and the area and the numbers of people associated with each countermeasure. Also it calculates the individual early (or acute) and chronic doses, the collective dose, and the health effects regarding the radiation exposure.

Before holding accident cost calculation of each accident sequence, the consequences which are able to be quantified and to be taken into consideration are determined. Consequences of the severe accidents to people can be divided into health effects, economic impacts and social impacts. Health effects include the health effects from radiation exposure and the psychological effects. Costs resulting from the radiation protection countermeasures are taken into account as economic impacts. The social impacts are difficult to deal with because they involve the responses of the human-being which make them specific to the accidents. In addition, it is very difficult to convert them to monetary values. The author decided to include only the cost resulting from harmful rumor as quantitative data of other social impacts is not available. Consequences of the severe accidents to the environment can be divided into on-site and off-site consequences. The on-site consequences can be represented by the increase in decommissioning cost and the off-site consequences can be quantified by summing up the costs for decontamination of the land contaminated by the released radioactive materials.

Then the results from the consequence analysis by OSCAAR, i.e., the expected values of the periods and the numbers of people involved in the radiation protection countermeasures and the collective dose of each severe accident sequence, are used as the input data to perform the calculation of the accident cost of each accident consequence. The ways to estimate the monetary values of each consequence are briefly explained below. The equations and the values of the parameters used for the calculation can be found in K Silva et al. [9].

Health Effects The cost regarding health effects from radiation exposure is estimated by a simple multiplication of the collective dose and the willingness to pay (WTP) per unit exposure. This is because the stochastic effects from the radiation exposure are supposed to be in linear relationship with the exposure dose according to the linear non-threshold hypothesis of ICRP [13]. The deterministic effects from the radiation exposure is not included because it is internationally recognized that full effort must be made to prevent the deterministic effects even though those measures can significantly increase other consequences (e.g. economic impacts) of the accident [14]. Therefore, there is no point to consider the deterministic effects together with other consequences. The psychological effect cost is estimated by summing up the compensations regarding psychological effects resulting from sheltering, evacuation and relocation. The unit value of the compensation [JPY/person-year] refers to the compensation in the Fukushima accident [15].

Economic Impacts Income losses, transportation costs, accommodation costs and capital utility losses of the sheltered, evacuated and relocated population are used to estimate the economic impacts of those countermeasures. Losses of income of people who could not work during the implementation of sheltering, evacuation and relocation are included into the cost estimations of all countermeasures. Transportation costs and accommodation costs are included in the case of evacuation and relocation. Capital utility losses are considered only in the relocation cost calculation. Food intake restriction cost is estimated by summing the losses of the agricultural and livestock products the six types of the agricultural and livestock products: milk, dairy products, meat, leaf vegetables, root vegetables and grains and the cost of waste management

Social Impacts The approximate value of the cost regarding damages by harmful rumor was taken from the report of the commission of management and financial survey of TEPCO [16].

Environmental Impacts The increase in decommissioning cost is estimated by multiplying the total electric power of the target reactor by the increase of decommissioning cost per unit electric power obtained from the report of the commission of management and financial survey of TEPCO [16]. The decontamination of the released radioactive materials is supposed to be done in the entire relocated area. Different decontamination techniques are chosen to suit different land use types. The decontamination cost consists of the total cost generated during the implementation of all decontamination techniques and the summation of the management cost of waste generated. The former is obtained by multiplying the target area of each decontamination technique by the costs generated during the implementation of the technique per unit area which includes the costs of the materials, equipment and labors spent, and sum them up. The latter is the product of the mass of waste generated (volume reduction by incineration is taken into account for burnable waste) and the unit cost for the radiation waste disposal.

All costs stated above are summed up to form the accident cost of each accident consequences, Finally, calculated accident cost of each accident sequence is averaged using their CFFs as a weighting factor.

$$AC = \frac{\sum_{p}' AC_p \times CFF_p}{\sum_{p} CFF_p} \tag{1}$$

where AC_p and CFF_p represent the accident cost and the CFF of the pth accident sequence, and AC represents the average accident cost.

2.2. Calculation Conditions

The methodology was applied to a virtual 1100 MWe boiling water reactor (BWR-5) which is located at the center of Tokai Research and Development Center (TRDC) of JAEA. Dominant severe accident sequences were selected, and the CFFs, release times, release duration times, and release ratios of those accident sequences were taken from the results of an open document of level 2 seismic PRA [17]. The radiation protection scenario was selected based on the recommendations of IAEA, ICRP and Nuclear Safety Commission of Japan (NSC) [18-20]. As TRDC is in Ibaraki Prefecture, the data of population, agricultural and livestock products and land use types were taken from the statistical data of Ibaraki Prefecture [21].

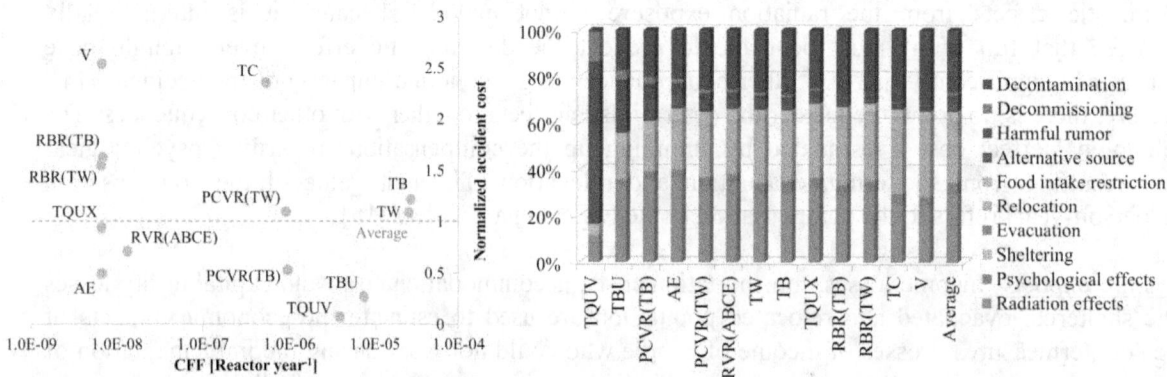

Fig. 1 (left) Normalized accident cost of each accident sequence [9]
Fig. 2 (right) Breakdowns of accident cost of all accident sequences and breakdown of average accident cost [9]

2.3. Results and Discussion

The normalized accident costs NAC_p of each accident sequence are shown with their CFFs in Fig. 1.

$$NAC_p = \frac{AC_p}{AC} \qquad (2)$$

Abbreviations, e.g., TB, TW, represent the accident sequences[*]. This figure shows both the occurrence probabilities (CFFs) and the consequences (accident costs) which are significant indicators to assess the risk of severe accidents in nuclear power plants. Many accident sequences with small CFFs, i.e., V, RBR(TB), RBR(TW), gave large accident costs. If only the CFF (or CDF) is used to indicate the risk, these accident sequences might be considered as insignificant due to their small CFFs. This implies that assessing only one indicator without assessing another may provide misleading information on risk of severe accidents.

Fig. 2 shows the breakdowns of accident costs of each accident sequence and of average accident cost which represent the relative sizes of each component of the accident costs. Accident sequences were sorted by their total accident cost in ascending order, and the breakdown of average accident cost is on the last bar. When the release is very small, e.g., TQUV, all components using constant values, i.e., alternative source cost, harmful rumor cost and decommissioning cost, dominate the accident cost. When the release is relatively small, e.g., PCVR(TB), AE, PCVR(TW) and RVR(ABCE), the radiation effect cost dominates the accident cost because the annual dose rates in most area are not high enough to trigger the relocation, and thus only limited area needs decontamination since the decontamination is assumed to be done only in the relocated area. When the release is moderate (TW, TB, TQUX, RBR(TB) and RBR(TW)), the radiation effect cost, the relocation cost and the decontamination cost are almost the same and dominate the accident cost since the relocated area (= decontamination target area) and the relocation period increase with the amount of source term. When the release is relatively large (TC, V), the relocation cost and the decontamination cost dominate the accident cost because the relocated area and the decontamination target area are significantly enlarged according to the increase of amount of source term while the increase of collective dose which determines the radiation effect cost is rather moderate. Breakdown of the average accident cost shows the similar trend to the accident sequences with moderate release. It can be concluded that the radiation effect cost, the decontamination cost and the relocation cost are the three components that dominate the accident cost. Therefore, measures to minimize these three costs without increasing one another or other costs have to be carefully considered in the decision makings in severe accident consequence management.

3. DISCUSSION ON DECONTAMINATION MODEL

3.1. Parameter Selection

First, All factors related to decontamination cost and the effects of decontamination, that may affect the accident cost were listed. These factors are qualitatively screened by selecting only factors that directly affect the three important cost components of the accident cost, i.e., radiation effect cost, relocation cost and decontamination cost. Selected factors are listed in Table 1. Then the authors carefully examined OSCAAR and identified 99 parameters, also listed in Table 1, to incorporate all selected factors into the accident cost calculation scheme.

---

[*] TB: Long-term loss of all AC power; TW: Loss of all decay heat removal function; TBU: Short-term loss of all AC power;
TQUV: Transient with loss of ECCS function; PCVR: Primary containment vessel rupture; TC: ATWS events
RBR: Reactor building rupture; RVR: Reactor vessel rupture; TQUX: Transient with loss of Depressurization
AE: LOCA with loss of ECCS injection; V: LOCA with loss of water injection

Table 1 Decontamination-related factors and parameters that affect the accident cost[1]

Factor	Parameter	Parameter No.
Factors/parameters that affect decontamination cost		
Determination of decontamination target area	Dose for decontamination target area setting [mSv/year]	1
Decontamination techniques used in each land use type	Fraction for application of each decontamination technique on roofs and walls of houses and buildings[2] [%] (2: B, 3: HPW)	2-3
	Fraction for application of each decontamination technique on gardens and playgrounds[2] [%] (4: RL, 5: RSS, 6: WLM, 7: RS, 8: CL)	4-8
	Fraction for application of each decontamination technique on agricultural areas[2] [%] (9: P, 10: RSS, 11: RS)	9-11
	Fraction for application of each decontamination technique on forests[2] [%] (12: RSF, 13: RS, 14: CL)	12-14
	Fraction for application of each decontamination technique on roads[2] [%] (15: SB, 16: CS, 17: W)	15-17
Unit cost of each decontamination technique	Unit costs of 12 decontamination techniques [JPY/m^2] (18: Determination of random number(s) used to determine the unit cost[3], 19: Random number to determine the unit cost for the case of same random number, 20: HPW, 21: B, 22: RS, 23: RL, 24: CL, 25: RSS, 26: WLM, 27: P, 28: RSF, 29: W, 30: SB, 31: CS)	18-31
Waste generated by each decontamination technique	Liquid and solid waste generated by each decontamination techniques[4] [m^3/m^2] (32: HPW (s), 33: HPW (l), 34: B (s), 35: B (l), 36: RS (s), 37: RL (s), 38: CL (s), 39: WLM (s), 40: RSF (s), 41: W (s), 42: W (l), 43: SB (s), 44: CS (s))	32-44
Waste management	Determination whether or not to include cost due to: Temporary waste storage (45), Waste transportation (47), Waste treatment (49), Interim storage (53), Waste disposal (55)	45, 47, 49, 53, 55
	Unit costs of: Temporary waste storage (46), Waste transportation (48), Liquid waste treatment (50), Solid waste treatment (incineration) (51), Solid waste treatment (classification and chemical process) (52), Interim storage (54), high level radioactive waste disposal (56), Disposal of controlled type waste (57) [JPY/m^3]	46, 48, 50-52, 54, 56-57
	Volume reduction rates for: Non-burnable solid waste (58), Burnable solid waste (59)	58-59
Factors/parameters that affect relocation cost		
Determination of decontamination target area	Dose for decontamination target area setting [mSv/year]	1
Determination of way of implication of each decontamination technique	Number of workers that can be involved in the decontamination work [man-year/year]	60
	Work speed of each decontamination technique [m^2/man-day] (61: HPW, 62: B, 63: RS, 64: RL, 65: CL, 66: RSS, 67: WLM, 68: P, 69: RSF, 70: W, 71: SB, 72: CS)	61-72
Dose reduction factors	Selection of data set of dose reduction factors	73
	Dose reduction factors for each decontamination technique [-] (74: HPW, 75: B, 76: RS, 77: RL, 78: CL, 79: RSS, 80: WLM, 81: P, 82: RSF, 83: W, 84: SB, 85: CS)	74-85
	Dose reduction factors for each land use type [-] (86: Houses, 87: Buildings, 88: Agricultural areas[5], 89: Forests, 90: Roads)	86-90
Occupational dose for workers involved with decontamination	Selection of range for calculation of occupational dose	91
	Ranges of average (92) and maximum (93) occupational dose calculation factors [-]	92-93
Period of staying in specific areas per day[6]	Period of staying in each land use type per day [hr] (94: Houses, 95: Buildings, 96: Gardens and playgrounds, 97: Agricultural areas, 98: Forests, 99: Roads)	94-99

[1] Following abbreviations represent 12 decontamination techniques, where HPW = High pressure (HP) water, B = Brushing, RS = Removing soil or covering with soil, RL = Removing, covering or harvesting lawn, CL = Cutting leaves and shrubs, RSS = Replacing soil with subsoil, WLM = Weeding or lawn mowing, P = Ploughing, RSF = Removing sediments and fallen leaves, W = Water, HP water or very HP water, SB = Sandblast or shotblast, CS = Cutting surface or resurfacing.

[2] The sums of the fractions of land use types are normalized to 100%, except for CL which can be applied in the area where other decontamination techniques has already been applied.

[3] Using same random number for all decontamination techniques or different random numbers for each decontamination techniques.

[4] (s) stands for solid waste and (l) stands for liquid waste.

[5] The same dose reduction factor is also used for gardens and playgrounds due to absence of data.

[6] The sum of periods of staying is normalized to 24 hours.

3.2. Model Description

3.2.1 Changes in decontamination cost estimation scheme

The decontamination cost is obtained by adding the total cost generated during the implementation of all decontamination techniques, to the summation of the management cost of waste generated, as is the case with Section 2. The author has improved the decontamination cost estimation scheme in order to include 99 parameters stated in Section 3.1 using information obtained from literatures and updates from the Fukushima accident [such as 22-24]. The detail of improvements of the decontamination cost estimation scheme is as follow:

(1) Decontamination target area is not the same as the relocated area, but is set based on the dose for target area setting,

(2) Decontamination techniques of each land use types are changed to match with the techniques selected in the Fukushima accident and in literatures,

(3) Distributions of fractions for application of each decontamination technique, unit costs of each decontamination technique, waste generated by each decontamination techniques, unit costs of each waste management step and, volume reduction rates are determined, and their values for each run are randomly selected from respective distributions,

(4) Costs from the entire procedure of waste management can be taken into account, and the inclusions of costs associated with respective steps of waste management to the accident cost calculation model are randomly determined,

(5) Volume reduction rate for non-burnable waste is also taken into account.

The total cost generated during the implementation of each decontamination technique for each land use type $DI_{l,t}$ [JPY] is calculated by

$$DI_{l,t} = F_{l,t} \times A_l \times U_{DI,t}, \text{ where } \sum_t F_{l,t} = 1. \tag{3}$$

$F_{l,t}$ stands for the fraction for application of the tth decontamination technique for the lth land use type [-], A_l for the total area of the lth land use type [m^2], and $U_{DI,t}$ for the unit implementation cost of the tth decontamination technique [JPY/m^2]. $F_{l,t}$s that possess no distribution, i.e., $F_{l,t}$s that do not appear in Table 2 as parameter number 2-17, are set to zero. On the other hand, the waste management cost of each decontamination technique for each land use types $WM_{l,t}$ [JPY] is estimated by

$$WM_{l,t} = F_{l,t} \times A_l \times \left[\{(WS_t + WL_t) \times X_{TS} \times U_{TS}\} + \{(WS_t + WL_t) \times X_{TR} \times U_{TR}\} + \right.$$
$$\left\{X_{WT} \times (WS_t \times U_{WT,WS_t} + WL_t \times U_{WT,WL_t})\right\} + \{X_{IS} \times VR_t \times WS_t \times U_{IS}\} + \tag{4}$$
$$\left. \{X_{WD} \times (WS_t \times VR_t \times U_{WD} + WS_t \times (1 - VR_t) \times U_{CWD})\}. \right.$$

Here, WS_t and WL_t are solid and liquid wastes generated by the tth decontamination technique per unit area [m^3/m^2] and VR_t is volume reduction rate for the tth decontamination technique. X is used to determine whether or not to include the respective step into the waste management cost (If yes, $X = 1$, if no $X = 0$.). U represents the unit cost of the respective waste management steps. Subscripts TS, TR, WT, IS, WD and CWD stand for temporary waste storage, waste transportation, waste treatment (volume reduction), waste interim storage, high level radioactive waste disposal and disposal of controlled type waste, respectively. The total decontamination cost is

$$DC = \sum_l \sum_t \left(DI_{l,t} + WM_{l,t} \right). \tag{5}$$

3.2.2 Changes in relocation cost estimation scheme

Relocation cost is estimated by summing the income losses, transportation costs, accommodation costs and capital utility losses, which is also the same as in section 2. The only difference is the estimation of relocation period. In previous studies, decontamination are supposed to be immediately done in the entire area where the dose is above the dose for decontamination target area setting (= dose level for the decision of return home), regardless the decontamination capacity. However, decontamination capacity can be limited by the number of workers that are prepared for the decontamination work. The number of workers that can be involved in the decontamination work N_{WK} [man-year/year] and the work speed of each decontamination technique WSP_t [m^2/man-day] are thus taken into account. The decontamination capacity DCP [m^2/year] can be estimated by

$$DCP = \sum_l \sum_t {}' F_{l,t} \times F_l \times N_{WK} \times WSP_t \times 365 \text{, where } \sum_l \sum_t \left(F_{l,t} + F_l \right) = 1. \tag{6}$$

F_l is the share of the lth area from the entire decontamination target area [-]. The values of $F_{l,t}$, N_{WK} and WSP_ts are randomly selected from respective distributions for each run. If the area where the dose is above the dose for decontamination target area setting is larger than the decontamination capacity, it will be reduced to the decontamination capacity. This will lengthen the relocation period and increase the relocation period, but will in turn reduce the radiation effect cost as the population is kept from the contaminated area for a longer time. Detail on calculation methodology is omitted as it is the same as in previous studies.

3.2.3 Changes in radiation effect cost estimation scheme

Radiation effect cost is the product of the collective dose (the sum of the collective dose of the population and that of the decontamination workers) and the WTP per unit exposure. As the dose reduction factors for each decontamination technique and for each land use type are introduced, the collective dose CD can be calculated by

$$CD = DR \times \left(CD_{POP,B} + CD_{OCP,B} \right), \tag{7}$$

$$DR = \sum_l \sum_t {}' F_{l,t} \times F_l \times DR_t \text{, where } \sum_l \sum_t \left(F_{l,t} + F_l \right) = 1 \tag{8}$$

when the set of dose reduction factors for each decontamination technique is used, and

$$DR = \sum_l {}' F_l \times DR_l \text{, where } \sum_l {}' F_l = 1 \tag{9}$$

when the set of dose reduction factors for each land use type is used. $CD_{POP,B}$ and $CD_{OCP,B}$ are the collective doses of the population and the decontamination workers before consideration of dose reduction factor [Sv], and DR is the average dose reduction factor [-]. The values of $F_{l,t}$, DR_ts and DR_ls are randomly selected from respective distributions for each run. In regard to the dose of decontamination workers, the occupational dose calculation factor OD [-] is introduced. The collective dose of decontamination workers before consideration of dose reduction factor can be estimated by

$$CD_{OCP,B} = \sum_d \sum_r \sum_y {}' X_{DC,d,r,y} \times OD \times D_{d,r,y} \times N_{WK,d,r,y}. \tag{10}$$

Here, $X_{DC,d,r,y}$ is used to indicate whether or not decontamination is done in the area represented by mesh (d,r) in the yth year (If yes, $X = 1$, if no $X = 0$.). $D_{d,r,y}$ and $N_{WK,d,r}$ are the annual dose and the number of decontamination workers in the area represented by mesh (d,r) in the yth year, respectively.

4. SENSITIVITY ANALYSIS

4.1. Elementary Effects Method

The authors performed a sensitivity analysis using the elementary effects method proposed by Morris [25] and revised by Campolongo et al. [26]. This method can identify: (1) parameters with large influence to the output and large extent of interactions with other parameters which require careful attention, and, (2) parameters with negligible influence of which the value can be fixed to constant.

In this method, we assume that the k-dimensional vector $\mathbf{X}$ of the model input has components X_i each of which can assume integer values in the set $\{0, 1/(p-1), 2/(p-1), ..., (p-2)/(p-1), 1\}$. This forms a k-dimensional p-level experimental region Ω ($k \times p$ matrix). For a given value $\mathbf{x}$ of $\mathbf{X}$, the elementary effect of the ith input parameter is defined as

$$d_i(\mathbf{x}) = \frac{y(x_1,...,x_{i-1}, x_i + \Delta, x_{i+1},...,x_k) - y(\mathbf{x})}{\Delta} \tag{11}$$

where Δ is a predetermined multiple of $1/(p-1)$, and $\mathbf{x} = (x_1, x_2, ..., x_k)$ is any selected value in Ω such that the transformed point $(\mathbf{x} + \mathbf{e}_i\Delta)$, where $\mathbf{e}_i$ is a vector of zeros but with one as its ith component, is still in Ω for each index $i = 1, ..., k$. In this study, accident cost y is the output of the model. The input is a 99-dimensional vector ($k = 99$), since there are 99 decontamination-related parameters to be examined. The number of levels p and Δ are set to 10 and 5/9, respectively. The number of runs r for each component X_i is set to 20. The way to determine p, Δ and r can be referred to in Saltelli et al. [27].

In each run, $\mathbf{x}$ is randomly selected from $\mathbf{X}$, $y(x_1,...,x_{i-1}, x_i + \Delta, x_{i+1},...,x_k)$ and $y(\mathbf{x})$ are then estimated, and the elementary effect $d_i(\mathbf{x})$ is consequently calculated. To calculate the accident cost y, all x_is are used as the percentile to pick up a value from the distribution of the ith parameter (the sequence of the parameters is defined in Table 1). After the rth run of the kth component X_k, the average of the absolute values of the elementary effects μ^*, and the standard deviation of the elementary effects σ, of each component X_i are calculated using

$$\mu^* = \sum_{i=1}^{r} |d_i|/r \text{, and} \tag{12}$$

$$\sigma = \sqrt{\sum_{i=1}^{r} (d_i - \mu)^2 / r} \text{, where } \mu = \sum_{i=1}^{r} d_i/r \text{.} \tag{13}$$

Both μ and μ^* can be used as an indicator of the importance of the parameter. μ^* is preferable to μ because $d_i(x)$s can give negative value and some effects may thus cancel each other out when computing the average if μ is used [22]. σ can be used to indicate the extent of interactions of the parameter with other parameters.

4.2. Determination of parameter distributions

Distributions of 99 parameters are formed base on the information obtained from literatures and updates from the Fukushima accident [such as 22-24]. Distributions of parameters of high importance

Table 2 Distributions of important parameters

No.	Parameter	Type of Distribution	Min.	Max.	Remarks
1	Dose of setting decontamination target area [mSv/year]	Discrete	1	20	4 annual dose rates (1, 5, 10 and 20) with same probability density (P(x) = 0.25).
55	Determination whether or not to include cost due to waste disposal	Discrete	0	1	[0, 0.5) = no/[0.5, 1) = yes.
60	Number of workers that can be involved in the decontamination work [man-year/year]	Uniform	5000	50000	Determined by the evaluator.
56	Unit cost of waste disposal [JPY/m^3]	Uniform	650000	3018000	
36	Waste generated by removing soil or covering with soil [m^3/m^2]	Uniform	0.000	0.079	

(parameters of which μ*s shown in Fig. 3a are the first to the fifth largest) are presented in Table 2 as examples.

4.3. Results & Discussion

The results of the sensitivity analysis, i.e., the μ*s and the σs of each parameter, are shown in Fig. 3a and 3b. Fig. 3a shows the overview for all parameters, where the graph is zoomed in in Fig. 3b to visualize parameters with small μ* and σ. The numbers in the graphs correspond to the parameter numbers in Table 1. It is observable from the figure that μ* correlates strongly with the σ, i.e., parameters with high importance tend to have large extent of interactions with other parameters. In this paper, the discussion will be done based only on μ* as it may be able to roughly represent the discussion on σ.

It is obvious from Fig. 3a that the dose of setting decontamination target area (1) is very influential to the accident cost as it determines the size of the decontamination target area. Fig. 3a also shows that parameters related to waste management also have very high importance since there are four waste management-related parameters (53, 55, 56 and 58) of which the μ*s and the σs are over 0.20. Very high μ*s and σs of 55 and 56 emphasize the importance of consideration of costs due to waste disposal

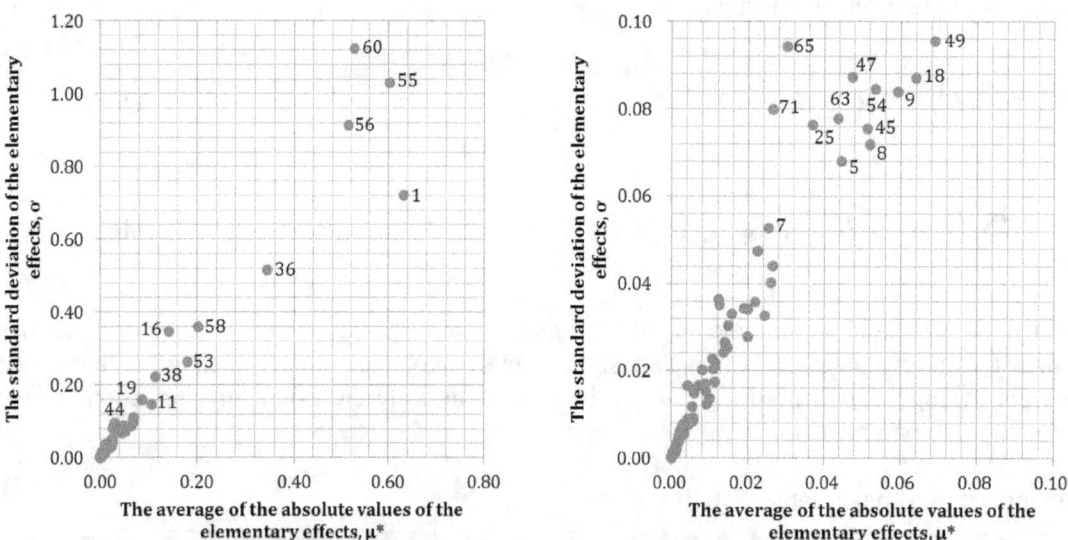

Fig. 3a (left) μ*s and σs of all parameters
Fig. 3b (right) Zoomed-up version of Fig. 3a to the region where $0 \leq \mu* \leq 0.10$ and $0 \leq \sigma \leq 0.10$
This figure shows all non-negligible parameters other than those shown in Fig. 3a

which is omitted in many earlier studies. This implies that in spite of inadequate information on accurate parameter values, it is important to consider the costs due to waste disposal in the estimation of accident cost, i.e., the estimation of the accident consequences. Another very important parameter is the number of workers that can be involved in decontamination work (60) as it directly affects the relocation period. The volumes of waste generated per unit area by decontamination techniques which generate a lot of waste (36, 38 and 44) also seem to be important as it influence the total amount of the waste. It is also observed that fractions for application of decontamination techniques with high unit cost (11 and 16) can be quite influential to the output. Lastly, the large μ^* of the parameter 19 implies that when the distributions of unit costs of all decontamination techniques are taken into account simultaneously, they may have large effect on accident cost. As for these parameters, more raw data collection is needed for parameters of which distributions are formed by limited number of data points. Further discussion on quality of the data collected or consultation with stake holders to determine the distributions on the specific values to represent the respective parameters may also be needed.

Fig. 3b shows that the influences of: fractions for application of many decontamination techniques (5, 7, 8 and 9), parameters that determine the unit cost of some decontamination techniques (18 and 25), waste management-related parameters other those stated above (45, 47, 49 and 54), and, work speeds of some decontamination techniques (63, 65 and 71), are not negligible (both μ^* and σ are over 0.05). Parameters that did not appear above are theoretically negligible, and can be fixed to constants in order to simplify the model and to reduce the calculation time. It is interesting that none of parameters that affect radiation effect cost, which was the largest component of accident cost in Section 2, are influential to accident cost. One possible reason is that much larger waste management cost and longer relocation period significantly increased the decontamination cost and the relocation cost, respectively, and the radiation effect cost became relatively smaller. However, it is to be noted that distributions of many radiation effect cost-related parameters, e.g., WTP per unit exposure, were not taken into account. Taking them into account may significantly increase the importance of radiation effect cost-related parameters to the accident cost. Similarly, other parameters with low μ^*s and σs must be carefully examined before fixing to a constant.

5. CONCLUSION

The calculation scheme of the accident cost, which is an index for severe accident consequence assessment, was introduced. The authors pointed out the needs of improvements of the calculation scheme, including data collection and further consideration of important parameters, and a global sensitivity analysis of the model. This study focused on the consideration of decontamination model. The decontamination model was formulated using decontamination-related parameters that directly affect the three important cost components: the decontamination cost, the relocation cost and the radiation effect cost. Distributions of all parameters were set, and a sensitivity analysis was performed to identify parameters with large influence to accident cost calculation and large extent of interactions with other parameters. Parameters with high importance tend to have large extent of interactions with other parameters. Parameters that are influential to the accident cost are: the dose of setting decontamination target area, a number of waste management-related parameters, the number of workers that can be involved in decontamination work, the volumes of waste generated per unit area by decontamination techniques which generate a lot of waste, the fractions for application of decontamination techniques with high unit cost, and the common random number when the same random number is used for calculation of unit costs for all decontamination techniques. Further studies, e.g., more raw data collection for some parameters, further discussion on quality of the data collected, or consultation with stake holders, may be needed for these parameters.

References

[1]   T. Homma et al., "*Radiological consequence assessments of degraded core accident scenarios derived from a generic level 2 PSA of a BWR*", JAERI-Research-2000-060, JAERI, 2000, Ibaraki.

[2]   J. C. Helton et al., "*Uncertainty and sensitivity analysis of chronic exposure results with the MACCS reactor accident consequence model*". Reliab. Eng. Syst. Saf., 50, 137-177, (1995).

[3]   T. Haste et al., "*MELCOR/MACCS simulation of the TMI-2 severe accident and initial recovery phases, off-site fission product release and consequences*", Nucl. Eng. Des., 236(10), 1099-1112, (2006).

[4]   NEA, "*Probabilistic risk criteria and safety goals*", NEA/CSNI/R(2009)16, 2009.

[5]   EC, "*ExternE: Externalities of energy. Methodology 2005 update*", 2005, Luxemburg.

[6]   S. Hirschberg et al., "*Severe accidents in the energy sector, First edition, Project GaBE: comprehensive assessment of energy systems*", PSI Bericht Nr. 98-16, Paul Scherrer Institut, 1998, Switzerland.

[7]   IAEA. "*Health and environmental impacts of electricity generation systems: procedures for comparative assessment*", Technical reports series No. 394, 1999, Vienna.

[8]   S. J. Park, "*Estimates of the economic consequences of a severe nuclear accident in Japan*", J. Natl. Econ., 191(3), 1-15, (2005).

[9]   K. Silva et al., "*Integration of direct/indirect influences of severe accidents for improvements of nuclear safety*", Proceedings of the 2012 20th international conference on nuclear engineering, 2012, California.

[10] K. Silva et al., "*Cost per severe accident as an index for severe accident consequence assessment and its applications*", Reliab. Eng. Syst. Saf., 123, 110-122, (2014).

[11] K. Silva et al., "*Estimation of cost per severe accident for improvement of accident protection and consequence mitigation strategies*", Proceedings of PSAM Topical Conference in Tokyo In light of the Fukushima Dai-ichi Accident, 2013, Tokyo.

[12] T. Homma et al., "*Uncertainty and sensitivity studies with the probabilistic accident consequence assessment code OSCAAR*", Nucl. Eng. Technol., 37(3), 245-258, (2005).

[13] ICRP, "*Low dose extrapolation of radiation-related cancer risk, Publication 99*", Elsevier, 2005.

[14] ICRP, "*Application of the Commission's recommendations for the protection of people in emergency exposure situations. Publication 109*", Elsevier, 2009.

[15] Committee for Nuclear Damage Compensation, "*Interim guideline for the determination of the scope of nuclear damage regarding accident on Fukushima Daiichi and Daini Nuclear Power Stations*", 2011, Tokyo.

[16] The Commission of Management and Financial Survey of TEPCO, "*Commission report*", 2011, Tokyo.

[17] J. Ishikawa et al., "*Systematic source term analysis for level 3 PSA of a BWR with Mark-II type containment with THALES-2 code*", JAERI-Research 2005–021, JAERI, 2005, Ibaraki.

[18] IAEA, "*Intervention criteria in a nuclear or radiation emergency. Safety series No. 109*", 2006, Vienna.

[19] ICRP, "*The 2007 recommendations of the International Commission on Radiological Protection. Publication 103*", Elsevier, 2007.

[20] NSC, "*Interim report of the concept for the revision of the radiation protection countermeasures of nuclear facilities*", 2012, Tokyo.

[21] Ibaraki Prefectural Government, "*The 2010 statistical yearbook of Ibaraki Prefecture*", 2010, Ibaraki.

[22] JAEA, "*Report of the results of the decontamination model projects*", 2012, Ibaraki.

[23] A. S. Nisbet et al., "Generic handbook for assisting in the management of contaminated inhabited areas in Europe following a radiological emergency, Version 2", EURANOS(CAT1)-TN(09)-03, 2010, UK.

[24] T. Yasutaka et al., "*A GIS-based evaluation of the effect of decontamination on effective doses due to long-term external exposures in Fukushima*", Chemosphere, (Accepted 29 June 2013).

[25] M. D. Morris, "*Factorial sampling plans for preliminary computational experiments*", Technometrics, 33(2), 161-174, (1991).

[26] F. Campologo et al., "*An effective screening design for sensitivity analysis of large models*", Environ. Model. Softw., 22, 1509-1518, (2007).

[27] A. Saltelli et al., "*Sensitivity Analysis in Practice, A Guide to Assessing Scientific Model*", Wiley, 2004, UK.

Safety of LPG rail transportation: influence of safety barriers

V. Busini*, M. Derudi, R. Rota

Politecnico di Milano, Department of Chemistry, Materials and Chemical Engineering "G. Natta",
Piazza Leonardo da Vinci 32, 20133 Milano, Italy

Abstract: The risk due to the road and rail transportation of liquefied petroleum gas (LPG) is well known. Severe scenarios were caused by road or rail accidents involving LPG pressurized tank cars. Consolidated approaches exist for the analysis, the prevention and the mitigation of risk due to the transportation of hazardous materials (HazMat) by road or rail. In Europe a specific regulation applies to the equipment used for the transport of HazMat and specific regulations apply to the qualification of equipment used for LPG transportation. Nevertheless, on June 29th, 2009, an extremely severe transportation accident involving LPG took place in the station of Viareggio, in Italy. A train carrying 14 tank cars of LPG derailed and several railcars overturned on the shunts in the Viareggio station. A tank was punctured, releasing its entire content that ignited causing an extended and severe flash-fire. The present study focused on the study of the effect of different parameters on the heavy gas cloud dispersion resulting from the accident, such as meteorological parameters and height of safety walls. It was found that, to be effective, the mitigation barriers must be carefully designed, with particular reference to their height with respect to the height of the heavy gas cloud.

Keywords: Hazardous Material Transportation, Consequence Analysis, CFD models, Heavy Gas Dispersion Modeling, LPG, Viareggio accident.

1. INTRODUCTION

Accidents involving the release of hazardous substances are a major concern in industrial and transportation risk assessment. They can lead to consequences with high magnitude as the formation of a hazardous cloud can spread over distances of kilometers and pose a risk to both human health and the environment. Urban areas are easily involved in the release of hazardous substances, not only because many industries are part of urban agglomerations as a result of the growth of cities, but also because the transport of substances by road and rail usually crosses the cities. The latter, though usually involves smaller amounts of substances, represents a serious hazard because the accident mitigation is less effective. In addition, the transport vehicles pass near highly vulnerable populated areas such as schools and hospitals. Such accidents in urban areas are therefore a highly dangerous scenario in terms of the magnitude of the consequences, worsen by the high population density present in these areas.

Urban areas are characterized by complex geometries because of the large number of buildings, of various shapes and sizes; these obstacles affect strongly the wind speed and direction because of the presence of trails, stagnant zones, recirculation and preferential pathways which may significantly complicate the scenario.

Accidental releases of hazardous substances have been studied since the '80s and have been investigated through the execution of tests of large spills and the development of numerical models. These models continue to be currently used for the study of the consequences of releases [1, 2], and some of them, such as DEGADIS, ALOHA, and UDM, are among the most widely used in engineering applications [3, 4]. These are lumped parameter models, usually one-dimensional and account for some physical phenomena using semi-empirical relationships whose parameters have been derived from field tests [5]. Since the experimental setup of these tests usually do not involve any obstacle, these models are able to provide reliable results in an open field, or when almost no obstacle occurs in the region covered by the cloud.

To analyze the effects of a multitude of obstacles with different geometries on the dispersion of the cloud, computational tools based on computational fluid dynamics (CFD) are necessary. This approach allows the complete three-dimensional analysis and forecasting velocity, temperature and

*valentina.busini@polimi.it

concentration fields in the domain of integration. Although this procedure may provide more detailed results, it requires a large amount of resources in terms of both computation time and skills of the analysts.

In previous works [6-8], the dispersion of the LPG cloud in Viareggio as it happened in 2009 accident has been simulated using a CFD model; the present work involves the study of the effect of different parameters on the heavy gas cloud dispersion, such as: atmospheric stability class, wind velocity, and presence of mitigation walls of different height.

2. MATERIALS AND METHODS

Computational fluid dynamics codes solve numerically the Navier–Stokes equations of motion (Eqs. (1) and (2)), the energy balance (Eq. 3) and the equation arising from the turbulence modeling [9].

$$\frac{\partial \rho}{\partial t} + \nabla \cdot (\rho \vec{v}) = 0 \tag{1}$$

$$\frac{\partial (\rho \vec{v})}{\partial t} + \nabla \cdot (\rho \vec{v} \vec{v}) = -\nabla p + \nabla \cdot (\bar{\bar{\tau}}) + \rho \vec{g} \tag{2}$$

$$\frac{\partial (\rho c_v T)}{\partial t} + \nabla \cdot (\rho \vec{v} c_p T) = \nabla \cdot (k_T \nabla T) \tag{3}$$

In the equation above, ρ is the density, t the time, v the velocity, p the pressure, τ the shear stress, g the gravity acceleration, c_v and c_p the specific heats, T the temperature and k_T the thermal conductivity.

In this work the k–ε model was used for representing the effects of the turbulence. This model introduces two additional transport equations for turbulent kinetic energy k (Eq. 4) and turbulent kinetic energy dissipation rate ε (Eq. 5), respectively [10]:

$$\frac{\partial (\rho k)}{\partial t} + \frac{\partial (\rho k v_i)}{\partial x_i} = \frac{\partial}{\partial x_j}\left[\left(\mu + \frac{\mu_t}{\sigma_k}\right)\frac{\partial k}{\partial x_j}\right] + G_k + G_b - \rho \varepsilon - Y_M + S_k \tag{4}$$

$$\frac{\partial (\rho \varepsilon)}{\partial t} + \frac{\partial (\rho \varepsilon v_i)}{\partial x_i} = \frac{\partial}{\partial x_j}\left[\left(\mu + \frac{\mu_t}{\sigma_\varepsilon}\right)\frac{\partial \varepsilon}{\partial x_j}\right] + C_{\varepsilon 1}\frac{\varepsilon}{k}(G_k + C_{\varepsilon 3}\,G_b) - C_{\varepsilon 2}\rho \frac{\varepsilon^2}{k} + S_\varepsilon \tag{5}$$

where v_i is the velocity component along x_i direction, μ the viscosity, μ_t the turbulent viscosity, G_k the shear stress-related turbulent kinetic energy production, G_b the buoyancy-related turbulent kinetic energy production, Y_M the compressibility-related kinetic energy production and S_k and S_ε are user-defined source terms. $C_{\varepsilon 1}$, $C_{\varepsilon 2}$, $C_{\varepsilon 3}$, σ_k, σ_ε and C_μ are empirical constants; in this work, Jones and Launder [11] values have been used.

The k–ε turbulence model was complemented with an Atmospheric Stability sub-Model (ASsM) [12], able to ensure the consistency of the CFD results with the Monin-Obukhov theory [8], by introducing a user-defined source term S_ε defined as (for neutral conditions):

$$S_\varepsilon(z) = \rho \frac{u_*^4}{z^2}\left[\frac{(C_{2\varepsilon} - C_{1\varepsilon})\sqrt{C_\mu}}{K^2} - \frac{1}{\sigma_\varepsilon}\right] - \mu \frac{2\,u_*^3}{K\,z^3} \tag{6}$$

Moreover, for the ground boundary conditions, the ASsM model uses a roughness constant, C_s, equal to 0.979 in the wall functions, and a surface shear stress, τ_w.

The domain is discretized through the use of a calculation grid allowing for transforming the partial differential equations into a system of algebraic equations.

The commercial package Fluent 12.1.2 [13] was used for all the computations, together with the boundary conditions summarized in Table 1.

.

Table 1: Summary of boundary conditions

Boundary	Type	Notes
wind Inlet boundary	velocity inlet	wind velocity, temperature and turbulence values for the wind inlet flux
wind Outlet boundary	pressure outlet	constant pressure outlet surface
top boundary	velocity inlet	wind velocity, tangential to the surface
ground boundary	wall	no slip boundary, roughness specification, fixed temperature; gas mass fraction was specified in the region of the pool
buildings walls	wall	flat and adiabatic
gas inlet boundary	mass flow inlet	mass flow, temperature and turbulence values for the gas flux

3. RESULTS

The meteorological condition at the moment of the accident in Viareggio were: stable class (F), wind speed 0.7 m/s, direction E-SE; the final footprint of LPG cloud, that is at 300 s after the release, is sketched in Figure 1 for the sake of comparison. For further details and for the comparison with the field evidences the readers can refer to previous works [6-8]. Because of the agreement found with the field evidences, we will refer in the following to this case as the "real case".

Figure 1: LFL (black) and UFL (red) footprints at 300 seconds

In this work, a neutral stability class with wind velocity of 5 m/s at 10 m (5D) was considered with two different wind directions: the actual wind direction of the Viareggio accident (E-SE) and the opposite one (W-NW). For these two wind direction three different domains were considered: the first one represents the domain of Viareggio as it was the day of the accident; the second one is the first one with a wall 2.5 m high on the east side of the station added; in the third domain two continuous walls 4.5 m high on both sides of the station (east and west) were added.

3.1. The real case with different atmospheric conditions

The results in Figure 2 are relative to the case with the actual wind direction of the Viareggio accident (E-SE) and show how a higher wind velocity, with respect to the real case shown in Figure 1, entails a higher dispersion of the LPG leading to a different and less widespread footprint of the cloud. The figure also shows that no upwind spreading of the cloud is present in this case: therefore, less houses are involved and they are located only on the east side.

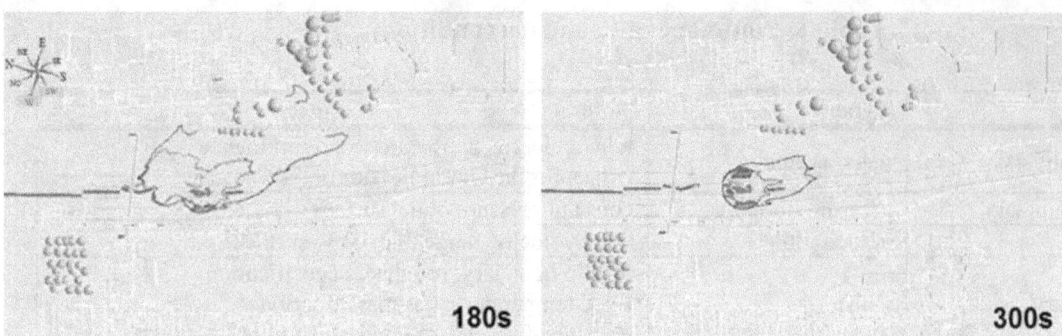

Figure 2: LFL (black) and UFL (red) footprints at 180 and 300 seconds for 5D E-SE atmospheric conditions

Figure 3, which is relative to the case with wind direction W-NW, shows how the already existing walls on the west side (2.3 meters high) are able to contain the cloud. Contrary to the previous case, we can see that the cloud extends only to the west side.

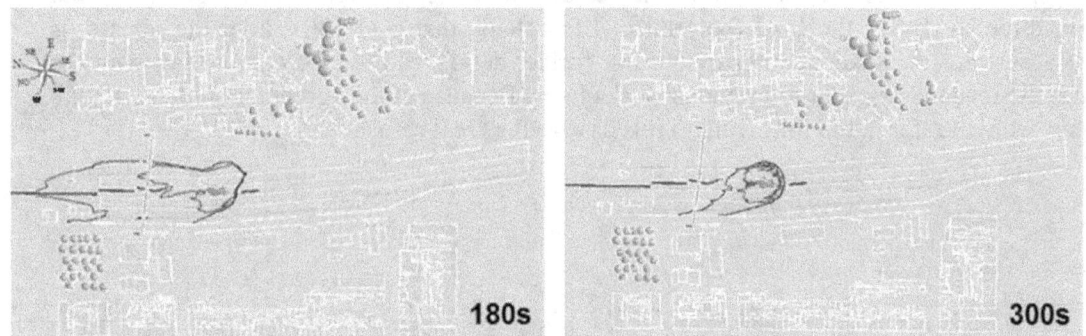

Figure 3: LFL (black) and UFL (red) footprints at 180 and 300 seconds for 5D W- SW atmospheric conditions

3.2 The case with 2.5 m high mitigation wall on the east side

Successively, a configuration with a mitigation wall 2.5 m high added on the east side was investigated. As evident from the results summarized in Figure 4, this mitigation wall is not able to contain completely the cloud that extends to the first to rows of houses.

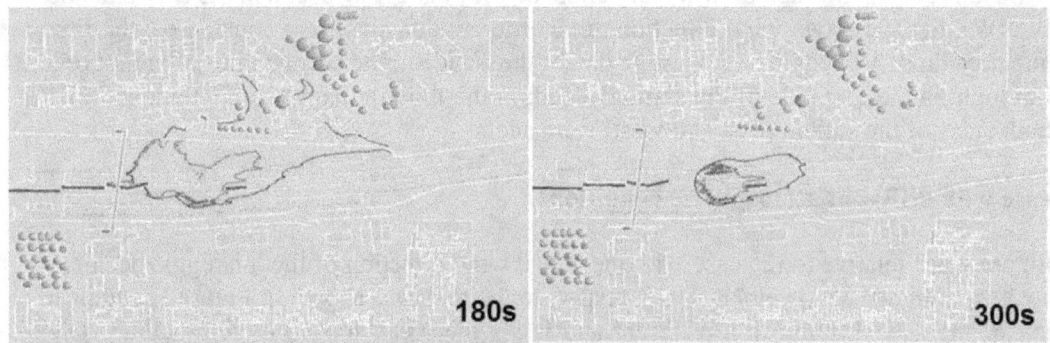

Figure 4: LFL (black) and UFL (red) footprints at 180 and 300 seconds for 5D E-NE atmospheric conditions and a wall 2.5 m high on the east side

The simulation performed for the W-SW wind direction (Figure 5) resulted in a footprint similar to that shown in Figure 3.

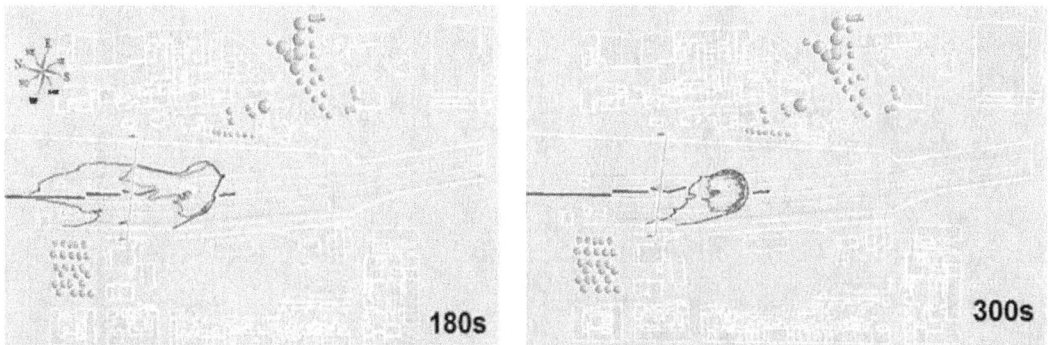

Figure 5 LFL (black) and UFL (red) footprints at 180 and 300 seconds for 5D W-SW wind atmospheric conditions and a wall 2.5 m high on the east side

The different behaviours of the cloud in the last two cases, that is the wall on the west side is effective while the one on the east side even though with the same height is not, arise from a canalization of the wind in the latter case, as shown in Figure 6.

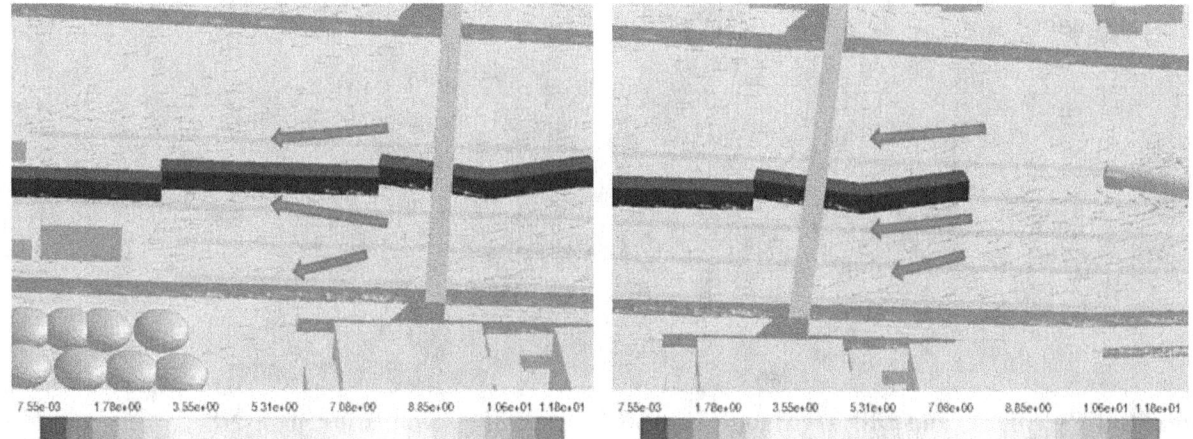

Figure 6: Velocity vectors along the train (left) and near the punctured tank (right) at 1 meter above the ground

We can see that the cloud is divided into two parts by rail tankers and thus channelled along the tracks following the line of rail tank cars; in this way a smaller portion of the cloud gets to the west wall.

3.2 The case with 4.5 m high mitigation walls on both sides

In the last configuration considered, a wall of 4.5 m was added on both sides. The height was chosen on the bases of the general criterion that a mitigation wall is effective in reducing the hazardous distance from a cloud dispersion when its height is comparable to that the cloud reaches without the mitigation wall [14]. Figure 7 shows that the height of the cloud in proximity of the mitigation wall position is about 4.5 m (as computed from the absolute height reported in the figure minus the altitude of Viareggio, which is equal to about 2.5 m on the sea level).

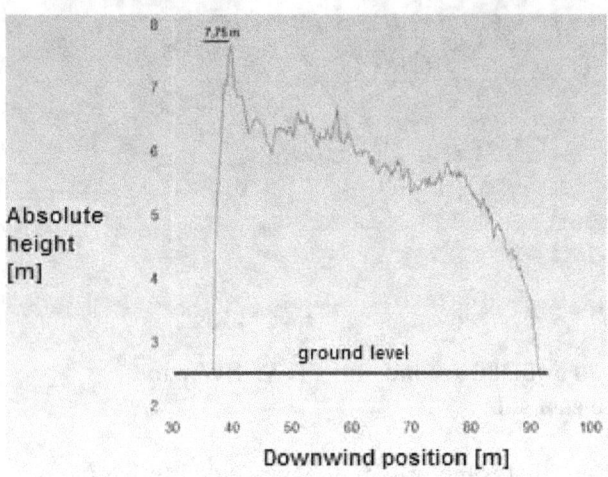

Figure 7: Downwind cloud height

With the presence of these walls, the cloud can be confined both on the east side (Figure 8) and on the west side (Figure 9).

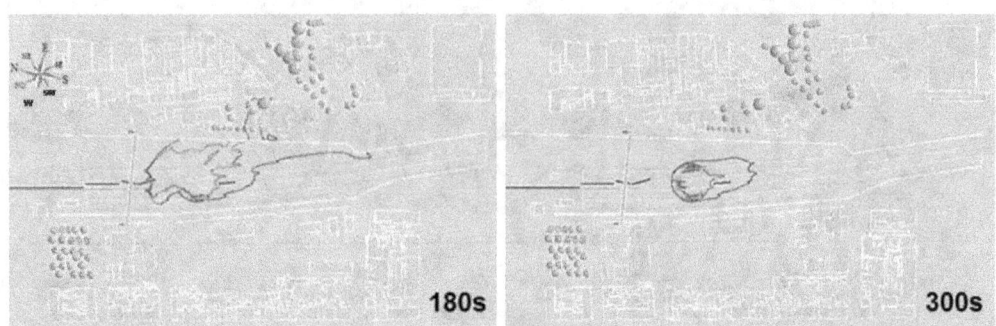

Figure 8: LFL (black) and UFL (red) footprints at 180 and 300 seconds for 5D E-NE atmospheric conditions and two walls 4.5 m high on both sides

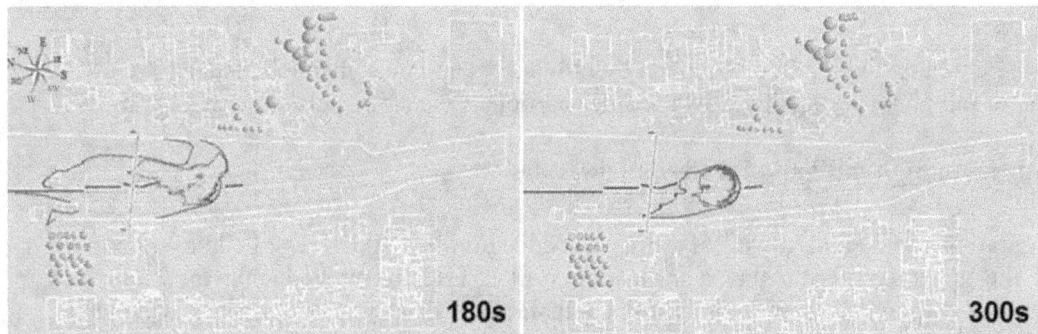

Figure 9: LFL (black) and UFL (red) footprints at 180 and 300 seconds for 5D W-SW atmospheric conditions and two walls 4.5 m high on both sides

Figure 9 also shows a peculiar behavior of the cloud footprint in the sketch at 180 s: the 4.5 m additional wall on the east side creates a special situation leading to the formation of a concentration area with the shape of "lagoon" near to the east wall. This behavior can be explained analyzing the velocity vectors on the ground and at 1 m height (Figure 10): they show that the wind has two main direction, one along the train and another along the wall.

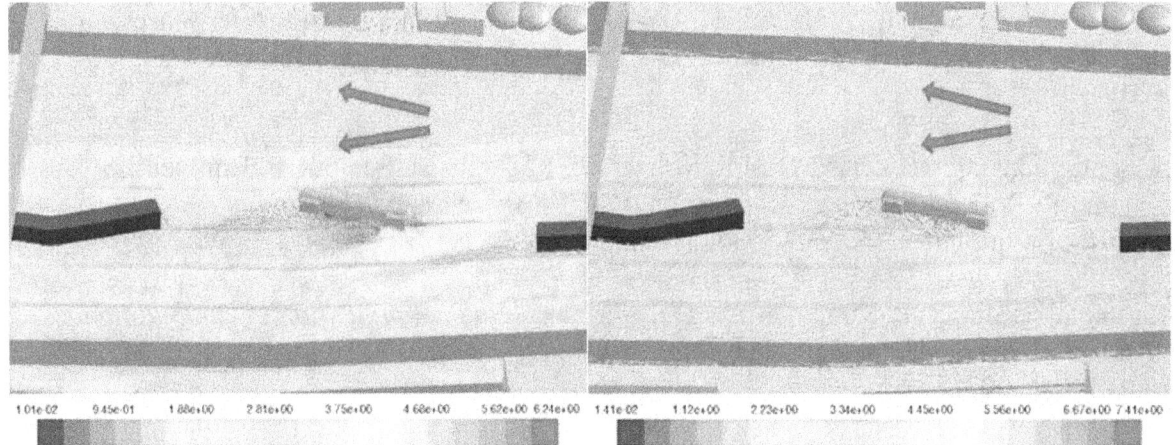

Figure 10: Velocity vectors at ground level (left) and 1 meter above the ground (right)

4. CONCLUSION

This work was aimed at studying the effect of different atmospheric conditions as well as the effectiveness of mitigation barriers of various heights on the dispersion of clouds produced by the release of LPG in urban areas.

As expected, a higher wind velocity together with a neutral atmospheric stability class entails a higher dispersion of the LPG leading to a different and less widespread footprint of the cloud, while the effect of introducing mitigation barriers is harder to predict.

The examined case studies derived from the real case of the accident of Viareggio considering two wind directions in three different geometrical domains: the first one represents the domain of Viareggio as it was the day of the accident, the second one is the same domain with a wall 2.5 m high on the east side of the railway station, while the third domain involves two continuous mitigation walls 4.5 m high on both sides of the railway station (namely, east and west sides). It has been found that in all the domains with the E-NE wind direction the cloud bypasses the mitigation barriers, while for the W-SW wind direction the cloud is always contained by the mitigation barriers thanks to a channeling of the cloud along the track.

These results clearly indicate the importance of carefully investigating through CFD tools the effect of introducing a mitigation barrier.

References

[1] M. Nielsen and S. Ott. "*A collection of data from dense gas experiments*", (1996).

[2] S. R. Hanna and R. E. Britter. "*Wind Flow and Vapor Cloud Dispersion at Industrial and Urban Sites*". 2002.

[3] A. Bernatik and M. Libisova. "*Loss prevention in heavy industry: risk assessment of large gasholders*", Journal of Loss Prevention in the Process Industries,17:271-278, (2004).

[4] D. R. Brook, N. V. Felton, C. M. Clem, D. C. H. Strickland, I. H. Griffiths and R. D. Kingdon. "*Validation of the Urban Dispersion Model (UDM)*", International Journal of Environment and Pollution,20:11-21, (2003).

[5] M. Fingas. "*The Handbook of Hazardous Materials Spills Technology*". 2002.

[6] V. Busini, M. Pontiggia, M. Derudi, G. Landucci, V. Cozzani and R. Rota. "*Safety of LPG Rail Transportation*", Icheap-10: 10th International Conference on Chemical and Process Engineering, Pts 1-3,24:1321-1326, (2011).

[7] G. Landucci, A. Tugnoli, V. Busini, M. Derudi, R. Rota and V. Cozzani. "*The Viareggio LPG accident: Lessons learnt*", Journal of Loss Prevention in the Process Industries,24:466-476, (2011).

[8] M. Pontiggia, G. Landucci, V. Busini, M. Derudi, M. Alba, M. Scaioni, S. Bonvicini, V. Cozzani and R. Rota. "*CFD model simulation of LPG dispersion in urban areas*", Atmospheric Environment,45:3913-3923, (2011).

[9] A. Luketa-Hanlin, R. P. Koopman and D. L. Ermak. "*On the application of computational fluid dynamics codes for liquefied natural gas dispersion*", Journal of Hazardous Materials,140, (2007).

[10] M. Pontiggia, M. Derudi, M. Alba, M. Scaioni and R. Rota. "*Hazardous gas releases in urban areas: Assessment of consequences through CFD modelling*", Journal of Hazardous Materials,176:589-596, (2010).

[11] G. S. W. Hagler, M. Y. Lin, A. Khlystov, R. W. Baldauf, V. Isakov, J. Faircloth and L. E. Jackson. "*Field investigation of roadside vegetative and structural barrier impact on near-road ultrafine particle concentrations under a variety of wind conditions*", Science of the Total Environment,419:7-15, (2012).

[12] V. Busini, M. Lino and R. Rota. "*Influence of Large Obstacles and Mitigation Barriers on Heavy Gas Cloud Dispersion: a Liquefied Natural Gas Case-Study*", Industrial & Engineering Chemistry Research,51:7643-7650, (2012).

[13] ANSYS Inc. ANSYS Fluent 12 User's guide. 2009.

[14] M. Derudi, D. Bovolenta, V. Busini and R. Rota. "*Heavy gas dispersion in presence of large obstacles: selection of modeling tools*", Industrial & Engineering Chemistry Research (10.1021/ie4034895).

Determination of Target Reliability Levels Based on Value to the Customer and Warranty Budgets

Michael Bartholdt[a*], Volker Schweizer[a] and Bernd Bertsche[a]

[a] University of Stuttgart, Stuttgart, Germany

Abstract: The method presented here serves to determine the system's as well as subsystems' target reliability levels combined. It centers the functions of the product to be developed and knows to weight requirements in line with the voice of the customer. Each subsystem's target reliability level is defined in accordance with its quantitative contribution to fulfilling the functions as desired by the customer. Statistically inevitable failures before the targeted product lifetime are often compensated by warranty and good-will expenditures. These costs are methodologically broken down and allocated to each subsystem purposefully in order to achieve the utmost customer satisfaction. By bringing together such costs per subsystem and its importance to the customer, reliability goals are obtained aligned with the value to the customer.

In contrast to most of the existing methods to define and allocate reliability goals (typically realized by two different methods), subsystem target reliability levels are defined at first. The system's reliability goal is then calculated by means of Boole-theory. Arbitrarily complex systems can be analyzed.

Keywords: Reliability Allocation, Reliability Optimization, Reliability Goal, Target Reliability Level, System Reliability

1. INTRODUCTION AND MOTIVATION

Reliability engineers are often called upon to make decisions on the product's goal reliability [1,2]. The question of how to meet this goal leads them to a reliability allocation problem. Which component needs to perform how reliably in order to meet the system's reliability goal and assure an economically competitive development at the same time?

A given product can be considered as a means of providing the functionality meeting the customer's needs. In these terms, the task of product development is to design the product so that it is able to perform its functions reliably. However, not all components of a given system need to be equally reliable. Neither do they contribute to a given function equally nor are the functions equally important. The omnipresent economical constraints in competitive environments require the development to allocate resources in line with the customers' requirements.

Product reliability has repeatedly been specified as the most important order winning criteria for complex products [3]. It seems reasonable to integrate the customers' voice into the determination of the product's reliability level.

In an effort to achieve the utmost customer satisfaction, statistically inevitable failures before the targeted product lifetime are compensated by warranty and good-will whose costs commonly make up about 10% of the company's turnover, i.e. about as much as the targeted profit margin [3]. To let reliability goal determination emerge from given warranty-related expenditures is therefore logical.

As will be discussed, the process of determining a system's goal reliability and allocating subsystem reliability goals is currently a two-step process. First, the system's goal is determined for which cost considerations often serve as the basis (cf. Chapter 2.1). The system's goal is then broken down into goals for each subsystem (cf. Chapter 2.2). Established methods do not merge the two steps into one.

Another reoccurring problem is the requirement of various factors which are empirically determined. Furthermore, the customer's point of view is hardly incorporated in neither the system's nor the subsystem's goal reliability determination. The customers' requirement regarding the functions of the product are not adequately taken into account either.

---

[*] michael.bartholdt@ima.uni-stuttgart.de

2. STATE-OF-THE-ART

Section 2.1 outlines current best practices for defining a system's reliability level. In sections 2.2., the most common methods for deriving target reliability levels of subsystems from the system's reliability goal are briefly reviewed in order to analyze what they have in common and what they lack, whereupon the problem description and motivation for the proposed method (cf. Chapter 3) is based.

2.1. Methods for defining system reliability goals under financial considerations

Different methods exist which aim at optimizing the relationship between costs and reliability levels. They help at defining economically sensible reliability goals. The most common methods for defining the system's reliability goal are listed in Table 1 [4].

Table 1: Overview of Approaches Optimizing Reliability Levels Under Financial Considerations

Approach by	Characterization
Selivanov	Product costs over the useful life (including maintenance costs)
Churchman	Maintenance costs with interest and acquisition costs
Brunner	Product costs under consideration of MTBF and the minimization of reliability levels
Köchel	Maximizing system reliability by means of redundancies
Kapur	Effort minimization algorithm
Kohoutek	Reliability costs as the sum of costs of development, manufacturing and warranty costs
Schnegas	Costs and reliability in the context of product definition

The approach by Selivanov considers the total costs of a technical product over its useful life and herein includes investment costs, costs proportional to the operating time as well as costs increasing with operation time (e.g. maintenance costs). This approach lacks the possibility to be used in early development stages as the two latter costs cannot be defined until well into the useful life.

Churchman takes into account initial costs as well as maintenance costs including interest. Here again, the connection to reliability is only given through maintenance costs, which have to be known. The reliability structure is not considered.

The approach by Brunner is based on finding the cost-optimal mean time between failures (MTBF) [5]. Considered costs are product costs in general as well as maintenance costs. The formula includes costs dependent on (desired) reliability levels. The determination of some needed factors requires expert knowledge which limits its application.

Köchel suggests a method to determine the financially optimal number of redundant identical subsystems. Apart from the redundancies, the product's reliability structure has to be serial. The method is suitable for early development stages.

Kapur's method requires an effort minimization algorithm which aims at minimizing the development effort by means of optimal component reliability levels. The definition of a threshold reliability level for each subsystem is the basis for deciding whether or not the individual reliability level needs to be improved. The reliability structure has to remain serial. Hartig [5] describes how such an effort minimization algorithm can be implemented and points out limits to the method.

Kohoutek's method is the only one which explicitly incorporates warranty costs. Costs of reliability also comprise costs of reliability design as well as costs of reliability in manufacturing, for which an empirical function needs to be defined. Practical applicability is limited due to the lack of the required quality function [4,5]. Furthermore, the product's reliability structure is not incorporated in the model.

Schnegas [4] describes a model according to which the material costs of a unit are calculated as the product of the volume and a cost-factor. Similarly, manufacturing costs are calculated. By comparing load levels to the geometry withstanding the load, reliability relevant information is obtained since stress levels relate to reliability and geometry relates to costs. [4,5,6]

2.2. Methods for allocating subsystem reliability goals

Two groups of methods can be distinguished: unweighted and weighted methods. Table 2 summarizes the most common methods.

Table 2: Overview of Approaches Apportioning Reliability Levels to Subsystems

Approach by/ name	Characterization
Equal Apportionment	Unweighted method. Each subsystem is assigned with the same reliability level.
AGREE	The importance of the subsystem to the system is considered in terms of its contribution to the total time of application.
ARINC	New failure rates are calculated based on pre-knowledge and the assumption that their share toward the system reliability remains the same.
Feasibility of Objectives	Empirically determined factors weigh the contribution of a subsystem to system reliability.
Karmiol	Empirically determined factors weigh the contribution of a subsystem to system reliability, consideration the subsystems' functional importance.
Bracha	Factors calculated based on partly empirically determined parameters weigh the contribution of a subsystem to system reliability.
Mettas	Based on the monetary effort to improve existing subsystem reliability levels, new goal reliability levels are derived for each subsystem.

Unweighted Method

The Equal Apportionment method allocates subsystem reliability targets assuming equal failure rates for each subsystem leading to the given system reliability level [7]. It is the most simple of all allocation methods. The system reliability needs to be known, the failure rates of all components are assumed to be constant (exponential failure distribution) and the system's reliability structure to be either completely serial or completely parallel.

Weighted Methods

Weighted Methods share the commonality that they are able to derive subsystems' reliability targets more accurately from the known system reliability level than the unweighted method described above due to using weighting factors. These factors stem from different scopes of consideration. The input information can be known or assumed. Weighted methods are more realistic and therefore have higher practical relevance [6,7]. Like the unweighted method, the weighted methods have in common that the system's reliability level (i.e. target) is known up front, based on which subsystems reliability levels (i.e. targets) are allocated.

The AGREE Method considers the total number of subsystems (N_{AGREE}) (1st tier) as well as the number of subsystems n_i within a subsystem (2nd tier) [6,7]. Since subsystems and sub-subsystems are considered in the calculation, the system's complexity is included in the derivation of the subsystems' reliability targets (1st tier). The importance factor E_{AGREE} incorporates the importance from a time input point-of-view for each subsystem, referenced to the total time of system application. All subsystems' failure distributions need to be exponential and the system's reliability structure to be serial.

The fundamental assumptions of the ARINC method are again, that the system is strictly serial as well as that all subsystem's failure distributions are exponential (i.e. have a constant failure rate) [7]. Additionally, the subsystems' failure rates need to be known (or assumed). Calculated are allowable failure rates λ_i' based on the system's new maximum allowable failure rate λ_{system} by means of weighting past failure rates λ_i of subsystems i to their sum (which equals the old system's failure rate). The per-cent contribution of the new failure rate to the new system's failure rate thus remains the same as before.

As for the Feasibility of Objectives method, each of the n subsystems' failure rate λ_i is the product of the system's failure rate with a factor, similarly to the ARINC method. The factor is the product of a set of four parameters which evaluate a given subsystem's complexity, state-of-the-art, operation time and environment qualitatively in the range of e.g. 1 to 10 requiring expert knowledge. This method also requires the system to consist of serially connected subsystems with constant failure rates [4].

The Karmiol method allows for describing serial systems where individual components can have identical redundancies. The failure distribution has to be exponential, meaning that the failure rate is constant. Similar to the Feasibility of Objectives method, different factors are used which influence the allocation of reliability goals to subsystems: complexity, criticality, state-of-the-art and operational profile (i.e. operation time and environmental exposure). Again, they are evaluated qualitatively in the range of 1 to 10. Additionally, a subsystem's relevance in terms of the number of its functions F_i is considered.

Bracha's method assumes that four factors influence the reliability allocation to subsystems: sublevel complexity, state-of-the-art, environmental conditions and operation time. Failure rates need to be constant again. Kececioglu describes in detail how to determine the factors a, c, e, and t which relate to each other [7]. The Bracha method is also able to cover systems in which individual subsystems have identical redundancies similar to the possibilities of the Karmiol method (and practically all of the methods discussed, e.g. by summarizing redundant parts into one by means of Boole-theory). [6,7]

Mettas' method [8] takes into account the effort (i.e. costs) for achieving certain reliability goals and allocates them accordingly. Subsystem reliability levels need to be known. The method can therefore not be applied for the development of new products unless the factors can be estimated in a substantiated way. In an effort to find the financial optimum, the sum of reliability related costs has to be minimized. The method requires some heavy assumptions to be made, e.g. factor f, which stands for the feasibility of increasing the reliability of subsystem i. It influences the relationship between costs and achievable reliability levels (cf. [8]). Some important aspects are not covered by this method either, such as the operation time or the importance to the customer [6].

3. PROBLEM DESCRIPTION

Different methods exist for determining the costs of reliability efforts and some are even able to link the financial effort with actual reliability goals for the system (such as the model of e.g. Brunner, Kapur). Only one method takes into account warranty costs as reliability related costs (i.e. Kohoutek). Many consider MTBF (Brunner) or maintenance frequencies (Churchman) as the reliability goal striven for. Some are too general (e.g. Selivanov) while others are too specific (e.g. Köchel). Most require significant levels of expert knowledge to determine the factors needed (e.g. Selivanov, Kapur, Kohoutek, Brunner) and most are limited in terms of reliability structures as they are able to cover only serial or only parallel structures but no combinations. The methods by Selivanov and Churchman only incorporate the product's reliability through maintenance costs. Others (Brunner, Kapur and Kohoutek) require parameters which are hard to determine and often have to be estimated.

Similarly, the methods for assigning subsystem reliability goals have significant limitations. Almost all of the most common methods as discussed above are limited to either strictly serial or strictly parallel systems. They allow individual components to be redundant with the same component. The application of practically all weighted methods requires considerable knowledge for estimating the factors needed, which limits their application, especially in early development stages. The Mettas method additionally requires the subsystem's current reliability levels to be known (which makes it an optimization method rather than an allocation method).

The method by Mettas is the only one which is able to cover complex system structures and as well as the only one considering costs directly as a factor influencing reliability allocation (cf. [8]). Yet it does not incorporate an "importance factor" representing the customers' point-of-view (such as the AGREE method) [6]. All but the method by Mettas are only able to cover (sub)systems with constant failure rates. The unweighted method of equal apportionment is generally considered too imprecise [6] but may help defining first reliability values for e.g. the Mettas method.

Based on these shortcomings of the existing methods, the method proposed in this paper offers a widely applicable approach to quantify the system's and subsystems' reliability goals based on the value to the customer while at the same time considering financial constraints from the manufacturer's point-of-view. Its goals therefore are the following:

- Accounting for the "voice of the customer" to assure designing the product that the customer actually wants, i.e. customer-driven target definitions.
- Merging the two steps for finding subsystem reliability goals into one step in order to facilitate the application.
- Reversing the process: first, subsystem reliability goals are to be defined first which then lead to the system's reliability level using the reliability structure (as this is more in line with the recommended process of product development where the product is developed based on the functions it is supposed to serve [7]).
- Putting the product's functions into the focus.
- Allowing for all kinds of system structures (i.e. combinations of serial and parallel).
- Incorporating the system reliability structure from early on as it has significant influence on economical resource allocation.
- Being independent from certain failure distributions (allow for non-constant failure rates).
- Assuring a straight-forward application without heavily relying on the estimation of parameters
- Applying target costing principles for defining the reliability goals instead of cost-counting principles: the benchmark for the desired reliability levels beside the customer is the market. The manufacturer should be able to scale efforts accordingly (i.e. the "voice of the manufacturer").

4. PROPOSED METHOD

In this chapter, the proposed method is described, starting with the assumptions taken as a basis. The flow chart (cf. Figure 1) will guide the reader through the method.

4.1 Assumptions

a. The product is regarded as a means of providing the functions the customer wants.
b. The maximum customer satisfaction is sought most economically.
c. Field usage of the product is considered the most critical phase to both failure behavior and ultimately customer satisfaction. The definition of target reliability levels is therefore linked to warranty budgets.
d. Budget allocation for the actual development of the product to the subsystems takes place in a separate step (not part of the proposed method; cf. e.g. [4,9]).
e. Customers' information on the desired functions is available or can be assessed.
f. The warranty budget and time, the product's parts list as well as the costs for rectifying a warranty claim are known.
g. The basic requirements for applying the Boole-theory are met.

4.2. Approach

Figure 1 illustrates the different steps of the proposed method which are described in this chapter. Each step as indicated in the flow chart is described below.

Step 1: The product's functions are defined and listed from the viewpoint of the customer and as input from development.

Step 2: The functions are ranked from the customer's point-of-view, e.g. by means of a pair-wise comparison, an evaluation grid or others. The result is a quantification of the product's functions' importance to the customer in percent.

Figure 1: Flow Chart of the Proposed Method

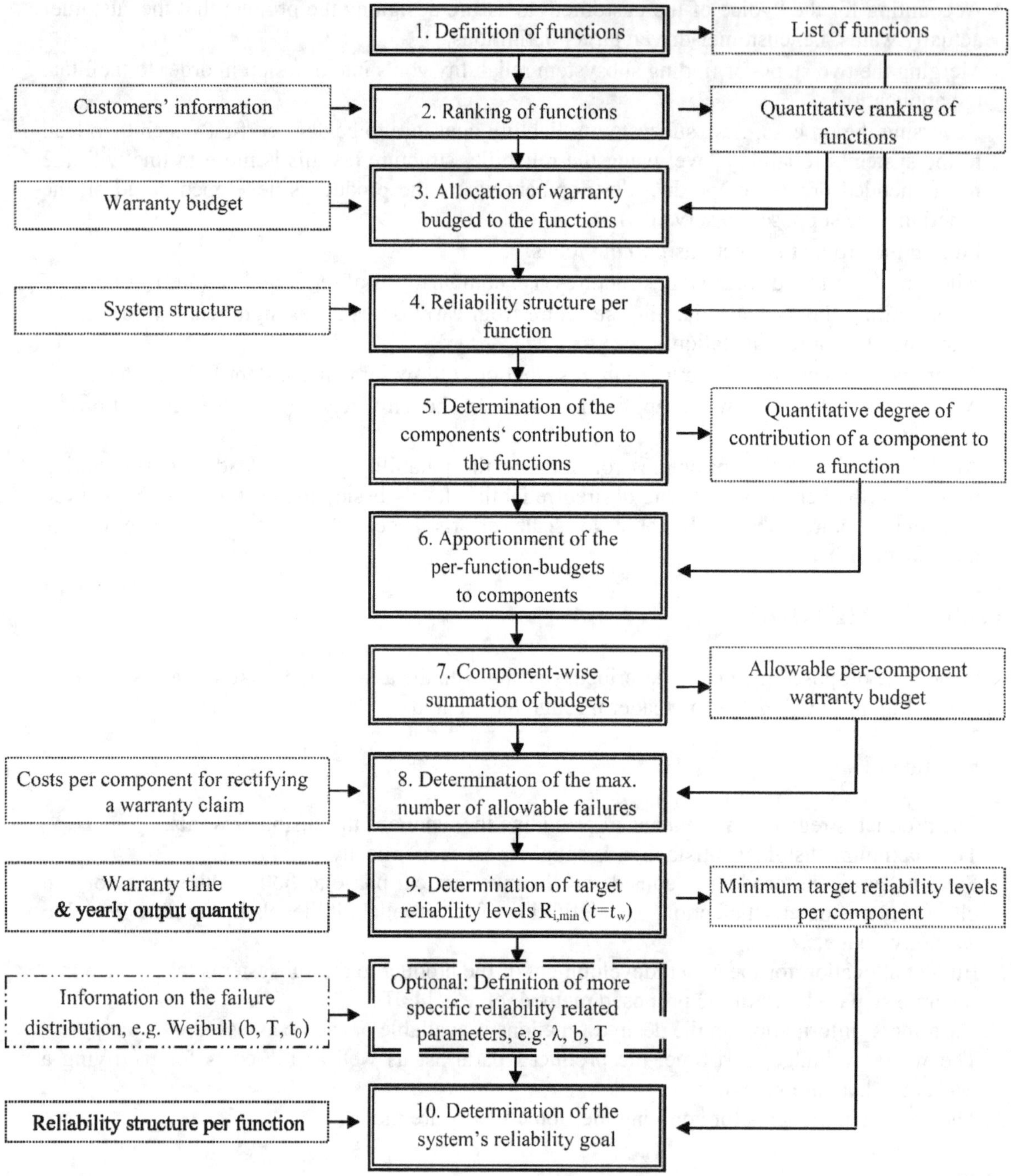

```
                              ┌─────────────────────┐        ┌─────────────────────┐
                              │ 1. Definition of    │───────▶│  List of functions  │
                              │    functions        │        └─────────────────────┘
                              └─────────────────────┘
                                        │                                │
┌─────────────────────┐       ┌─────────────────────┐        ┌─────────────────────┐
│ Customers' information│─────▶│ 2. Ranking of       │───────▶│ Quantitative ranking│
└─────────────────────┘       │    functions        │        │   of functions      │
                              └─────────────────────┘        └─────────────────────┘
                                        │                                │
┌─────────────────────┐       ┌─────────────────────┐
│  Warranty budget    │──────▶│ 3. Allocation of    │◀───────────────────┘
└─────────────────────┘       │    warranty budged  │
                              │    to the functions │
                              └─────────────────────┘
                                        │
┌─────────────────────┐       ┌─────────────────────┐
│  System structure   │──────▶│ 4. Reliability      │◀───────────────────┐
└─────────────────────┘       │    structure per    │
                              │    function         │
                              └─────────────────────┘
                                        │
                              ┌─────────────────────┐        ┌─────────────────────┐
                              │ 5. Determination of │        │ Quantitative degree │
                              │    the components'  │───────▶│ of contribution of a│
                              │    contribution to  │        │ component to a      │
                              │    the functions    │        │ function            │
                              └─────────────────────┘        └─────────────────────┘
                                        │                                │
                              ┌─────────────────────┐
                              │ 6. Apportionment of │◀───────────────────┘
                              │    the per-function-│
                              │    budgets to       │
                              │    components       │
                              └─────────────────────┘
                                        │
                              ┌─────────────────────┐        ┌─────────────────────┐
                              │ 7. Component-wise   │───────▶│ Allowable per-      │
                              │    summation of     │        │ component warranty  │
                              │    budgets          │        │ budget              │
                              └─────────────────────┘        └─────────────────────┘
                                        │                                │
┌─────────────────────────┐   ┌─────────────────────┐
│ Costs per component for │──▶│ 8. Determination of │◀───────────────────┘
│ rectifying a warranty   │   │    the max. number  │
│ claim                   │   │    of allowable     │
└─────────────────────────┘   │    failures         │
                              └─────────────────────┘
                                        │
┌─────────────────────────┐   ┌─────────────────────┐        ┌─────────────────────┐
│ Warranty time           │──▶│ 9. Determination of │───────▶│ Minimum target      │
│ & yearly output quantity│   │ target reliability  │        │ reliability levels  │
└─────────────────────────┘   │ levels R_{i,min}    │        │ per component       │
                              │ (t=t_w)             │        └─────────────────────┘
                              └─────────────────────┘
                                        │
┌─────────────────────────┐   ┌─────────────────────┐
│ Information on the       │──▶│ Optional: Definition│
│ failure distribution,   │   │ of more specific    │
│ e.g. Weibull (b, T, t_0)│   │ reliability related │
└─────────────────────────┘   │ parameters, e.g.    │
                              │ λ, b, T             │
                              └─────────────────────┘
                                        │
┌─────────────────────────┐   ┌─────────────────────┐
│ Reliability structure   │──▶│ 10. Determination of│◀───────────────────
│ per function            │   │ the system's        │
└─────────────────────────┘   │ reliability goal    │
                              └─────────────────────┘
```

Step 3: The importance levels are multiplied with the given warranty costs accounted for by product development, e.g. defined by means of target costing (benchmarking) or experience gained from previous products. The outcome is the warranty budget to be allocated to a given function in order to ensure the customer's requirements toward the function regarding reliability.

Step 4: A function is usually realized by several components interacting together. The system's parts list (considered as given) is mirrored according to the list of functions, e.g. by means of a matrix using a value managerial approach (cf. [10]). It can now be quantitatively determined how a component contributes to one or several given functions.

Step 5: Based on a component's position in the function-specific reliability structure, values are assigned to the different components function-wise. The value is one if the components are all in series. If the system is composed of e.g. two components which are redundant to each other, the influence on the function is 0.5 (50%) each, if three components are redundant, the influence is 0.33 (33%) each etc.

For a system where component 1 is serial to two components 2 and 3 which are parallel (cf. Figure 2), the value would be 0.5 (50%) for component 1 and 0.25 (25%) for component 2 (cf. Table 3).

Figure 2: Exemplary Reliability Structure for a Function

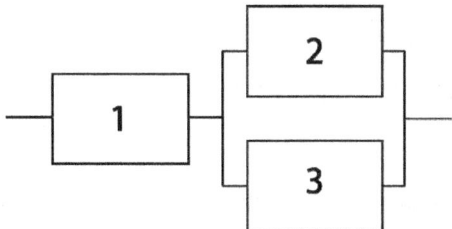

Table 3: Exemplary Determination of the Components' Contribution to the Function

	Component			
	1	2	3	Sum of the influence
Influence on the function (absolute value)	1	1/2	1/2	2
Influence (per-cent value)	0.5 (1/2=0.5 i.e. 50%)	0.25 (((1/2)/2) = 25%)	0.25 (25%)	1 (100%)

Step 6: Multiplying the percent value with the warranty budget allocated to the function (as determined in step 2) results in the warranty budget allocated to individual components in order to reliably fulfill the function of interest.

Step 7: The budgets apportioned to each component function-wise are now summed up component-wise resulting in the allowable per-component warranty budget.

Step 8: The allowable warranty budget of a component is divided by the expected total costs for rectifying one warranty claim (i.e. replacing the component etc.) filed by a customer. This data is considered as a given. These costs are known in case of a predecessor product or can be estimated in relation to its complexity. The result is the number of financially allowable failures per component (during the same time the warranty budget was defined for, e.g. 1 year).

Step 9: The reliability goal for the component at the end of the warranty-time can be determined by dividing the total number of units still intact (cf. Step 8) by the total number of components in the field (i.e. as planned).

Optional step: The failure rate could be assumed to be constant over the product's lifetime. In case the failure rate cannot be assumed to be constant, the Weibull distribution can serve to describe the (planned) failure behavior, for which additional information is required on its parameters such as the shape or scale parameter.

Step 10: By means of the system reliability structure, the system's goal reliability can be calculated applying Boole-theory formula.

5. ILLUSTRATIVE EXAMPLE

All steps of the methods are quantitatively summarized in Table 5.

Suppose a product consisting of seven components (*i*=1,2,3…7) provides three functions which have been identified by the manufacturer in accordance with the customer (step 1).

Step 2: The functions are ranked by means of a pair-wise comparison (cf. Table 4), a common method for prioritizing alternatives [11]. Reading from left to right, the values have the following meaning: 1= less important than; 2=as important as; 3= more important than.

Table 4: Ranking of Functions by Pairwise Comparison

	Function 1	Function 2	Function 3	Sum	Rank	Importance
Function 1	-	3	3	6	1.	50%
Function 2	1	-	3	4	2.	33%
Function 3	1	1	-	2	3.	17%

Step 3: Suppose management has decided to allocate the warranty budget over the warranty time of 2 years to be €100,000. The allocated warranty budgets per function result at €50,000 for function 1, €33,000 for function 2 and €17,000 for function 3.

Step 4: Find the reliability structure per function. In this example, the structures are found to be as illustrated in Figure 3. Such a reliability structure can have scalable detail which makes the method applicable at early or later stages of development. Please note that not all components contribute to all functions. The structures can be arbitrarily complex.

Figure 3: Reliability Structure per Function of the Exemplary Product

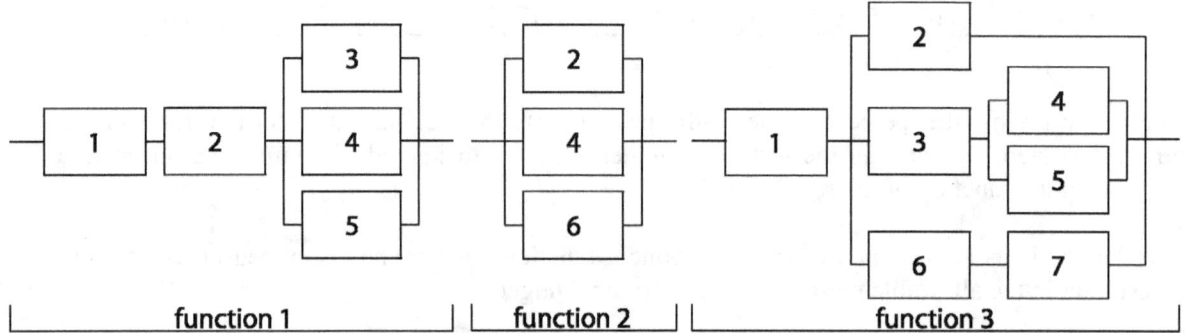

Step 5: The individual component's contribution to the function is analyzed similar to the example in chapter 4.2. Such a correlation often stems from the House of Quality or as part of a value managerial analysis [10]. Please note that in function 3, the same influence factor needs to be allocated to components 6 and 7 since they are in series, i.e. they are equally important. Components 4 and 5 combined represent the same importance as component 3. As they are redundant to each other, their individual importance is half as great as that of component 3 (cf. Example of chapter 4.2.). Please find all calculated contributions in Table 5.

Step 6: By multiplying the per-cent value with the allowable warranty budget of each function, each component's share is derived. Component 1 was found to contribute 33% toward the fulfillment of function 1. Therefore, 0.33·€50,000 = €16,667 are allocated to that component for the matter of function 1.

Step 7: The per-component-and-function budgets are added up component-wise resulting in the budget total per component which can be interpreted as the maximum allowable warranty budget to be spent on the given component. Component 2 for instance contributes to all of the fictional functions in this example. The allowable warranty budget for component 2 sums up to €16,667 + €11,000 + €2,125 = €29792.

Step 8: In order to determine the maximum number of allowable failures, information is needed on the per-component costs for rectifying a warranty claim, e.g. the replacement costs for each component. In case of an existing predecessor product, this amount would be known from e.g. field data. If such a predecessor product does not exist (as is the case of a new product development), the costs need to be estimated. The replacement costs for component 1 are in this example assumed to be €100, for component 2 to be €500 etc.

The maximum allowable number of failures per component is then calculated by dividing the per-component budget total by the component's replacement costs. For component 1, the number of allowable failures is thus €23,042/ €100 = 230.4, i.e. 230.

Step 9: Reliability can be empirically expressed as the number of items still intact divided by the total number of items [3,9]. The planned yearly output quantity is 5,000 units, i.e. 10,000 units per two years. The minimum reliability for component 1 at the end of the design period (2 years) is $R_{1,min}(t=2\text{years}) = (10,000 - 230)/10,000 = 0.9770 = 97.70\%$. Component 1's minimum reliability after 2 years has thus been derived to be $R_{1,min} = 97.70\% = R_1(t_w)$ representing its reliability goal.

Optional step: If the applicant of the method supposes a constant failure rate over the warranty time t_w (as is the case for exponential failure distributions, i.e. Weibull shape parameter $b = 1$), it can be easily calculated. For constant failure rates, the reliability at time t_w can be calculated as follows [3,9]:

$$\lambda = \frac{-\ln(R(t_w))}{t_w}$$

The resulting failure rate for component 1 over two years is $\lambda_1 = -(\ln(0.9770))/2 = 11.634 \cdot 10^{-3}$ (1/year), representing the maximum allowable constant failure rate.

If the applicant of the model is dedicated to determining a Weibull failure distributions with a shape parameter $b \neq 1$, two alternatives are briefly discussed below, which all use $R_{i,min}(t_w)$ as a given input.

a. If the lifetime of the component T is required to have a certain (minimum) value (or can be estimated e.g. based on predecessor versions of the component or has been defined as a strategic goal), the component-specific (maximum) Weibull shape parameter b can be calculated by [3]

$$b = \frac{\ln(-\ln(R_{i,min}))}{\ln(\frac{t_w}{T})},$$

where $R_{i,min}$ is the reliability goal for the individual component at the end of the warranty time t_w.

b. If on the other hand the shape parameter b of the component is known, i.e. the Weibayes method is applicable [12], the component's lifetime T can be found by [3]

$$T = (-\ln(R_{i,min}))^{-\frac{1}{b}} \cdot t_w .$$

In case the failure distribution of the component was to be described best by a shape parameter $b = 1.5$, the resulting lifetime T is obtained to $T=24.54$ years. The situation is illustrated in the Weibull plot of Figure 4. If a combination of b- and T- values for a distribution crosses the red area, it is to be regarded as inadmissible as it will lead to an inferior reliability. Distributions crossing the critical region have in common, that an elevated amount of failures occur at a very early stage before t_w, which

Figure 4: Weibull Plot of Different Failure Distributions

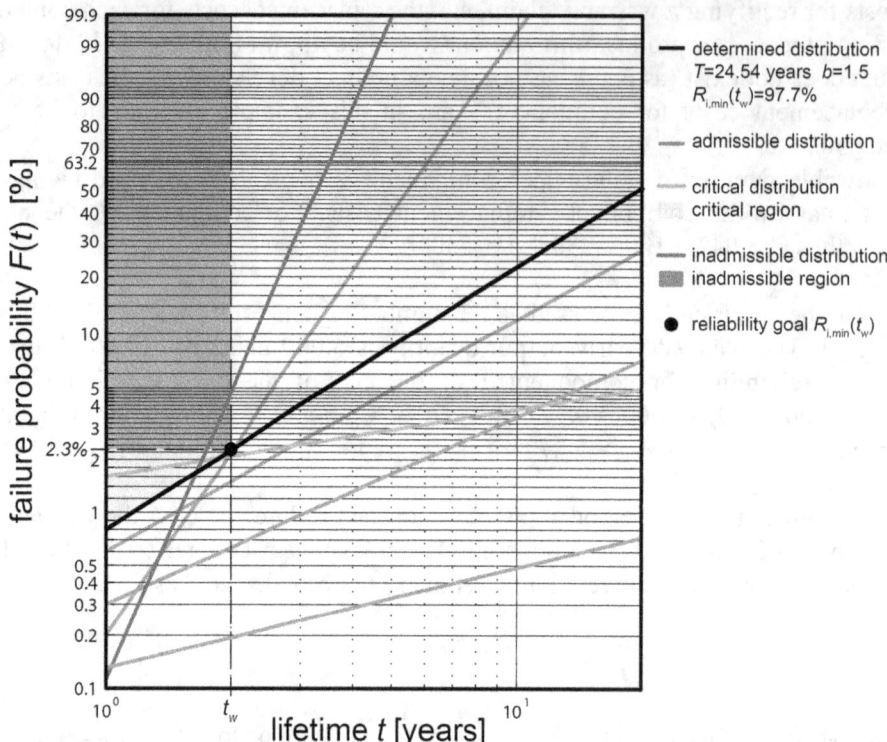

is financially unfavorable from the manufacturer's point-of-view. Furthermore, it may lead to particularly disappointed customers.

Step 10: The system's reliability goal R_{System} can now be calculated applying Boole-theory [3]. In order to do so, the system's reliability structure needs to be known, which in the assumed case equals the serial connection of all functions (cf. Figure 3). The reliability structure expressions per function are as shown in Table 5 where their values are also indicated:

Table 5: Reliability Formula and Values for the Illustrative Example

	Reliability formula	Reliability
Reliability for function 1, R_I:	$=R_1*R_2*(1-(1-R_3)(1-R_4)(1-R_5))$	97.1236%
Reliability for function 2, R_{II}:	$=1-(1-R_2)(1-R_4)(1-R_6)$	99.9999%
Reliability for function 3, R_{III}:	$=R_1*[1-(1-R_2)(1-(R_3*(1-(1-R_4)(1-R_5)))(1-(1-R_6R_7)))]$	97.6986%
Reliability for the system R_{System}:	$=\{R_1*R_2*(1-(1-R_3)(1-R_4)(1-R_5))\}$ $*\{1-(1-R_2)(1-R_4)*(1-R_6)\}$ $*\{R_1*[1-(1-R_2)(1-(R_3*(1-(1-R_4)(1-R_5)))(1-(1-R_6R_7)))]\}$	94.8883%

In case the costs for achieving the individual component's target reliability level as derived in the method presented here are to be estimated, the method my Mettas [8] could be used for instance.

Table 6: Stepwise Determination of the Reliability Goal

		Component							Total
		1	2	3	4	5	6	7	
Function 1 (€50,000)	Influence (absolute value)	1	1	1/3	1/3	1/3	0	0	3
	Influence (per-cent value)	1/3 (33%)	1/3 (33%)	1/9 (11%)	1/9 (11%)	1/9 (11%)	0	0	1 (100%)
	Budget share	€16,667	€16,667	€5,555	€5,555	€5,555	0	0	€50,000
Function 2 (€33,000)	Influence (absolute value)	0	0.33	0	0.33	0	0.33	0	1
	Influence (per-cent value)	0	0.33 (33%)	0	0.33 (33%)	0	0.33 (33%)	0	1 (100%)
	Budget share	0	€11,000	0	€11,000	0	€11,000	0	€33,000
Function 3 (€17,000)	Influence (absolute value)	1	1/3	1/3	1/6	1/6	1/3	1/3	8/3
	Influence (per-cent value)	3/8 (37.5%)	1/8 (12.5%)	1/8 (12.5%)	1/16 (6.25%)	1/16 (6.25%)	1/8 (12.5%)	1/8 (12.5%)	1 (100%)
	Budget share	€6,375	€2,125	€2,125	€1,062.5	€1,062.5	€2,125	€2,125	€17,000
Budget total		€23,042	€29,792	€7,680	€17,617.5	€6,617.5	€13,125	€2,125	€100,000
Repl. costs		€100	€500	€300	€1000	€750	€800	€100	€3,750
Max. failures		230	59	25	17	8	16	21	n/a
$R_{i,min}(t_w)$		97.70%	99.41%	99.75%	99,83%	99,92%	99.84%	99.79%	n/a
Failure rate (over t_w)		$11.634 \cdot 10^{-3}$	$2.959 \cdot 10^{-3}$	$1.252 \cdot 10^{-3}$	$0.851 \cdot 10^{-3}$	$0.400 \cdot 10^{-3}$	$0.801 \cdot 10^{-3}$	$1.051 \cdot 10^{-3}$	n/a

6. CONCLUSION

The method presented here as summarized by means of the flow chart in Figure 1 is able to derive subsystem target reliability levels and thereafter the system's reliability target based on value to the customer and warranty budgets. The procedure starts off by putting the customer requirements toward the functions of the product in the focal point since the product to be developed is persistently perceived as a means of fulfilling the functionality the customer wants. In order to assure the maximum customer satisfaction, statistically inevitable failures during the granted warranty time are compensated for by the manufacturer. Allocating allowable warranty budgets to each subsystem based on its importance toward fulfilling the system's functions ultimately results in the allowable number of failures per warranty time, whereupon reliability relevant parameters can be derived. This results in a reliability allocation as much in line with the voice of the customer as possible.

Contrasting with existing methods, the one presented here is able to combine the determination of a system reliability target with breaking it down to subsystem reliability targets. Also, the method does not depend as heavily as e.g. the method ARINC, AGREE, the Feasibility of Objectives or the method by Karmiol, Bracha and Mettas on empirically defined factors. This enhances the applicability of the method significantly. Furthermore, the method can be applied in early development stages of either a new product development or of design changes as long as the system structure per function can be determined. How encompassing a function is regarded is scalable. The method is also applicable to arbitrarily complex systems.

References

[1] J. Juskowiak, V. Schweizer, M. Stohrer, B. Bertsche. *"Reliability Growth Model in Early Design Stages"*, RAMS Proceedings, 2013RM-220, (2013)

[2] V. Schweizer, J. Juskowiak, M. Bartholdt, F. Jakob, P. Zeiler, B. Bertsche. *"Modification and Application of an S-shaped Reliability Growth Model"*, ESREL Proceedings, p. 3029-3035, (2013)

[3] B. Bertsche. *"Reliability in Automotive and Mechanical Engineering"*, Springer, 2008, Heidelberg, Germany

[4] B. Bertsche, P. Göhner, U. Jensen, W. Schinköthe, H.-J. Wunderlich. *"Zuverlässigkeit mechatronischer Systeme"*, Springer, 2009, Heidelberg

[5] J. Hartig. *"Kostenrelative Zuverlässigkeitsoptimierung in der Konzeptphase maschinenbaulicher Produktentwicklung"* (English title: *Cost-related Reliability Optimization in Early Design Stages of Product Development)*, Rostock University, 1996, Rostock, Germany

[6] D. Kirschmann. *"Determination of Extended Reliability Goals in the Context of Product Development"*, Institute of Machine Components, 2012, Stuttgart, Germany

[7] D. Kececioglu. *"Reliability Engineering Handbook"*, DEStech Publications, Inc., 2002, Lancester, USA

[8] A. Mettas, ReliaSoft Corporation. *"Reliability Allocation and Optimization for Complex Systems"*, RAMS Proceedings, 2000RM-87, (2000)

[9] A. Birolini. *"Reliability Engineering: Theory and Practice"*, Springer, 2004, Heidelberg, Germany

[10] DIN EN 12973 *"Value Management"*, Deutsches Institut für Normung e.V., 2002

[11] L. L. Thurstone. *"A Law of Comparative Judgment"*, Psychological Review, p. 274-286, 1927, Washington, USA

[12] R.B. Abernethy. *"The New Weibull Handbook"*, R.B. Abernethy, 2000, North Palm Beach, USA

How to integrate correctly hardware common cause failures in frequency calculations?

Hervé Brunelière[a*], Monica Rath[a], and Wenjie Qin[a]
[a] AREVA NP SAS, Paris La Défense, France

Abstract: Hardware common cause failures are generally the highest contributors in the I&C systems reliability and availability studies.

Comparisons of results from calculations of frequency of spurious actuations by a safety system or frequency of failures of a control system with operation feedback of such failures show that the frequency calculations are often overestimated. This is due to the use of « classic » common cause failure parameters.

This is mainly explained by the fact that, for these undesired events, failures are generally not hidden ones and are then detected within few hours. Then, for common cause failures that are not simultaneous, the first failure is often repaired before the second one appears.

This over conservatism can lead to inappropriate design choices like addition of redundancies or interlocks to minimize the frequency of an undesired event based on a calculation that does not reflect the real situation. This is then a concern for a designer and for a utility to limit as far as possible the impact of this over conservatism.

One solution is to consider only independent failures in frequency calculations. In this case, the result is underestimated as simultaneous common cause failures that are possible and credible are not considered in the result. Then, the risk is not to implement some necessary measures in the design due to over optimistic results.

The paper will discuss possible solutions to handle these types of failures in calculations based on real cases.

Illustrations will be based on a typical architecture of an I&C system based on Teleperm XS platform similar to the ones currently implemented in nuclear power plants.

The paper will also integrate discussions on relevance of the different methodologies including no consideration of CCF at all, degraded CCF factors values and possibilities of extrapolation. These methodologies will be compared based on their impact on calculation results and the consistency with operational experience.

Keywords: CCF, frequency, methodology, failure mode

1. GOALS

The goal of the paper is to make a status of AREVA NP SAS work in progress for the improvement of consideration of common cause failures in frequency calculations. Illustration example is based on a typical and theoretical architecture of an I&C system based on Teleperm XS (TXS) platform similar to the Protection Systems currently implemented in nuclear power plants.

Relevance of the different methodologies is preliminarily assessed. These methodologies are mainly compared based on their impact on calculation results and on the consistency with operational experience.

2. CONTEXT

2.1. Typical CCF methodology

In European countries, Common Cause Failures (CCF) are generally calculated using extended beta factors methodology.

The probability of failure due to CCF of k out of m identical components is assessed using following equation:

$$Q_m^k = \beta_m^k Qt \ (1)$$

Where Qt is the probability of failure of one component

Typical values for β_n^n are:

$$\beta_2^2 = 0{,}05 \ (2)$$
$$\beta_3^3 = 0{,}02 \ (3)$$
$$\beta_4^4 = 0{,}01 \ (4)$$

These data are mostly used for projects in Europe. They were even previously given in European Utility Requirements in the Probabilistic Safety Assessment chapter but were removed in last version.

This method has following properties:

- It distinguishes partial and total loss of components that are subject to CCF.
- Probability of CCF is proportional to probability of failure of one component.
- Application of the method assumes that all CCF happen simultaneously.
- The commonly used values for these extended beta factors are quite high.

The two last properties are consequence of the fact that most calculations are made for assessing probability of failure on demand of redundant mitigation means. Then the most significant contributors are Common Cause Failures of non self-monitored failures that are only detectable during periodic tests. "Standard" beta factors take then into account accumulation of faults between two periodic tests, i.e. failures revealing in a typical interval of several months. Their adequacy to failure modes that are detectable within few hours is challengeable.

2.2. Increasing need for frequency calculations

The context is evolving in the nuclear industry:
- Focus is more and more made on availability aspects in parallel to safety concerns.
- Robustness of the preventive line (all means that control the plant in order to avoid any event that could lead to occurrence of incidents and accidents) is more and more important.

It is then necessary to focus on undesired events like failure of control functions, occurrence of a postulated initiating event or spurious actuation of safety functions. These events are of course quantitatively evaluated by calculating their frequencies.

3. ILLUSTRATION EXAMPLE

In order to show concretely how the subject raised by this paper is of importance, it is chosen to make an illustration example based on a theoretical architecture of a Protection System based on Teleperm XS platform. Frequency of spurious Reactor Trip actuation is assessed with different assumptions on CCF values.

3.1. System architecture and other main assumptions

Figure 1: Simplified architecture of the theoretical Teleperm XS based Protection System (one division)

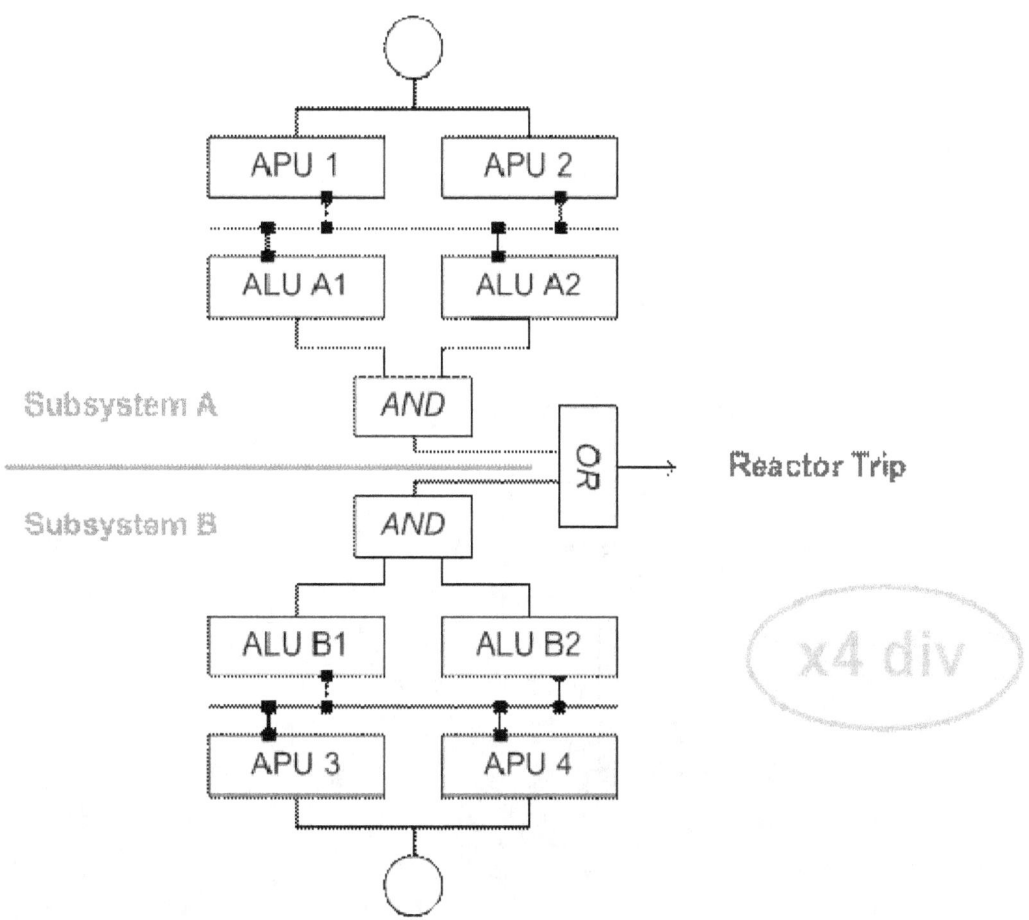

Reactor Trip functions are implemented in the Acquisition and Processing Units (named APUx) and Acquisition and Logic Units (named ALUs) of the four divisions. In each APU, the result of each threshold is transferred in the ALUs of the four divisions. The ALUs are performing 2-out-of-4 logic between redundant signals from APUs. If threshold result(s) have a faulty status, the voting logic is degraded as follows:

Table 1: Degradation of voting logics inside the theoretical Teleperm XS based Protection System

Number of faulty inputs	Voting logic
0	2-out-of-4
1	2-out-of-3
2	1-out-of-2
3 or 4	Actuation

In every division, the two ALUs of each sub-system are generating de-energized orders, which are combined with a functional AND logic (electrical OR). For reactor trip order of one division, the signals coming from both sub-systems are combined with a functional OR logic (electrical AND).

The functional structure is given in Figure 2.

Figure 2: Logical implementation of Reactor Trip functions in the theoretical Teleperm XS based Protection System (one division)

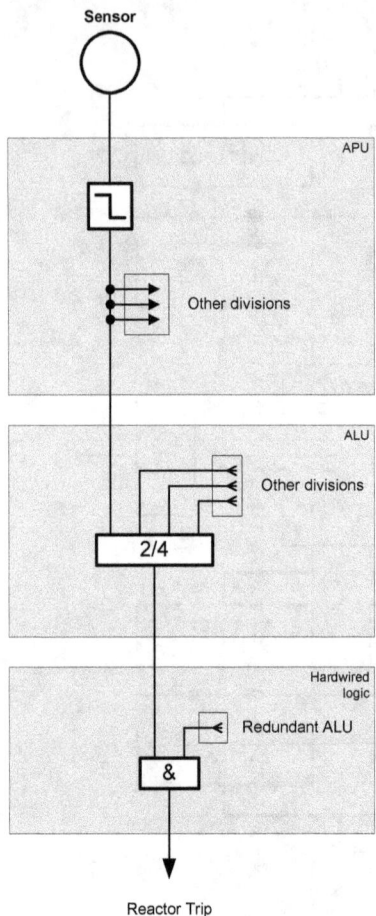

Main assumptions related to the calculations are given hereafter:

- Sensors, actuators and support systems are not considered. Only automation system failures are taken into account.

- Every spurious failure is assumed to be detected and repaired in an interval of less than 8 hours (time between occurrence of the failure and start up of the new or repaired I&C module).

- Software failures are not considered here as their potential for CCF is to be considered separately for hardware failures and their integration is not useful for the aim of this paper.

3.2. Results with conservative beta factors

3.2.1. Results with conservative beta factors

When calculating their frequencies with common cause factors as described in section 2.1:

- Frequency of spurious reactor trip due to the theoretical Protection System is 2.4E-05 per hour which corresponds to 2.1E-01 per year, i.e. one reactor trip every 4,8 years only due to hardware failures.

- Main contributors to the result are Common Cause Failures of components (that are assumed to happen simultaneously).

3.2.2. Results without assuming CCF

The model for assessing the frequency of a spurious reactor trip signal by hardware failures of the same theoretical Protection System is now updated to eliminate any potential for CCF. Only independent failures are possible. This is in the logic of a probabilistic interpretation of IEC 62340 standard [1] in which CCF have to be assumed only at the moment of the demand.

The same calculation now shows significantly different conclusions:

- Frequency of spurious reactor trip due to the theoretical Protection System is 4.6E-07 per hour which corresponds to 4.0E-03 per year, i.e. one reactor trip every 250 years only due to hardware failures.
- The main contributors are independent failures of two components, the second failure occurring before the repairing of the first one, i.e. between 0 and 8 hours after first failure.

3.2. Conclusion from these assessments

As the second result is 52 times lower than the first one, it shows that this strategy for assessing CCF turns to be the key point of the correct frequency evaluation.

The frequency calculations in the first case are overestimated. This is mainly due to the use of the so called "classic" Common Cause Failure parameters and the consideration of simultaneous failures.
For these undesired events, failures are generally not hidden ones and are then detected within a few hours. Then, most of the time, the first failure is repaired before the second one appears.
This over conservatism can lead to unnecessary design choices like addition of redundancies or interlocks to minimize the frequency of an undesired event. This is then a concern for a designer and for a utility to limit this over conservatism at a maximum.

The frequency calculations in the second case are underestimated. Indeed, there remains a potential that common cause failures happen in a very short interval. This may be the case for failures corresponding for example to shutdown of systems or systems stopping to proceed. This is less credible for spurious operations of functions if designed fall-back position is non actuation of safety function, as "fail to 1" failure modes in high-quality I&C systems generally have a very low potential to happen by common cause.
This under conservatism leads to over optimistic results and, depending on what is calculated, gives a bad representation of the safety or the availability of the plant.

3.3. What can we do?

With such an amplitude in the obtained results, it appears necessary to find a way to calibrate the model. Actions are ongoing to see how benefits from operational experience can be used for the purpose of this calibration.

If it is decided to stick to the use of beta factors methodology, it appears necessary to lower the values of CCF parameters that are used in frequency calculations.

3.4. Some examples of results with lower beta values

3.4.1. Example 1

A proposal of revaluation of beta factors is:

$$\beta av_n^n = \beta_n^n / 10 \ (5)$$

The idea is that most CCF are not simultaneous. Then it seems achievable to prove that less than 10% of them happen in an interval of 8 hours

Tuned values for βav_n^n (n from 2 to 4) are then:

$$\beta av_2^2 = 5E - 03 \quad (6)$$
$$\beta av_3^3 = 2E - 03 \quad (7)$$
$$\beta av_4^4 = 1E - 03 \quad (8)$$

When calculating their frequencies with such common cause factors:

- Frequency of spurious reactor trip due to the theoretical Protection System is 2.8E-06 per hour which corresponds to 2.4E-02 per year, i.e. one reactor trip every 41 years only due to hardware failures.
- The resulting main contributors are still Common Cause Failures of components happening in a short interval.

3.4.2. Example 2

As βav_n^n corresponds to the percentage of cases where, if one component fails in a group of n identical components, the (n-1) other ones would fail in the few hours, it can be assumed that the higher is the n value, the higher is the factor between typical beta factors and revaluated beta factors used for frequency assessments.

A proposal is for n = 2 to 4.

$$\beta av_n^n = \beta_n^n / (5 + n) \quad (9)$$

This seems reasonable at least when n is not too high.

Tuned values for βav_n^n are then:

$$\beta av_2^2 = 7,14E - 03 \quad (10)$$
$$\beta av_3^3 = 1,25E - 03 \quad (11)$$
$$\beta av_4^4 = 5,56E - 04 \quad (12)$$

When calculating their frequencies with such common cause factors:

- Frequency of spurious reactor trip due to the theoretical Protection System is 6,2E-06 per hour which corresponds to 5,4E-02 per year, i.e. one reactor trip every 19 years only due to hardware failures.
- The resulting main contributors are still Common Cause Failures of components happening in a short interval.
  - In 76% of the cases, 2 failures happen in this interval
  - In 24% of the cases, at least 3 failures happen in this interval

4. CONCLUSION

This paper discusses possible management of hardware CCF in frequency calculations. It is based on a typical I&C system with a design which is similar to systems implemented in nuclear power plants. In order to be an example of calculation that can be compared to operational experience, spurious reactor trip by a Protection System is taken as example.

Result with classic beta factors appears too conservative. One solution is to assume that all failures are independent. In this case, the result is underestimated. The factor 52 between the results of both scenarios shows the importance of this parameter.

This paper also introduces work to assess relevance of different methodologies to model as adequately as possible common cause hardware failures. These methodologies can easily be compared based on their impact on calculation results (lower or higher frequency).

At this stage, it can be assumed that the final chosen solution could be:

- Use of degraded CCF factor values compared to "classic ones".
- Use of corrective factors or functions to adapt the preliminary calculated results according to similar scenarios from operational experience.

Future work is to assess the results compared to:

- Uncertainties that can be assumed (does this methodology give reliable and trustable results or not?).
- Consistency with operational experience.
- Level of confidence that the final result is still conservative (is it possible to defend this calculation in front of a safety authority?).

Data from NUREG ([2]) is also analyzed in parallel.

The recommendations that will come from this work may be different according to the different applications and to the different components.

Additionally, there are systems where some components of a type are in operation and other ones are in standby. Components in standby are actuated in case of failure of first ones. The subject of potential Common Cause Failure between these components that are identical, but for which initial operating conditions are different, also needs to be addressed.

References

[1] IEC 62340 - Nuclear power plants – Instrumentation and control systems important to safety – Requirements for coping with common cause failure (CCF) – December 2007
[2] NUREG/CR-6268 - Common-Cause Failure Database and Analysis
System: Event Data Collection, Classification, and Coding

The Basic Idea of Quantitative Model of Reactor Protection System Considering Stochastic Process

Hitoshi Muta[a]

[a] Tokyo City University, Tokyo, Japan

Abstract: In nuclear power plants such as ABWR and the latest PWR, digital instrumentation and control system have been installed increasingly to reactor emergency shutdown system which is one of the important safety functions. However, it has been found that it is difficult to model the digital equipment reliability in probabilistic risk assessment (PRA). And some of issues such as taxonomies of failure modes have been studied in the international framework, OECD/NEA/WGRisk task group called DIGREL.

In this paper, the reactor trip actuation failure event logics and frequencies resulting from the multiple failures and the demand following the initiating event are analyzed qualitatively and quantitatively. This paper presents the example of the reliability analysis of the digital Reactor Protection System (RPS) considering stochastic process, the approach given by this paper will be applicable to establish the PRA model of digital RPS of the actual nuclear power plant.

Keywords: DI&C, RPS, Self-diagnostic Function; Stochastic Process; Markov Transition Model.

1. INTRODUCTION

Several PRA studies to model the digital safety system have been done so far. For example, in ABWR PRA conducted in Japan [1][2][3][4][5], conventional Fault Tree Analysis (FTA) technique was used in reliability analysis of the digital RPS. However, it has been found that it is difficult to model the digital equipment reliability in probabilistic risk assessment (PRA) using conventional FTA technique because FTA cannot simulate precisely state transition among various states and functions of digital equipment.

OECD/NEA CSNI WGRisk set up the task group DIGREL to develop the basis of reliability analysis method of the digital safety system and now discussing about several issues related to quantitative modeling including the taxonomies of digital system failure modes [6][7][8]. Although conventional FTA technique has been applied to reliability analysis of the digital RPS so far, introducing more dynamic approach is essential to properly assess the effects of repair or manual shutdown operation following detection of faults by self-diagnostic function. However, there is few PRA including dynamic reliability models of DI&C system currently, and dynamic approach is considered to be still in a trial stage.

In this paper, the reactor trip actuation failure event logics and frequencies resulting from the multiple failures and the demand following the initiating event are analyzed qualitatively and quantitatively. Then the comparison is made between the method obtained in this paper and FTA technique to clarify the difference.

[Notations and definitions]

P_i: probability of the state in i

λ_M [1/hr]: initiating event frequency (the probability of demand of the RPS per unit time at time t, given the RPS is not actuated at time t)

λ[1/hr]: constant hardware failure rate

λ_D [1/hr]: constant detected hardware failure rate

λ_U [1/hr]: constant undetected hardware failure rate

R [1/hr]: constant restart rate of the reactor (the probability of transfer per unit time at time t, given the system is in shutdown state at time t)

m [1/hr]: constant renewal rate of the reactor (the probability of renewal per unit time at time t, given the system is in ATWS at time t)

ω_C [1/hr]: ATWS frequency per unit calendar time in the steady state

$\omega_C{}^*$ [1/hr]: the part of ATWS frequency ω_C caused by the independent hardware failure and the demand

$\omega_C{}^{**}$ [1/hr]: the part of ATWS frequency ω_C caused by the common cause hardware failure and the demand

ω_R [1/hr]: ATWS frequency per unit reactor operational time in the steady state

$\omega_R{}^*$ [1/hr]: the part of ATWS frequency ω_R caused by the independent hardware failure and the demand

$\omega_R{}^{**}$ [1/hr]: the part of ATWS frequency ω_R caused by the common cause hardware failure and the demand

2. The Digital Reactor Protection System Description

The RPS is one of the most important functions to control reactivity, actuate reactor trip in an emergency situation and maintain the reactor in safe state. This accident sequences are defined as the anticipated transient without scrum (ATWS) event in PRA study and are very important to core damage risk [1]-[5]. Figure 1 is a simplified event tree that is focused on accident sequences of ATWS event. If neither does the RPS actuate nor do control rods insert successfully, ATWS event will occur. Generally, the RPS is typically consisted of multiple channels such as "1-out-of-2 twice" or "2-out-of-4" configuration, including several devices and complicated connections [9]. However, to make the discussion easier, the RPS is hypothetically assumed to be composed of two independent channels that are regarded as simple 1-out-of-2 configuration as shown in Figure 2. Here, reactor trip is actuated by two solenoid valves A & B opening in case that 1-out-of-2 channel of the RPS is activated with trip signal from a sensor.

The following postulates are put on the RPS described above:

a) a hardware failure and a software failure are considered,

b) for the RPS hardware failure mode, the division level failure which is being discussed in DIGREL [6] is applied in this paper, so it is possible to be considered that hardware failure is defined as loss of function of a channel of the RPS including sensor, PI/O, MUX, DTM, TLU and OLU,

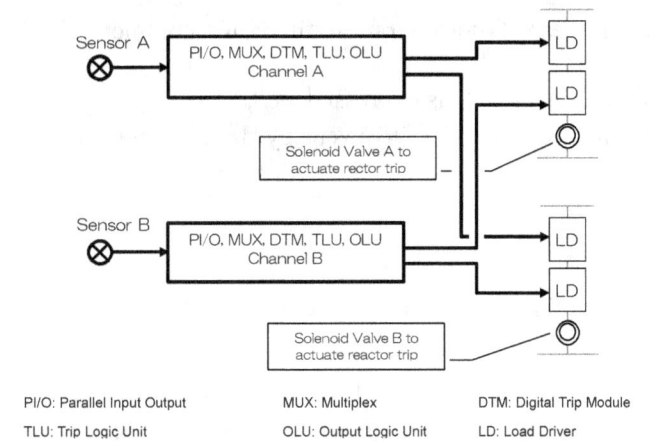

PI/O: Parallel Input Output   MUX: Multiplex   DTM: Digital Trip Module

TLU: Trip Logic Unit   OLU: Output Logic Unit   LD: Load Driver

Figure 1. Outline of Simplified Digital Reactor Protection System (1-out-of-2 Configuration)

c) to avoid the reliability model too much complicated and to be focused on clarifying the validity of this approach, the failures of LDs, solenoid valves and control rod drive (CRD) system out of the logic circuits are assumed to be negligible,

d) the hardware failure mode in a channel of the RPS is classified into the common cause failure and the independent failure,

e) the common cause failure and the independent failure are respectively classified into the detected failure and the undetected failure,

f) the detected fault is detected instantly by a self-diagnostic function which runs continuously,

g) the undetected fault can be detected by a surveillance test which is executed at an interval of T,

h) to make discussion easier, the software fault is not considered in this paper,

i) a plant personnel starts to repair the failed channel of the RPS after detection of single hardware fault by self-diagnostic function or surveillance test,

j)   a plant personnel makes the reactor shutdown immediately after a detection of hardware fault of both channels by self-diagnostic function or surveillance test,

k)   a duration of shutdown operation is T_{SD},

l)   the reactor returns to the initial state after the shutdown state and restarts, and

m)   the reactor is renewed after the state of core damage following ATWS sequence.

The following statistical assumptions are made:

i)   the initiating events and the failures of the RPS occur statistically-independently and randomly,

ii)   the start of the initiating event can be modeled by the exponential distribution with the demand rate of λ_M,

iii) the reactor returns from shutdown state to the initial state by the constant transfer rate of R,

iv)   the hardware fault in a channel of the RPS that is according to the postulate b), can be modeled by the exponential distribution with the total constant hardware failure rate of λ, which can be divided into a detected hardware fault and an undetected hardware fault.

v)   a detected hardware fault can be modeled by the exponential distribution with the constant hardware failure rate of λ_D,

vi)   an undetected hardware fault can be modeled by the exponential distribution with the constant hardware failure rate of λ_U,

vii) a repair of detected hardware fault can be modeled by the exponential distribution with the constant restoration rate of μ_R, which can be approximated as *(1/MTTR)*,

viii) a repair of undetected hardware fault can be modeled by the exponential distribution with the constant restoration rate of μ_R, which can be approximated as *(1/MTTR)*,

ix)   a shutdown operation can be modeled by the exponential distribution with the constant transition rate of μ_{SD}, which can be approximated as *(1/T_{SD})*,

x)   a mean failure-duration time of "single undetected hardware fault" can be approximated as *(T/2)*, a mean failure-duration time of either "independent double undetected hardware fault" or "superposition of independent undetected hardware fault and independent detected hardware fault" can be approximated as *(T/3)* and a mean failure-duration time of "common cause undetected hardware fault" can be approximated as *(T/2)*,

xi)   the ratio of the common cause hardware failure to λ_D and λ_U is expressed as β,

xii) the reactor is renewed after core damage event following ATWS event by the constant renewal rate of m.

3. Analyses of ATWS event frequency

In this section, the process of ATWS event is analysed for the reactor equipped the RPS defined in the previous section.

3.1. *Core damage event logics and the fault tree*

In general, ATWS event could occur through one of the following sequential logics:

(A) an initiating event occurs in fault of both channels of the RPS, or

(B) both channels failure of the RPS occurs in a demand state after an initiating event.

It is obvious that those two logics are mutually exclusive. The two sequential logics for the core damage events are developed by the FTA technique with the priority AND-gate as shown in Figure 3 and Figure 4.

The top event, ATWS event (E-top), occurs when the logic of either "the demand and common cause hardware fault (E1)", "the demand and independent multiple hardware faults (E2)" or "the demand and the software fault (E3)" is true.

The first logic E1 becomes true when either Sequence #1-1 or Sequence #1-2 is true. Sequence #1-1: if a common cause hardware failure occurs in the demand state resulting from an initiating event, then ATWS event occurs. However, since the RPS is a stand-by system and initiating events must be prior to the actuation of the RPS, it is not necessary to consider this sequence. Sequence #1-2: if an initiating event occurs in the common cause hardware fault resulting from a common cause hardware failure, then ATWS event occurs.

The second logic of E2 is classified into the sequences #2-1 through #2-6. Sequence #2-1: if the independent hardware failure of channel B occurs in the state resulting from a demand after an initiating event and an independent hardware fault of channel A, then ATWS event occurs. The other sequences can be expressed in the same manner such that #2-2 is "Demand>channel B failure>channel A failure", #2-3 is "channel A failure >Demand> channel B failure", #2-4 is channel B failure>Demand> channel A failure. However, these sequences are not needed being considered because of the reason same as excluding the sequence #1-1. Although one of the RPS channel is fault for the sequences #2-3 and #2-4, they can be treated same as #2-1 and #2-2 because one channel is enough to actuate the RPS.

The last logic is a software failure. Since this study doesn't focus precise method, a software failure is not considered furthermore. Although there is a probability that any combination of E1 and E2 occur simultaneously, it is considered that the probability is sufficiently small to be negligible based on the rare event approximation. This means that these three ATWS event logics E1 and E2 can be treated separately,

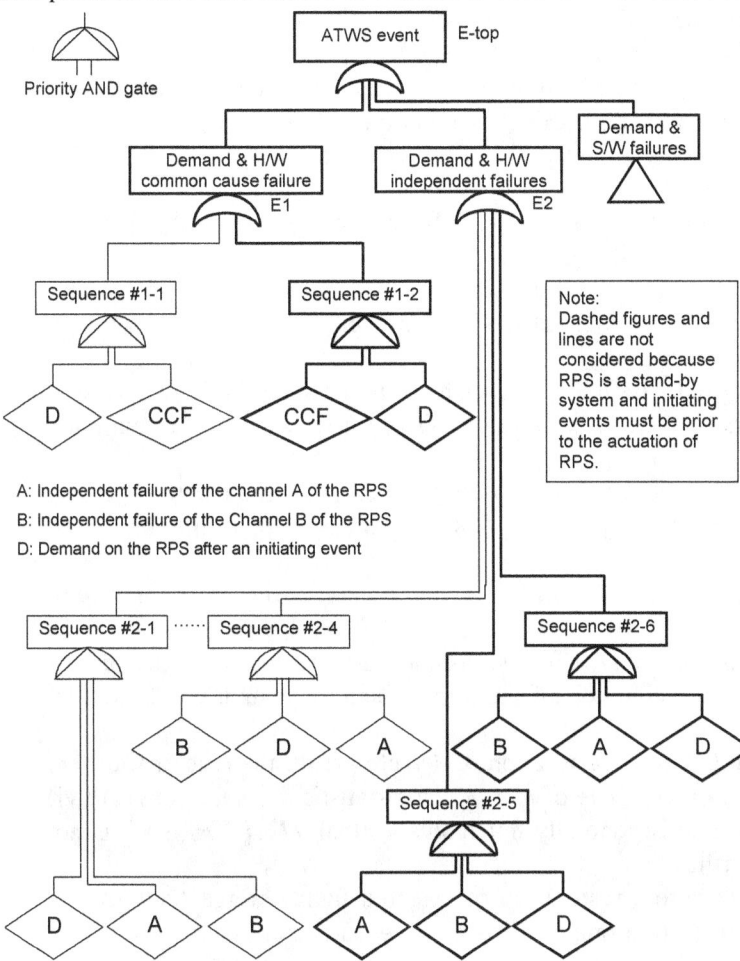

Figure 2 Outline of Simplified Digital Reactor Protection System (1-out-of-2 Configuration)

and the ATWS frequency can be approximated by the summation of these ATWS event frequencies of these three logics in the following discussion. In the next section, ATWS event caused by independent hardware faults are analyzed as an example.

3.2. ATWS event caused by independent hardware faults

This can be modeled by a state-transition diagram as shown in Figure 5 based on the postulates a) through n), statistical assumptions i) through xiv) and sequential logics (A) and (B). Definitions of states are as follows:

State A: Normal state, there is no demand and faults,

State B: Shutdown state, a plant is not in an operation but in a safe state,

State C: One of the channels is in an undetected fault, but there is no demand,

State D: One of the channels is in a detected fault, but there is no demand,

State E: Both of the channels are in undetected faults, but there is no demand,

State F: One of the channels is in an undetected fault and the other is in a detected fault, but there is no demand,

State G: Both of the channels are in detected faults, but there is no demand, and

State H: Core damage state caused by ATWS.

The transitions and their transition rates are:

1) if a demand after an initiating event occurs in Normal State A, the reactor transfers from A to Shutdown State B, this occurs by the transition rate of λ_M; see statistical assumption ii),

2) if an undetected failure occurs in one of the channels in A, the reactor transfers from A to State C, this occurs by the transition rate of $2(1-\beta)\lambda_U$, in which the RPS is being repaired after a detection of undetected fault by surveillance test; see statistical assumption iv), vi) and xi),

3) if a detected failure occurs in one of the channels in A, the reactor transfers from A to State D, this occurs by the transition rate of $2(1-\beta)\lambda_D$, in which the RPS is being repaired after a detection of detected fault by self-diagnostic function; see statistical assumption iv), v) and xi),

4) the reactor returns from B to A, this occurs by the transition rate of R; see statistical assumption iii),

5) if the RPS is restored in C, the reactor transfers from C to A, this occurs by the transition rate of $\{1/(T/2+MTTR)\}$; see statistical assumption viii) and x),

6) if a demand after an initiating event occurs in C before restoration of undetected fault, the reactor transfers from C to B, in which one of the channels of the RPS is being actuated, this occurs by the transition rate of λ_M; see statistical assumption ii),

7) if another undetected failure occurs in C before restoration of undetected fault, the reactor transfers from C to State E, this occurs by the transition rate of $(1-\beta)\lambda_U$; see statistical assumption iv), vi) and xi),

8) if another detected failure occurs in C before restoration of undetected fault, the reactor transfers from C to State F, this occurs by the transition rate of $(1-\beta)\lambda_D$; see statistical assumption iv), v) and xi),

9) if the RPS is restored in D, the reactor transfers from D to A, this occurs by the transition rate of $(1/MTTR)$; see statistical assumption vii),

10) if a demand after an initiating event occurs in D before restoration of detected fault, the reactor transfers from D to B, in which one of the channels of the RPS is being actuated, this occurs by the transition rate of λ_M; see statistical assumption ii),

11) if another undetected failure occurs in D before restoration of detected fault, the reactor transfers from D to State F, this occurs by the transition rate of $(1-\beta)\lambda_U$; see statistical assumption iv), vi) and xi), however, since this transition rate is generally much smaller than $(1/MTTR)$, so it is not necessary to consider this transition further,

12) if another detected failure occurs in D before restoration of detected fault, the reactor transfers from D to State G, this occurs by the transition rate of $(1-\beta)\lambda_D$; see statistical assumption iv), v) and xi), however, this transition rate is generally much smaller than $(1/MTTR)$, so it is not necessary to consider this transition further,

13) if a plant personnel makes the reactor shutdown successfully in E before a demand after a detection of simultaneous undetected fault of both channels by surveillance test, the reactor transfers from E to B, this occurs by the transition rate of $\{1/(T/3+MTTR)\}$; see statistical assumption ix) and x),

14) if a demand occurs in E before completion of shutdown operation, the reactor transfers from E to state H, this occurs by the transition rate of λ_M; see statistical assumption ii),

15) if a plant personnel makes the reactor shutdown successfully in F before a demand after a detection of simultaneous detected fault and undetected fault of each channel by self-diagnostic function and surveillance test, the reactor transfers from F to B, this occurs by the transition rate of $\{1/(T/3+MTTR)\}$; see statistical assumption ix) and x),

16) if one of the channels of the RPS is restored in F after a detection of detected fault by self-diagnostic function, the reactor transfers from F to C, this occurs by the transition rate of $(1/MTTR)$; see statistical assumption vii),

17) if a demand occurs in F before completion of shutdown operation, the reactor transfers from F to H, this occurs by the transition rate of λ_M; see statistical assumption ii),

18) if a plant personnel makes the reactor shutdown successfully in G before a demand after a detection of simultaneous detected fault of both channels by self-diagnostic function, the reactor transfers from G to B, this occurs by the transition rate of $(1/T_{SD})$; see statistical assumption ix). However, the probability in state G is quite small, because the transition rate from G to B is negligible as explained in 12), so it is not necessary to consider this transition,

19) if a demand occurs in G, the reactor transfers from G to H, this occurs by the transition rate of λ_M; see statistical assumption ii), however, the probability in state G is quite small as explained as 12) and 18), so it is not necessary to consider this transition, and,

20) The reactor transfers from H to A by a renewal of the reactor, this occurs by the transition rate of m; see statistical assumption xii).

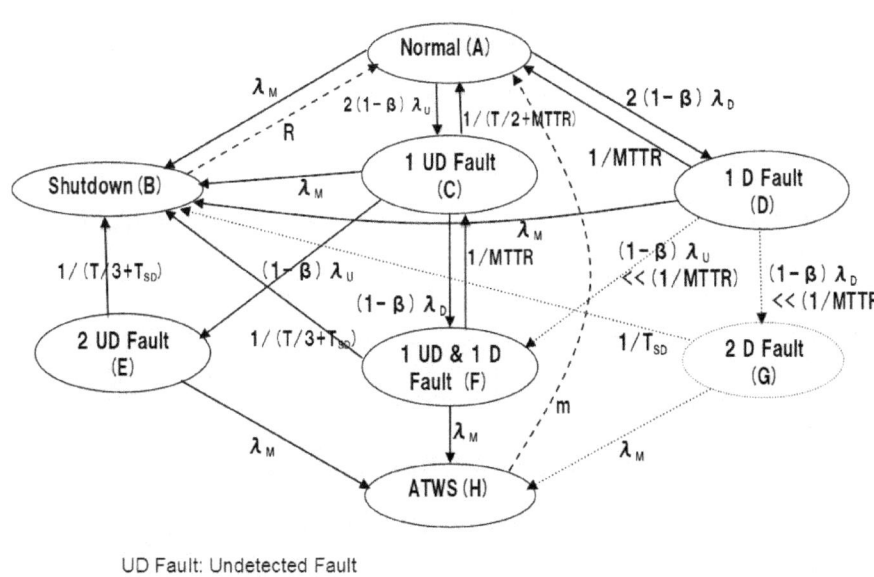

UD Fault: Undetected Fault

D Fault: Detected Fault

Figure 3. A state-transition diagram of ATWS event for independent hardware failures

Because of the above descriptions 11), 12), 18) and 19), it is not necessary to consider state G and transitions from D to F, from D to G, from G to B and from G to H. The state-transition diagram presents the following simultaneous equations in a steady state:

$$P_A + P_B + P_C + P_D + P_E + P_F + P_H = 1 , \tag{1}$$

$$\{\lambda_M + 2(1-\beta)\lambda_U + 2(1-\beta)\lambda_D\} \cdot P_A = R \cdot P_B + \frac{1}{\frac{T}{2}+MTTR} \cdot P_C + \frac{1}{MTTR} \cdot P_D + m \cdot P_H , \tag{2}$$

$$R \cdot P_B = \frac{1}{\frac{T}{3}+T_{SD}} \cdot P_E + \frac{1}{\frac{T}{3}+T_{SD}} \cdot P_F + \lambda_M \cdot P_D + \lambda_M \cdot P_C + \lambda_M \cdot P_A , \tag{3}$$

$$\{\frac{1}{\frac{T}{2}+MTTR} + \lambda_M + (1-\beta)\lambda_U + (1-\beta)\lambda_D\} \cdot P_C = 2(1-\beta)\lambda_U \cdot P_A + \frac{1}{MTTR} \cdot P_F , \tag{4}$$

$$(\lambda_M + \frac{1}{MTTR}) \cdot P_D = 2(1-\beta)\lambda_D \cdot P_A , \tag{5}$$

$$(\lambda_M + \frac{1}{\frac{T}{3}+T_{SD}}) \cdot P_E = (1-\beta)\lambda_U \cdot P_C , \tag{6}$$

$$(\frac{1}{\frac{T}{3}+T_{SD}} + \frac{1}{MTTR} + \lambda_M) \cdot P_F = (1-\beta)\lambda_D \cdot P_C , \tag{7}$$

and

$$m \cdot P_H = \lambda_M \cdot (P_E + P_F) . \tag{8}$$

Based on Figure 5 and the definition, the ATWS frequency caused by the independent hardware fault and the demand per unit calendar time in the steady state, ω_C is given as

$$\varpi_C{}^* = \lambda_M \cdot P_E + \lambda_M \cdot P_F (= m \cdot P_H) . \tag{9}$$

From equations (1) through (9), $\omega_C{}^*$ becomes

$$\varpi_C{}^* = \frac{(\lambda_M \cdot X_1 + \lambda_M \cdot X_2)}{1 + X_1 + X_2 + X_3 + X_4 + \dfrac{1}{R} \cdot \left(1 + \lambda_M (X_3 + X_4) + \dfrac{X_1 + X_2}{\dfrac{T}{3} + T_{SD}} \right) + \dfrac{\lambda_M}{m} \cdot (X_1 + X_2)}$$
. (10)

Here,

$$X_1 = \frac{(1-\beta) \cdot \lambda_U}{\lambda_M + \dfrac{1}{\dfrac{T}{2} + MTTR}},$$

$$X_2 = \frac{(1-\beta) \cdot \lambda_D}{\lambda_M + \dfrac{1}{MTTR} + \dfrac{1}{\dfrac{T}{3} + T_{SD}}},$$

$$X_3 = \frac{\dfrac{1}{\dfrac{T}{2} + MTTR} + \lambda_M + (1-\beta)(\lambda_D + \lambda_U) - \dfrac{1}{MTTR} \cdot X_2}{2(1-\beta) \cdot \lambda_U}$$

and

$$X_4 = \frac{2(1-\beta) \cdot \lambda_U \cdot X_3}{\lambda_M + \dfrac{1}{MTTR}}.$$

Since $\varpi_R{}^*$ which is defined as the ATWS frequency caused by the independent hardware fault and the demand per unit reactor operational time in the steady state, can be obtained by normalized by operational time, $\varpi_R{}^*$ is given as

$$\varpi_R{}^* = \frac{\varpi_C{}^*}{1 - P_B - P_H}.$$
(11)

Moreover, equation (11) can be rewritten as

$$\varpi_R{}^* = \frac{\lambda_M \cdot (X_1 + X_2)}{1 + X_1 + X_2 + X_3 + X_4}.$$
(12)

It is easily found that equation (12) does not include the parameter of R and m. This means that $\varpi_R{}^*$ is not affected by the value of R and m (>0). Namely, because P_B and P_H become null when $R \to \infty$ and $m \to \infty$ (see equation (10)), equation (12) is equal to equation (10) in which $R \to \infty$ and $m \to \infty$. In addition, the reciprocal of $\varpi_R{}^*$ is equal to the meantime from the state A to H, because the ATWS frequency of the reactor is equal to the reciprocal of mean time from the initial state to ATWS in the steady state.

3.3. *ATWS event frequency*

For ATWS event caused by common cause hardware faults can be analyzed in the same manner described in the section 3.2. Therefore, ATWS event frequency per unit calendar time is

$$\omega_C = \omega_C{}^* + \omega_C{}^{**},$$
(13)

and, ATWS event frequency per unit reactor operational time is

$$\omega_R = \omega_R{}^* + \omega_R{}^{**}.$$
(14)

4. CONCLUSION

Digital devices have been realizing advanced functions such as complicated control or self-diagnostic. On the other hand, the method based on conventional FTA technique has become difficult to analyse the effects of recovery or shutdown operation following detection of faults by the diagnostic function appropriately.

This paper shows that, taking account of the relationship among the RPS failures, demand after the initiating event, detection of RPS fault by self-diagnostic or surveillance tests, repair of the RPS components and plant shutdown operation by the plant operators as a stochastic process, the ATWS event can be modelled by the event logic fault tree and state-transition diagrams assuming the hypothetical 1-out-of 2 digital RPS. Then, the ATWS event frequency is formulated base on the state-transition diagrams. Because introducing more dynamic approach is essential to properly assess the effects of repair or manual shutdown operation following detection of faults by self-diagnostic function specific to the digital safety system as shown in this paper.

Thus the approach given by this paper will be applicable to establish the PRA model of the digital RPS of the actual nuclear power plant. For simplicity, this paper assumes simplified 1-out-of-2 configuration RPS. However, the approach given by this paper will be applicable to the analysis of actual RPS equipped 1-out-of-2 twice or 2-out-of-4 configuration.

References

[1] Nuclear Power Engineering Corporation (NUPEC). [The Report of Establishment of Level 1 PSA Method of ABWR Plant at Power Operation (1998)]. Japan: Nuclear Power Engineering Corporation; 1999. INS/M98-26. [Japanese]

[2] Nuclear Power Engineering Corporation (NUPEC). [The Report of Establishment of Level 1 PSA Method of ABWR Plant at Power Operation (1999)], Japan: Nuclear Power Engineering Corporation; 2000, INS/M99-29. [Japanese]

[3] Nuclear Power Engineering Corporation (NUPEC). [The Report of Establishment of Level 1 PSA Method for Internal Events at Power Operation = Reliability Analysis of Digital Reactor Protection System = (2002)]. Japan: Nuclear Power Engineering Corporation; 2003, INS/M02-29. [Japanese]

[4] Japan Nuclear Energy Safety Organization (JNES). [The Report of Establishment of Level 1 PSA Method for Internal Events at Power Operation = Improvement of Reliability Analysis of Digital Reactor Protection System (PWR) =]. Japan: Japan Nuclear Energy Safety Organization; 2007, JNES/SAE07-029. [Japanese]

[5] Japan Nuclear Energy Safety Organization (JNES). [The Report of Establishment of Level 1 PSA Method for PWR Plants at Power Operation = 3-Loop PWR Plant with/without Accident Management Countermeasures=]. Japan: Japan Nuclear Energy Safety Organization; 2009, JNES/SAE08-013. [Japanese].

[6] Stefan Authen, Jan-Erik Holmberg. Reliability Analysis of Digital Systems in a Probabilistic Risk Analysis for Nuclear Power Plants. NUCLEAR ENGINEERING AND TECHNOLOGY. 2012 June; VOL.44 NO.5.

[7] Ewgenij Piljugin, Stefan Authén, Jan-Erik Holmberg. Proposal for the Taxonomy of Failure Modes of Digital System Hardware for PSA. Paper presented at: 11th International Probabilistic Safety Assessment and Management Conference & The Annual European Safety and Reliability Conference; 2012 June; Helsinki, Finland.

[8] Tsong-Lun Chu, Meng Yue, Wietske Postma. A Summary of Taxonomies of Digital System Failure Modes Provided by the DigRel Task Group. Paper presented at: 11th International Probabilistic Safety Assessment and Management Conference & The Annual European Safety and Reliability Conference; 2012 June; Helsinki, Finland.

[9] Japan Nuclear Energy Safety Organization (JNES). [The Report of Improvement of Reliability Model of Digital Reactor Protection System]. Japan: Japan Nuclear Energy Safety Organization; 2010, JNES/SAE10-013. [Japanese]

[10] T Shimodaira, Y Sato and K Suyama. [Estimation of hazardous event rate for repairable 1-out-of-2 safety-related systems based on state transition models]. Trans of the institute of Electronics, Information and Communication Engineers. 2005 August; Vol.J88-A. No.8: P.962-973. [Japanese]

A quantitative software testing method for hardware and software integrated systems in safety critical applications

Hai Tang[a], Lixuan Lu*[a]

[a] University of Ontario Institute of Technology, Oshawa, ON, Canada

Abstract: Most of today's Safety Instrumented Systems (SIS) are hardware and software integrated systems. In these systems, failures can occur in both hardware and software. Hardware failures and their effects have been studied extensively in the literature. However, the methods and results dealing with hardware failure are not directly applicable for software reliability modeling, due to the difference of nature between hardware and software. This is especially of concern when the SIS is used for safety critical applications. In this paper, a hardware and software integrated reliability model is proposed to model the reliability of the integrated system. The requirement on software reliability is then determined based on the hardware reliability and the requirement on the Safety Integrity Level (SIL) of the integrated system. Following this, a Bayesian stopping rule is used to determine the minimal number of successful software runs, in order to provide a certain level of confidence that the reliability requirement on the software is achieved.

Keywords: Probabilistic risk assessment (PRA)/probabilistic safety assessment (PSA), Safety instrumented systems (SIS), Hardware and software integrated reliability model, Bayesian stopping rule, Software reliability demonstration test (SRDT).

1. INTRODUCTION

Most of today's Safety Instrumented Systems (SIS) are hardware and software integrated systems, used in safety critical applications, such as in nuclear power plant control, chemical processes and machine guarding. They are used to detect hazardous events and perform safety-related functions to reduce the risk of human injury and fatality. Although there have been many methods proposed in literature for probabilistic risk assessment (PRA)/probabilistic safety assessment (PSA) of instrumentation and control systems, most of the analyses are geared towards hardware failures and their effects. Quantitative reliability modeling and analysis of hardware and software integrated systems still pose great challenges, especially in the situation when the a quantitative software reliability assessment is taken into consideration.

Software and hardware fail differently. This is caused by many fundamental aspects. Hardware is the physically connected components that perform or support a function within the system. It normally has a time-related failure rate because of the hardware wear out process. Hardware failure occurs when some form of stress exceeds the associated strength of the product. On the other hand, software is a collection of instructions which enables a controller to perform a specific task on the hardware platform, and software cannot work alone without hardware. Software failure mechanisms are different from hardware failures in that all software failures are inadvertently designed into the system [1]. Software does not wear out as hardware, thus software failure is time-independent.

When building reliability models for hardware and software integrated systems, there are many factors to consider. First of all, due to the fundamental difference between hardware and software failure mechanisms, it is obvious that the methods developed to model the hardware reliability are not suitable for software reliability modeling, and vice versa. Software failures are the design defects highly related to design process management and developers' skills and experience. It is not easy to quantitatively incorporate all these factors with existing reliability models, although many software

*Contact Author: Lixuan.Lu@uoit.ca

reliability models have been developed for reliability assessment of different phases of the software lifecycle [2, 3]. All of these models have their own weakness, and none of them can cover the whole software lifecycle. Moreover, general purpose software reliability models always require failure data during the development, testing and operating phase of software lifecycle, which can be difficult to obtain for safety critical software. Software used in SIS requires an extremely high reliability, whose failure is very rare or may be never found from the software testing.

An integrated reliability model is developed in this paper to analyze the Safety Integrity Level (SIL) of an integrated SIS, in which the time-independent software reliability model is coupled with the time-related hardware model to calculate the average probability of failure on demand (PFDavg). A reliability demonstration test is designed with Bayesian stooping rules, which gives a proof with certain level of confidence that the software is able to respond successfully when a SIS is required to perform a safety function on demand.

2. SAFETY INTEGRETY LEVEL

2.1. Average Probability of Failure on Demand

IEC 61508 [4] is an international standard for functional safety of systems. It specifies the requirement on the entire safety lifecycle management of a SIS, including system design, development and certification stages. According to the standard, the safety analysis procedure requires a quantitative determination of the risk reduction by using the safety system as a protection layer to the Equipment Under Control (EUC).

Safety Integrity level (SIL) is used as a measure of the risk reduction achieved by the safety system when a hazardous event occurs in the EUC. The probability of a SIS failing dangerously is the Probability of Failure on Demand (PFD). This represents the risk that the safety system fails to perform safety functions as designed when a demand for the safety function occurred. Assuming a single channel safety system has a dangerous failure rate λ_D, the PFD is the probability of a dangerous failure and is shown in Eq. (1):

$$PFD(t) = 1 - e^{-\lambda_D t} \tag{1}$$

Safety systems usually have an extremely low failure rate, which allows the PFD to be calculated by the approximated Eq. (2) in real applications.

$$PFD(t) = \lambda_D t \tag{2}$$

PFD average (PFDavg) is the average of probability of failure on demand, which is defined in Eq. (3):

$$PFDavg = \frac{1}{T} \int PFD(t) dt \tag{3}$$

2.2. Safety Integrity Level (SIL)

The *PFDavg* quantitatively represents the probability that a safety system fails performing the designed safety function when a demand occurs. The SIL is an index number uesed in real applications to indicate system risk reduction levels according to system *PFDavg* values. As shown in Table 1, four SILs, ranking form SIL 1 to SIL 4, are specified to show different ranges of system PFDavg. Systems with a higher SIL have a better risk reduction to the EUC. A lower *PFDavg* corresponds to a higher SIL, which means the system has a higher probability to perform a safety function correctly when a hazardous event occurs. Safety systems usually use redundancy on the component level or system level to achieve a certain SIL.

Table 1: Safety Integrity Level

Safety Integrity Level	Probability of failure on demand (PFDavg)	Risk Reduction factor (RRF)
1	10^{-1} to 10^{-2}	10 to 100
2	10^{-2} to 10^{-3}	100 to 1000
3	10^{-3} to 10^{-4}	1000 to 10000
4	10^{-4} to 10^{-5}	10000 to 100000

3. SOFTWARE RELIABILITY

3.1. Input Domain Based Model

Software plays a more and more important role in modern control systems, due to the fact that most control functions generated by the programmable logic controllers are software based functions. Safety functions performed by SIS require an interaction between hardware and software, in which the software processes the reading of the sensors and provides an output as the control signal to the EUC. SIS will fail into a dangerous condition if the software has an incorrect response to the hazardous event when the SIS is expected to perform a safety function.

Software reliability is a quantitative measurement of how well the software system work according to the design specifications. Unlike hardware, software does not wear out over time. There are many software reliability models in the literature to model software reliability in different stages of the software lifecycle. Reliability growth models are widely used in industrial applications. They provide a rudimentary estimation on system reliability and can be used to support project management. However, the assumptions made to use these models are still questionable and the characteristics of the software under evaluation are insufficiently accounted for. In addition, the techniques used to estimate the software reliability require a failure history. This can be difficult to obtain for safety critical systems. Because software in these systems normally ha a much higher reliability level than general purpose software and it rarely or never fails during testing [5].Input domain based models are more advantageous over the reliability growth models when modeling the reliability of safety critical software, because the method does not rely on empirical assumptions and the result is acquired through a direct statistical approach according to the probability of successful execution in software testing processes.

In an input domain based model [6], n inputs are randomly selected from the input data set E=(E_i: i=1, 2, ... , N), where E_i is the subset of software input domain. The inputs are sampled with the input distribution of operational profile P=(P_i: i=1, 2, ... , N); where P_i is the probability of choosing E_i as the input. If f failures are found by the execution of n inputs, then the software reliability can be estimated as:

$$R = 1 - f/n \tag{4}$$

When using the input domain based model, a prior knowledge of the system operational profile is required for testing purposes. The operational profile is a quantitative characterization of how the software is used in the real applications. A step by step approach to develop the operation profile for software testing can be found in [7]. For safety critical software test, the operational profile could be developed according to the operational history of a plant or by expert knowledge. If the input distribution of the software is unknown, then the operational profile can be developed by assuming a uniform input distribution over the software input domain.

3.2. Application of Input Domain Based Model for Safety Critical Software

Numerically, if no fault is detected from a software test, then the reliability of software can be estimated as 1. However, a realistic software product will never achieve 100% reliability unless all combinations of all possible inputs are tested and they all give the correct results as expected. To achieve this goal, an infinite number of tests are required. This is obviously impossible to use in real applications due to the time and cost for testing. There must be a stopping rule for software test to determine the minimal number of successful runs to provide a certain level of confidence that the software achieves its reliability goal.

In this paper, a hardware and software integrated reliability model is developed to calculate the requirement of software reliability when the software and software are integrated as a system to meet a probability of failure on demand for a certain SIL. Afterwards, a Bayesian stopping rule for safety critical software testing is applied to calculate the minimal executions before the termination of the test.

4. HARDWARE AND SOFTWARE INTEGRATED RELIABILITY MODEL

4.1. Hardware and Software Integrated System

An integrated reliability model is developed to model the system reliability of hardware and software integrated systems. The hardware subsystem is the physically interconnected devices including sensors, controllers and actuators. The software subsystem is the programs running on the hardware platform which enable the system to perform a specific task. The difficulties of reliability modeling of hardware and software integrated systems are caused by the natures of how hardware and software fail. Hardware fails over time. The reliability of hardware usually follows time-related distributions, such as exponential distribution or Weibull distribution. On the other hand, software failures are time-independent. Software does not wear out. The failures of software are the errors designed into the systems. When doing quantitative reliability assessment of hardware and software integrated systems, a model is needed to couple the time-related hardware reliability model and time-independent software reliability model together.

4.2. Integrated Reliability Modeling

To have the SIS respond successfully when a safety function is on demand, the hardware should work in a failure-free condition and the software should process the output properly as designed and give a correct control command to the hardware. Failures in either hardware or software could cause a failure on demand for the SIS.

With a dangerous failure rate $\lambda_{D,hw}$, the probability of failure on demand of hardware ($PFD_{hw}(t)$) can be calculated by Eq. (5) below:

$$PFD_{hw}(t) = 1 - e^{-\lambda_{D,hw}t} = \lambda_{D,hw}t \tag{5}$$

Software reliability R_{sw} is a time-independent value and it can be calculated by Eq. (6):

$$R_{sw} = 1 - F_{sw} \tag{6}$$

where F_{sw} is the probability of failure of the software. The reliability target of the software (R_{sw}) needs to be calculated from the integrated system reliability model as the reliability goal for software testing.

Use PFD_{sys} to represent the probability of failure on demand of the integrated system when the hardware and software are working together to respond to a safety function on demand. The system will fail when "hardware fails" OR "software fails". Therefore, the probability of failure on demand of the integrated system is calculated as shown in Eq. (7):

$$PFD_{sys}(t) = PFD_{hw}(t) \cup F_{sw} = PFD_{hw}(t) + F_{sw} - PFD_{hw}(t) \times F_{sw} \qquad (7)$$

The average probability of fail on demand of the hardware and software integrated system ($PFDavg_{sys}$) can be calculated as Eq. (8):

$$PFDavg_{sys} = \frac{1}{T} \int PFD_{sys}(t)dt \qquad (8)$$

Substituting Eq. (7) into Eq. (8), Eq. (9) is obtained:

$$PFDavg_{sys} = \frac{1}{T} \int (PFD_{hw}(t) + F_{sw} - PFD_{hw}(t) \times F_{sw})dt \qquad (9)$$

where T is usually referred to as the interval of proof test, which brings the system back to its original state after operating continuously for a period of time. Since the software reliability is time-independent, F_{sw} can be treated as a constant value when doing the integration, thus Eq. (9) is rearranged as:

$$PFDavg_{sys} = \frac{1}{T} \int PFD_{hw}(t)dt + F_{sw} - F_{ws} \times \frac{1}{T} \int PFD_{hw}(t)dt \qquad (10)$$

Substituting Eqs. (6) and (8) into Eq. (10) gives:

$$PFDavg_{sys} = PFDavg_{hw} + (1 - R_{sw}) - (1 - R_{sw}) \times PFDavg_{hw} \qquad (11)$$

4.3. Software Reliability Requirement

The reliability of the software can be calculated by solving Eq. (11), which yields:

$$R_{sw} = \frac{1 - PFDavg_{sys}}{1 - PFDavg_{hw}} \qquad (12)$$

To meet a given requirement on SIL n (n=1, 2, 3, 4), the $PFDavg$ of the hardware and software integrated system must be lower than the maximal allowed average probability of failure on demand ($PFDavg_{sys,max,n}$) for that SIL. Thus, the software reliability must be higher than the critical software reliability ($R_{sw,critical,n}$), which is the software reliability value that makes the $PFDavg_{sys}$ equals to $PFDavg_{sys,max,n}$. According to Eq. (12), the software reliability requirement of a SIL n hardware and software integrated system is:

$$R_{sw,critical,n} = \frac{1 - PFDavg_{sys,max,n}}{1 - PFDavg_{hw}} \qquad (13)$$

5. BAYESIAN STOPPING RULE FOR SOFTWARE TESTING

5.1. Bayesian theory for software testing

After the software reliability requirement for the system is determined, the Software Reliability Demonstration Test (SRDT) is performed to verify that the software meets the reliability requirement with an acceptable degree of confidence. A stopping rule for the software reliability testing is required to make a decision of whether the test should be continued or terminated. When testing a safety critical system, the stopping decision could be made based on Bayesian analysis of the testing results [8].

There are two possible results for each test: fail or pass. If the SIS performs correctly when a safety function is on demand, the test is passed. If the system fails to response properly for a safety function demand, the test is failed. Assuming the value of probability of failure for software is p, the number of failures f from n tests follows a Binomial distribution:

$$P(F = f) = C_f^n \, p^f (1-p)^{n-f} \tag{14}$$

A prior conjugate is used to represent the changes of the parameter of interest p with extra information gathered from the test. The conjugate distribution follows a Beta(a,b) distribution:

$$f(p) = \frac{p^{a-1}(1-p)^{b-1}}{B(a,b)} \tag{15}$$

Where $B(a,b)$ is the Beta function, and a and b are parameters chosen by the assessor to represent the prior knowledge of the parameter p. casein the case where, there is no information available for selecting a and b, a uniform prior can be used with $a=b=1$.

If f failures are found from n tests, the posterior distribution of p is Beta$(a+f,b+n-f)$:

$$f(p \mid f,n,a,b) = \frac{p^{a+f-1}(1-p)^{b+n-f-1}}{B(a+f,b+n-f)} \tag{16}$$

For uniform prior, Eq. (16) becomes:

$$f(p \mid f,n,1,1) = \frac{p^f (1-p)^{n-f}}{B(1+f,1+n-f)} \tag{17}$$

5.2. Stopping rule based on Bayesian decision theory

If F_{sw} is used to represent the maximum probability of failure for software, and C is used to represent the confidence level of the test result, the reliability requirement for a software test could be expressed as:

$$P(p < F_{sw}) \geq C \tag{18}$$

To meet the requirement on reliability and confidence, the smallest value of successful execution without failure n_1 is the smallest value of n which satisfies Eq. (19):

$$\int_0^{F_{sw}} \frac{(1-p)^n}{B(1,1+n)} dp \geq C \tag{19}$$

If one failure occurs after s_1 ($s_1 < n$) executions, the posterior distribution for p becomes:

$$f(p \mid 1, s_1, 1, 1) = \frac{p(1-p)^{s_1-1}}{B(2, s_1)} \tag{20}$$

This is the new prior distribution of p for the following testing stage. The posterior distribution after n_2 failure free tests are observed is:

$$f(p \mid 1, s_1 + n_2, 1, 1) = \frac{p(1-p)^{s_1+n_2-1}}{B(2, s_1 + n_2)} \tag{21}$$

For a given requirement on F_{sw} and C, the smallest value n_2 for failure free execution in the following stage of test after the fault is fixed can be solved from Eq. (22):

$$\int_0^{F_{sw}} \frac{p(1-p)^{s_1+n_2-1}}{B(2, s_1 + n_2)} dp \geq C \tag{22}$$

To continue this process, if the jth failure occurs on the s_jth test executions, the number of failure free execution for the next test stage (n_{j+1}) can be computed by solving the general Eq. (23):

$$\int_0^{F_{sw}} \frac{p^j (1-p)^{\sum_{i=1}^{j} s_i + n_{j+1} - j}}{B(j+1, \sum_{i=1}^{j} s_i + n_{j+1} - j + 1)} dp \geq C \tag{23}$$

In real applications, a simplified stopping rule developed below can be used to reduce the calculation for the process as described above. If j failures occurred, let the total number of executions until n_{j+1} failure free tests observed in the $j+1$th stage of test be N.

$$N = s_1 + s_2 + \cdots + s_j + n_{j+1} \tag{24}$$

N is the minimum number of executions required to successfully demonstrate the required level of reliability with predetermined acceptable confidence level. Regardless of when these failures are happened during the test, this test process can be treated equivalently as a single test process in which j failures are observed out of N executions. To meet the requirement of software reliability, the total executions N that contains j failures should be the minimum value of N which satisfies Eq. (25):

$$\int_0^{F_{sw}} \frac{p^j (1-p)^{N-j}}{B(j+1, 1+N-j)} dp \geq C \tag{25}$$

According to the analysis above, the stopping rule for the software testing only depends on the total number of executions and the total number of failures out of these executions. Given a reliability requirement F_{sw} and C, the stopping rule for reliability test can be calculated before the test is carried out.

6. NUMERICAL EXAMPLE

This part of paper will give a numerical example to demonstrate the application of the methodology developed above.

Assuming a hardware and software integrated system is used as an emergency shutdown system for a chemical plant to stop the process in EUC when a hazardous event occurs. The dangerous failure rate λ_D for the hardware system is evaluated as 1.6×10^{-7} /h and proof testing will perform every year (8760 hours), which brings the safety instrumented system back to its original state. According to the hazard analysis of the chemical plant, a SIL 3 is required for the emergency shutdown system.

The software reliability requirement for the given SIL can be calculated from the SIL requirement and hardware failure rate. From Eq. (5), the probability of failure on demand of the hardware subsystem is:

$$PFD_{hw}(t) = 1.6 \times 10^{-7} t \tag{26}$$

The $PFDavg_{hw}$ of hardware is the integral average of hardware PFD over the time of proof testing interval T, and is calculated as below:

$$PFDavg_{hw} = \frac{1}{T} \int PFD_{hw}(t)dt = \frac{1}{T} \int_0^T \lambda_D t \, dt = \frac{\lambda_D T}{2} = \frac{1.6 \times 10^{-7} \times 8760}{2} = 7 \times 10^{-4} \tag{27}$$

For a SIL 3 application, the maximal allowed $PFDavg_{sys}$ for the system is 10^{-3}, thus the software reliability requirement for the system is calculated from equation (13) as:

$$R_{sw,critical,3} = \frac{1 - PFDavg_{sys,max,3}}{1 - PFDavg_{hw}} = \frac{1 - 10^{-3}}{1 - 7 \times 10^{-4}} = 0.9997 \tag{28}$$

The probability of failure for software is therefore:

$$F_{sw} = 1 - R_{sw,critical,3} = 1 - 0.9997 = 3 \times 10^{-4} \tag{29}$$

A uniform prior distribution is used for the Bayesian analysis, where $a=b=1$. If a 99% confidence on the software reliability is required by the software test, the minimal number of executions with j failures is calculated by solving Eq. (30) below based on Eq. (25).

$$\int_0^{3 \times 10^{-4}} \frac{p^j (1-p)^{N-j}}{B(j+1,1+N-j)} dp \geq 0.99 \tag{30}$$

This equation is solved by using numerical method. Based on different numbers of system failures, the solution of Eq. (30) gives the stopping rule for the system safety test. The results are shown in Table 2. As can be seen, when the number of failures during test increases, the total number of required executions increase as well, in order to be 99% confident that the required software reliability level is achieved.

7. CONCLUSION

In this paper, a quantitative method is developed to determine the minimum number of testing required on software in hardware and software integrated systems, given the required Safety Integrity Level (SIL) on the overall system and hardware failure data. Due to the fundamental differences between hardware and software failure mechanisms, hardware and software are usually analyzed separately in practice by using distinct methodologies. Hardware failures are typically time-related with a certain failure rate λ , while the software failures are time-independent, since errors are inevitably designed into the final software product. To better model hardware and software integrated systems for

Table 2: Stopping rule for software testing

Numbers of failures, j	Total number of executions, N
0	15347
1	22124
2	28016
3	33479
4	38677
5	43690
6	48563
7	53328
8	58003
9	62604
10	67143
11	71627
12	76063

reliability analysis and to provide a practical method for reliability assessment, a reliability model for hardware and software integrated systems is developed in this paper to model the effect of both hardware and software on system safety and to produce a quantitative assessment result. Using this model, the requirement for software reliability is first determined based on hardware reliability and system SIL. A Software Reliability Demonstration Test (SRDT) based on a Bayesian stopping rule is then used to calculate the minimum number of executions that is required to prove the reliability requirement is achieved. One advantage of the method proposed in this paper compared with existing methods is that it couples both hardware and software reliability within the system. In addition, unlike the software reliability growth models based on empirical assumptions, the method proposed here measures the software reliability via the input domain based model, with which the result of the assessment could come directly from the observation of the testing. Finally, as shown in the numerical example, this method is practical and easy to use in real life applications.

References

[1]   William M. Goble, *"Control systems safety evaluation and reliability"*, International Society of Automation, 2010.

[2]   C.V. Ramamoorthy and F.B. Bastani, *"Software Reliability—Status and Perspectives"*, IEEE Transactions on Software Engineering, vol.SE-8, pp.354-371, (1982).

[3]   A.L. Goel, *"Software Reliability Models: Assumptions, Limitations, and applicability"*, IEEE Transactions on Software Engineering, vol.SE-11, pp. 1411-1423, (1985).

[4]   International Electrotechnical Commission, *"IEC 61508 Second Edition: Functional Safety of Electrical/Electronic/Programmable Electronic Systems"*, (2010).

[5]   A. Pasquini, E. De Agostino and G.D. Di Marco, *"An input-domain based method to estimate software reliability"*, IEEE Transactions on Reliability, vol. 45 , no. 1, pp. 95-105, (1996).

[6]   E. Nelson, *"Estimating software reliability from test data"*, Microelectronics Reliability, vol. 17, pp. 67-73, (1978).

[7]   J.D. Musa, *"Operational profile in software-reliability engineering"*, IEEE Software, vol. 10, pp. 14-32, (1993).

[8]   B. Littlewood and D. Wright, *"Some conservative stopping rules for the operational testing of safety critical software"*, IEEE Transactions on Software Engineering, vol. 23, pp. 673-683, (1997).

OECD/NEA WGRISK task on failure modes taxonomy for digital I&C – DIGREL

Abdallah Amri[a], Stefan Authén[b], Herve Bruneliere[c], Gilles Deleuze[d], Gabriel Georgescu[e], Jan-Erik Holmberg[f*], Man Cheol Kim[g], Keisuke Kondo[h], Ming Li[i], Ewgenij Piljugin[j], Wietske Postma[k], Jiri Sedlak[l], Carol Smidts[m], Jan Stiller[j], and Nguyen Thuy[d]

[a]OECD/NEA, Paris, France
[b]Risk Pilot AB, Stockholm, Sweden
[c]AREVA, Paris, France
[d]EDF R&D, Paris, France
[e]Institut de Radioprotection et de Sûreté Nucléaire, Paris, France
[f]Risk Pilot AB, Espoo Finland
[g]Chung-Ang University, Seoul, Korea
[h]Nuclear Regulation Authority, Japan
[i]United States Nuclear Regulatory Commission, USA
[j]Gesellschaft für Anlagen- und Reaktorsicherheit, Germany
[k]Nuclear Research and consultancy Group, the Netherlands
[l]ÚJV Řež, Husinec - Řež, Czech Republic
[m]Ohio State University, USA

Abstract: The OECD/NEA CSNI Working Group on Risk Assessment (WGRisk) has set up a task group called DIGREL to develop a taxonomy of failure modes of digital components for the purposes of probabilistic risk analysis (PRA). The failure modes taxonomy is based on a failure propagation model and a definition of five levels of abstraction: 1) system level, 2) division level, 3) I&C unit level, 4) I&C unit modules level, 5) basic components level. This structure corresponds to a typical reactor protection system architecture. The failure propagation model consists of the following elements: fault location, failure mode, uncovering situation, failure effect and the end effect. These concepts are applied to define the relationship between a fault in hardware or software modules (module level failure modes) and the effect on I&C units (I&C unit level failure modes). The purpose of the taxonomy is to support PRA, and therefore focuses on high level functional aspects rather than low level structural aspects. This focus allows handling of the variability of failure modes and mechanisms of I&C components. It reduces the difficulties associated with the complex structural aspects of software in redundant distributed systems.

Keywords: Probabilistic risk analysis, digital I&C, failure mode, taxonomy

1. INTRODUCTION

Digital protection and control systems appear as upgrades in older Nuclear Power Plants (NPP), and are commonplace in new NPP. To assess the risk of NPP operation and to determine the risk impact of digital systems, there is a need to quantitatively assess the reliability of the digital systems in a justifiable manner. Due to the many unique attributes of digital systems (e.g., functions are implemented by software, units of the system interact in a communication network, faults can be identified and handled online), a number of modelling and data collection challenges exist, and international consensus on the reliability modelling has not yet been reached.

In 2007, the OECD/NEA CSNI directed the Working Group on Risk Assessment (WGRisk) to set up a task group to coordinate an activity in this field. One of the recommendations was to develop a

---

* jan-erik.holmberg@riskpilot.se

taxonomy of failure modes of digital components for the purposes of Probabilistic Risk Assessment (PRA) [1]. This resulted in a follow-up task group called DIGREL. An activity focused on development of a common taxonomy of failure modes is seen as an important step towards standardised digital I&C reliability assessment techniques for PRA. Standard technological equipment of NPP process systems, like pumps, are either in the running or standby mode. On the opposite, computer based systems are typically always in the running mode – the difference in the modes is that they process different sets of input parameters and consequently solve different branches of algorithms. The need of specific taxonomy establishment is hence obvious.

The paper will give an overview of the digital I&C failure modes taxonomy, which will be published as an OECD/NEA working report in 2014. Results presented here should be considered preliminary proposals and not as the task group consensus thoughts.

2. GENERAL APPROACH

2.1. Introduction

A failure modes taxonomy is a framework of describing, classifying and naming failure modes associated with a system. One of the main uses of digital I&C failure modes taxonomy is to support the performance of reliability analyses and to unify the operational experience data collection of digital I&C systems. In the work of the DIGREL task, needs from PRA have guided the definition of the taxonomy, meaning, e.g., that I&C system and its failures are studied from their functional significance point of view.

In PRA, a failure modes taxonomy is applied in the systems analysis, including the performance of FMEA (failure modes and effects analysis) and the fault tree modelling. In PRA, the definitions for the failure modes and the related level of details of abstraction in the fault tree modelling can be kept in a high level as long as relevant dependencies are captured and reliability data can be found.

The DIGREL task has taken advantage from recent and on-going R&D activities carried out in the member countries in this field. This knowledge has been merged by inviting experts in the field to contribute to the activity. Example taxonomies have been collected from the member countries, and analysed, and the conclusions from the taxonomy examples and workshop discussions have been taken into account when considering principles for the taxonomy [2]. This material showed some variety in the handling of I&C hardware failure modes, depending on the context where the failure modes have been defined. Regarding the software part of I&C, failure modes defined in NPP PRAs have been simple – typically a software CCF failing identical processing units.

The taxonomy has been developed jointly by PRA and I&C experts which have slightly different views and needs on defining the failure modes. The PRA experts' perspective follows the needs of PRA modelling in order to capture relevant dependencies and to find justifiable reliability parameters. I&C experts are focused on failure mechanisms and their recovery means, e.g. verification and validation (V&V) measures. An important aspect in the development of the taxonomy has been for PRA and I&C experts to define the "meeting point" for the two perspectives. The "meeting point" means both agreeing on common terminology and defining the issues and levels of abstraction which the taxonomy shall address.

2.2. Requirements for the taxonomy

The development of a taxonomy is dependent on the overall requirements and prerequisites since they will set boundary conditions e.g. for the needed level of detail of hardware components and for the structure of the failure modes. A different set of requirements may result in a different taxonomy. The following targets for the taxonomy have been defined by the task group:

- Defined unambiguously and distinctly

- Forms a complete/exhaustive set, mutually exclusive failure modes
- Organized hierarchically
- Data to support the taxonomy should be available
- Analogy between failure modes of different components
- The lowest level of abstraction of the taxonomy should be sufficient to pinpoint existing dependencies of importance to PRA modelling
- Supports PRA practice, i.e. appropriate level for PRA, and fulfil PRA requirements/conditions
- Captures defensive measures against fault propagation and other essential design features of digital I&C.

2.3. Example system

The taxonomy is focused on the reliability analysis of the reactor protection system, which reduces the scope of failure modes and failure effects considerably. This limitation can be justified by several arguments. Firstly, there is a general consensus that protection systems (reactor trip & Engineered Safety Feature Actuation Systems (ESFAS)) shall be included in PRA, while control systems can be treated in a limited manner. Secondly, the system architecture and the mode of operation of protection systems versus control systems are different, which creates quite different basis for the reliability analysis and modelling. Thirdly, the I&C of the control systems is versatile having both on demand and continuous functions and they do not necessarily have a redundant structure. Even if the taxonomy is focused on the protection systems, it can be useful for control systems, too.

A representative fictive digital protection system example has been developed to be used as a reference in the application and demonstration of the taxonomy. Though there are technical differences between solutions provided by different vendors, many of the features of protection systems are similar for all vendors. Therefore the example is considered representative enough for the failure modes taxonomy purposes

The simplified model takes into account the following:

- Typical architecture of digital I&C systems performing safety functions of the Reactor trip system (RTS) and the ESFAS, jointly called hereafter Reactor Protection System (RPS) functions
- Typical hardware components of the digital I&C platforms
- Typical operation modes of the RPS: ready to actuate a safety function on demand (maintenance, testing, etc. modes are not considered)
- Typical means and features for failure detection and recovery
- Typical majority voting for actuation of RTS and ESFAS functions.

The example system implements I&C safety functions (of Category A according to [3], of Class 1E according to [4]). The overall system architecture describes its organisation in terms of divisions and I&C units.

The architecture of a digital I&C system is established primarily by hardware (e.g. analog and digital circuit boards/modules, units, cabinets) and their communication paths (e.g. direct wired connections, network communications, signal distribution boards). The architecture determines essentially the propagation paths of the probable failures of the hardware and of the software.

The example system is organised into two separated subsystems A and B, which are based on the same I&C platform, but implement functions that are diverse. The two subsystems do not exchange information, and for shared actuators, their outputs are fed to simple hardwired logic to determine system-level outputs.

The overall example system architecture can thus be summarised as shown in Figure 1. Each subsystem is itself organised into four redundant divisions, each division of subsystem being composed of different types of I&C units, namely:

- Acquisition and processing units (or APUs): these units acquire process-related information from sensors, and perform calculations to determine the division outputs.
- Voting units (or VUs): these units receive the results determined by the APUs of their division and subsystem and for which voting is required.
- Data Communication Units (or DCUs): these units allow APUs and VUs to communicate with one another. DCUs are integral parts of APUs and VUs, and therefore not shown in the figure below.

Figure 1: Example protection system architecture.

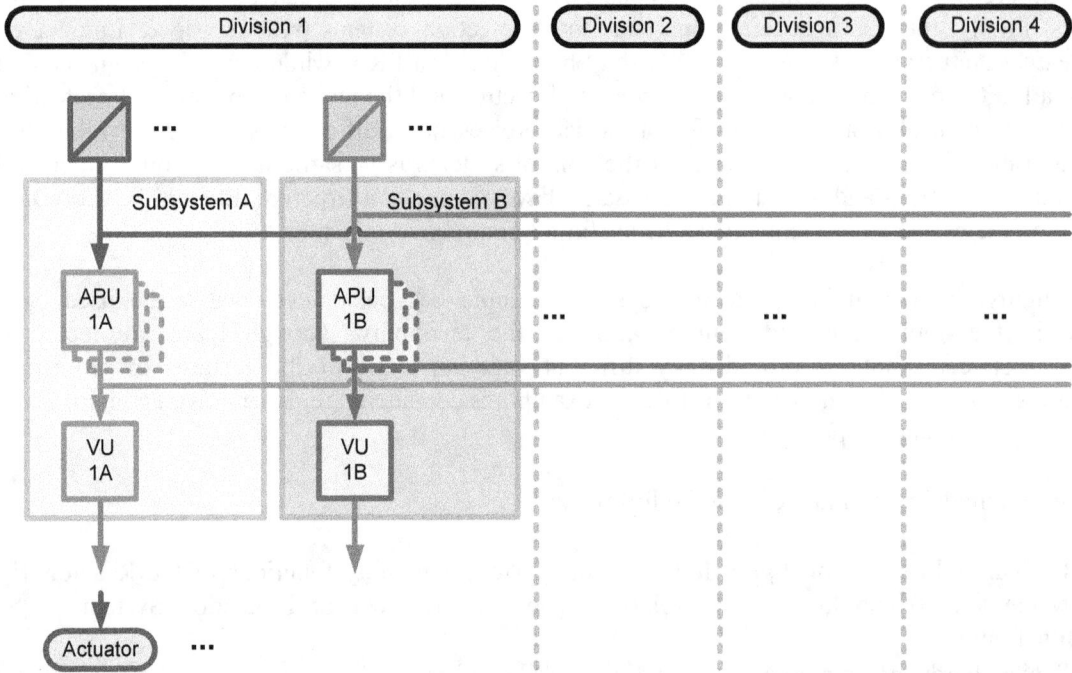

2.4. Levels of abstraction

The DIGREL task group has defined five levels of abstraction for the consideration of failure modes. The levels were identified when comparing examples of taxonomies used by different organisations [2], and are:

1) system level (complete reactor protection system),
2) division level,
3) I&C unit level,
4) I&C unit modules level,
5) basic components level.

This structure corresponds to a typical reactor protection system architecture, which is the scope of the DIGREL work. To handle complexity at the level of system, division and I&C units, failure modes are considered as much as possible only from the functional point of view. No significant distinction is made between hardware or software aspects at these levels. At the module and basic component levels, the taxonomy differentiates between hardware and software related failure modes.

3. FAILURE MODES TAXONOMY FOR DIGITAL I&C

3.1. Basic principles

The main approach is to define failure modes hierarchically and functionally. Failure modes are considered both from top-down and bottom-up perspective. The top-down structuring starts from the failure modes of the actuator functions (e.g. a pump fails to start, a valve fails to open), identifies associated I&C functions and continues down to units, modules and even to basic components, if so wished.

In the bottom-up view the failure modes are defined at low level of abstraction (typically at the module level or the basic component level) and then the failure effects are considered at the higher level. The result is a set of mappings between failure modes and effects between two levels of hierarchy. The PRA practitioner has to choose suitable level of abstraction for each individual PRA and its application.

The taxonomy is developed using a specific conceptual model of failure and failure propagation. The important elements of the failure model are:

- fault location,
- failure mode,
- uncovering situation,
- failure effect,
- end effect.

These concepts are applied, in particular, to define the relationship between a *fault* in hardware or software modules (module level *failure modes*) and the *end effect* on I&C units (I&C unit level failure modes). In the analysis, a fault is postulated in a hardware or software module (*fault location*). For hardware modules, different *failure modes* are explicitly defined. Software module *failure modes* are directly associated with the *failure effect*. *Uncovering situation* describes when, where and how the module failure is significant at the I&C unit level. A taxonomy of generic *failure effects* is defined to provide a simple but exhaustive way to categorise the effect of wrong output in a module.

The *end effect* describes the final propagation of the failure, taking into consideration all these elements of the failure model. In this consideration, a distinction can be made between the "maximum possible end effect", when fault tolerance design (FTD) is not effective or does not exist, and the "most likely end effect", assumes that FTD features are present and effective. FTD is effective only when the fault is detected by online monitoring, which is one of the uncovering situation categories.

A comprehensive description of the failure model can be found in [5], and is illustrated in Figure 2. Failure propagation is the path from a "locally" postulated fault to a system or plant level end effect, and it is dependent on the "context", which defined by the "plant condition", "initiating event" and "activation conditions". The propagation can be considered at different levels of abstraction following the I&C architecture. The most interesting part for PRA modelling is though the propagation between module and I&C unit levels.

Figure 2: Failure model [5].

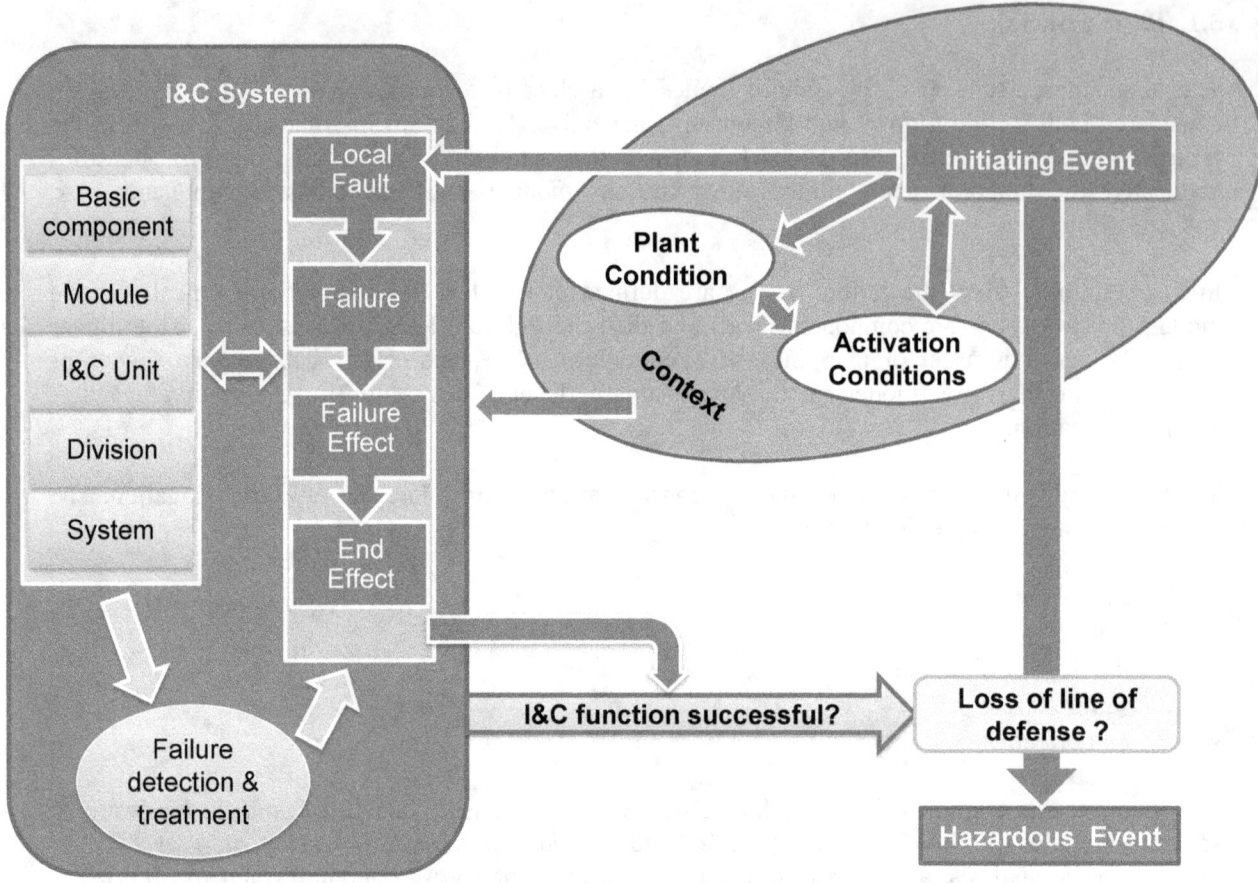

Figure 2: Failure model [5].

This general approach has been developed in the course of the DIGREL project. Its applicability and usefulness need to be assessed in further research efforts.

3.2. Failure mode taxonomy at System and division levels

Practically, the safety-related function of the system is defined as the generation of safety-related actuation signal in a predefined time interval only when required. Since the "division" designates the division of the protection system which is responsible for controlling the actuators in the corresponding division, the function of a division is the same as for a system. Thus, the failure modes in the division level are similar with those of the system level, which are

- failure to actuate the function (including late actuation),
- spurious actuation.

3.3. Failure mode taxonomy at I&C unit and module levels

The key part of the digital I&C failure modes taxonomy is in the I&C unit and module levels where the fundamental functionality of the system can be discussed, e.g., the defensive measures against faults. It is practical to keep these two levels together in the taxonomy since the meaning is to define the relation between failure modes of an I&C unit and the modules.

In the analysis, the existence of faults is postulated in the modules (hardware or software), and the question is to determine 1) how the unit is affected and 2) how other units that communicate with the defected unit are affected. In order to answer to these questions, the following issues need to be defined:

- The fault location: In which hardware or software module the fault is located?
- Failure effect:
  - *Fatal, ordered failure*: generation of outputs ceases, outputs are set to specified, supposedly safe values. Halt/abnormal termination of function with clear message.
  - *Fatal, haphazard failure*: generation of outputs ceases, outputs are in unpredictable states. Halt/abnormal termination of function without clear message.
  - *Non-fatal, plausible behaviour*: I&C runs with wrong results that are not evident. An external observer cannot determine whether the I&C unit or the hardware module has failed or not.
  - *Non-fatal, non-plausible behaviour*: I&C runs with evidently wrong results. An external observer can decide that the I&C unit or the hardware module has failed.
- Uncovering situation:
  - *Online detection*. Covers various continuous detection mechanisms.
  - *Offline detection*. E.g. periodic testing, and also other kind of periodic controls which can be credited in PRA.
  - *Revealed by demand*.
    - Latent failure, revealed by demand. A failure is present that is not detectable by online or offline mechanisms (test independent failure).
    - Failure triggered by demand. A specification error causes a failure on demand in an unexpected context.
  - *Revealed by spurious actuation*. The activation of the fault triggers spurious actuation before any FTD has time to take place. This situation covers two variants:
    - Spurious actuation due to functional failure, incl. voting logic
    - Spurious actuation due to failure of detection mechanism.

The combination of fault location, failure effect, uncovering situation together with the fault tolerant design (FTD) of the system are usually sufficient to determine the functional end effect in the I&C unit (APU or VU). Determination must be done case by case and is the essential part of the failure analysis.

An important issue is that it is neither necessary nor reasonable to assume all possible combinations, which considerably reduces the number of relevant failure modes (see Table 1). Fatal haphazard failures are not considered in this analysis, because here it is assumed that modules of the reactor protection system do not fail in an unknown state. Fatal failures are ordered and are detected by online detection or by spurious effect.

Non-fatal failures are more dangerous since any uncovering situation may be possible. In case of non-plausible behaviour, failure is detected by online detection or by spurious effect. "Plausible behaviour" refers to the case where the failure is not detected by online detection.

Table 1: Relevance of the combinations of local effects and detection situations

Failure effect	Uncovering situation				
	Online detection	Offline detection	Revealed by spurious action	Latent revealed by demand	Triggered by demand
Fatal, ordered	R	NR	R	NR	R
Fatal, haphazard	NR	R	R	R	R
Non-fatal, plausible behaviour	NR	R	R	R	R
Non-fatal, non-plausible behaviour	R	NR	R	NR	R

R: Combination relevant for further analysis of end effects
NR: Combination not relevant for the analysis of the effects. Non-relevance is due to logical considerations.

In the analysis of functional impacts on I&C units, we distinguish between the impact on a single I&C unit and impact on multiple I&C units. The latter is especially important when analysing the impacts of software faults (systematic fault in the design). From a single I&C unit point of view, the following functional failure modes can be considered

- Loss of all functions (outputs) of the I&C unit,
- Loss of a specific function,
- Spurious output (one function),
- Spurious output (all functions).

The above list is not exhaustive, and, e.g., for voting units the functional end effect may be more complex (e.g. degraded voting logic). Diesel load sequencer is also an example of a rather complex I&C function, for which a large number of failure modes may be assumed (but it can be sufficient to model only few of them in PRA).

In the example I&C architecture (Figure 1), the following end effects of a failure can be assumed:

- Failure of one function (or more) in one subsystem,
- Failure of one function (or more) in only one division in one subsystem,
- Failure of one function (or more) in both subsystems,
- Failure of one set of redundant APUs/VUs,
- Failure of multiple sets of redundant APUs/VUs in only one subsystem,
- Loss of one subsystem,
- Failure of multiple sets of redundant APUs/VUs in both subsystems,
- Loss of one subsystem and of one or more sets of redundant APUs in the other subsystem,
- Loss of both subsystems.

At the module level, a distinction is made between the treatment of hardware and software related failure modes. The taxonomy report [5] includes comprehensive list of hardware module failures which are associated with the failure effect and uncovering situations, and this is sufficient to determine the functional impact on I&C units. For instance, one of the failure modes of the processor module is that the processor stops execution of code, which is a fatal ordered failure and is detected by the online detection. The functional impact is loss of all APU/VU functions according to FTD. See [5] for more examples.

The approach for software modules is to successively postulate a single software fault in each software module regardless of the likelihood of such faults, and to determine the maximum possible extent of the failure, regardless of the measures taken by design or operation to limit that extent. The following list of software modules are considered:

- Operating system (OS). This module controls the overall functioning of the I&C unit (APU/VU/DCU), and is the same for all the units of all divisions or of all subsystems of the example system.

- Elementary functions (EFs). These modules provide readily useable standard (library) functions such as Boolean logic, mathematical functions or delays. They are the same for all units of the example system. However, an important difference with respect to the OS is that a specific APU/VU will use only a specific subset of all available EFs.

- Application-specific software modules (AS). These modules implement specific I&C functions in APUs and VUs. Homologous APUs (resp. VUs) in redundant divisions have the same sets of AS modules.

- Functional requirements specification modules (FRS). These are virtual software modules associated with application functions. There is one such module per application function required of an APU or a VU. The purpose to consider FRS modules is to allow the representation of errors in functional requirements specifications, which by operating experience has been shown to be not uncommon [6].

- Data communication software (DCS). This module implements the data communication protocol. It is part of the platform software, and all DCUs of the example system have the same DCS.

- Data link configuration (DLC). This module specifies the nodes that can be part of a given network, and the data messages that can be exchanged between the nodes of the network. The two subsystems use different networks, and therefore the DLCs of their DCUs are different.

- SW in COTS-modules (Commercial off-the-shelf). These modules are specific pieces of software present in hardware modules in APU, DCU, VU or any other module of the system (e.g. power supply) other than OS and AS. The implementation in software belongs to a commercial company, and the source code is not freely nor publicly available. It is restricted from use, such as modification or V&V, for the end user.

4. CONCLUSION

Due to the many unique attributes of digital systems, a number of modelling and data collection challenges exist, and international consensus has not yet been reached regarding their modelling in PRA. Currently in PRA, computer-based systems are mostly modelled by using simple approaches, and the primary goal is to model dependencies (I&C systems' support systems). There is a general consensus that protection systems shall be included in PRA, while control systems can be treated in a limited manner, depending on the importance on the plant safety.

The objective of OECD/NEA DIGREL task was to develop a failure mode taxonomy for reliability assessment of digital I&C systems for use in PRA. The I&C failure mode taxonomy has been developed to support modelling and quantification efforts. It will also help define a structure for data collection and to review PRA studies.

The proposed failure mode taxonomy has been developed by first collecting examples of taxonomies provided by the task group organisations. This material showed some variety in the handling of I&C hardware failure modes, depending on the context where the failure modes have been defined. Regarding software part of I&C, failure modes defined in nuclear power plant PRAs have been simple — typically a software related CCF failing identical processing units.

The failure modes taxonomy is based on a failure propagation model and a definition of five levels of abstraction: 1) system level, 2) division level, 3) I&C unit level, 4) I&C unit modules level, 5) basic components level. This structure corresponds to the typical reactor protection system architecture, which was the scope of the taxonomy researched in this document. The failure propagation model consists of the following elements: fault location, failure mode, uncovering situation, failure effect and the end effect. These concepts are applied to define the relationship between a fault in hardware or software modules (module level failure modes) and the effect on I&C units (I&C unit level failure modes). To handle complexity, at the level of system, division and I&C units, failure modes are considered as much as possible only from the functional point of view. No significant distinction is made between hardware or software aspects at these levels. At the module and basic component levels, the taxonomy differentiates between hardware and software related failure modes.

This approach has been developed in the course of the DIGREL project. Its applicability and usefulness need to be assessed in further research efforts. The purpose of the taxonomy is to support PRA, and therefore focuses on high level functional aspects rather than low level structural aspects. This focus allows handling of the variability of failure modes and mechanisms of I&C components. It reduces the difficulties associated with the complex structural aspects of software in redundant distributed systems.

Acknowledgements

Opinions presented are those of the authors and not necessarily those of the organisations of the authors or of other members of the DIGREL task group.

Contributions from the WGRISK/DIGREL task group members are acknowledged. The Finnish and Swedish work has been financed by NKS (Nordic nuclear safety research), SAFIR2014 (The Finnish Research Programme on Nuclear Power Plant Safety 2011–2014) and the members of the Nordic PSA Group: Forsmark, Oskarshamn Kraftgrupp, Ringhals AB and Swedish Radiation Safety Authority. NKS conveys its gratitude to all organizations and persons who by means of financial support or contributions in kind have made the work presented in this paper possible. Work by GRS has been funded by the German Federal Ministry for the Environment, Nature Conservation, Building and Nuclear Safety

References

[1] *"Recommendations on assessing digital system reliability in probabilistic risk assessments of nuclear power plants"*, NEA/CSNI/R(2009)18, OECD/NEA/CSNI, 2009, Paris.
[2] W. Postma, T.-L. Chu, M. Yue, *"Observations and discussion from the taxonomies of digital system failure modes provided by the DIGREL task group"*, ANS PSA 2013 International Topical Meeting on Probabilistic Safety Assessment and Analysis, Columbia, SC, September 22–26, 2013, on CD-ROM, American Nuclear Society, 2013, LaGrange Park, IL. Paper 91.
[3] *"Nuclear power plants — Instrumentation and control important to safety — Classification of instrumentation and control functions"*, IEC 61226, ed. 3.0 International Electrotechnical Commission, 2009, Geneva.
[4] *"IEEE Standard for Qualifying Class 1E Equipment for Nuclear Power Generating Stations"*, IEEE Std. 323-2003, Institute of Electrical and Electronics Engineers, 2003.
[5] *"Failure modes taxonomy for reliability assessment of digital I&C systems for PRA"*, Report prepared by OECD/NEA Working Group RISK Task group DIGREL, draft March 2014.
[6] *"Estimating Failure Rates in Highly Reliable Digital Systems"*, EPRI 1021077, Electric Power Research Institute, 2010, Palo Alto (limited distribution).

A Component-based Approach for Assessing Reliability of Compound Software

Monica Lind Kristiansen[a]*, Bent Natvig[b], and Harald Holone[c]
[a]Department of Informatics, Østfold University College, Halden, Norway
[b]Department of Mathematics, University of Oslo, Oslo, Norway
[c]Department of Informatics, Østfold University College, Halden, Norway

Abstract: Predicting the reliability of software systems based on a component approach is inherently difficult, in particular due to failure dependencies between the software components. This paper describes a component-based approach for assessing reliability of compound software, where failure dependencies between software components are explicitly addressed. This is done by finding accepted upper bounds for probabilities that pairs of software components fail simultaneously and then by including these into the reliability models. To find these accepted upper bounds, the approach applies principles of Bayesian hypothesis testing on simultaneous failure probabilities. In addition, the restrictions imposed on the simultaneous reliabilities and failure probabilities by the marginal reliabilities and failure probabilities are taken into account. To illustrate the approach, we use an example based on mobile positioning systems for backtracking. This is for instance used to help people with dementia to find their way home if they get lost.

Keywords: Failure dependencies, Component-based approach, Bayesian hypothesis testing, Mobile positioning systems.

1. INTRODUCTION

Before computerized systems can be used in any kind of critical applications, evidences that these systems are dependable are required [1]. This is desirable for most systems, but essential for systems which affect human safety and welfare, e.g. mobile positioning systems, patient monitoring systems, fly-by-wire systems, railway signal systems etc. Considering that most computerized systems are built as a structure consisting of several software components, of which some might have been pre-developed and used in other contexts, there is a need for methods for assessing reliability of compound software [1].

Although several different approaches to construct component-based software reliability models have been proposed in, among others, Cortellessa and Grassi [2], Gokhale and Trivedi [3], Gokhale [4], Palviainen et al. [5], Goseva-Popstojanova and Trivedi [6], Hamlet [7,8], Krishnamurthy and Mathur [9], Krka et al. [10], Kuball et al. [11], Popic et al. [12], Reussner et al. [13], Singh et al. [14], Trung and Thang [15], Vieira and Richardson [16], and Yacoub et al. [17], most of these approaches tend to ignore failure dependencies between software components [18–20]. This topic is discussed more thoroughly in, among others, Cortellessa and Grassi [2], Dai et al. [21], Gokhale and Trivedi [4], Guo et al. [22], Littlewood et al. [23], Lyu [1], Nicola and Goyal [24], Popic et al. [12], Popov et al. [25], and Tomek et al. [26].

In our research, we have developed a component-based approach for assessing reliability of compound software in which failure dependencies between software components are explicitly addressed [27].

---

* Corresponding author, monica.kristiansen@hiof.no
[1] Software systems consisting of multiple software components.

The idea is to find accepted upper bounds for probabilities that pairs of software components fail simultaneously and then include these into the reliability models. To find these accepted upper bounds, the approach applies principles of Bayesian hypothesis testing [28] on simultaneous failure probabilities. This includes taking into account the restrictions the marginal reliabilities and failure probabilities put on the simultaneous reliabilities and failure probabilities.

In Section 2, the motivation of our research is given. This is followed up by a description of our component-based approach in Section 3. This section also describes the assumptions of the approach and the direct restrictions on the simultaneous probabilities. In addition, some general rules for selecting the most important component dependencies are given. These rules are based on the concepts of data-parallel and data-serial components, which are defined in Section 3.3. In Section 4, the component-based approach is illustrated with an example from positioning services on mobile phones for geofencing and backtracking. A summary of the results, some conclusions and ideas for further work are given in Section 5.

2. BACKGROUND

Our research started by analyzing two interesting papers written by Cukic et al. [29] and Smidts et al. [30]. These papers present a Bayesian hypothesis testing approach for finding upper bounds for failure probabilities of single software components. The authors' idea is to complement testing with available prior information regarding the software components so that adequate confidence can be obtained with a feasible amount of testing.

In their approach, the null hypothesis (H_0) and the alternative hypothesis (H_1) are specified as: $H_0 : \theta \leq \theta_0$ and $H_1 : \theta > \theta_0$, where θ_0 is a probability in the interval $(0, 1)$ representing the upper bound for the failure probability θ of a software component. The upper bound θ_0 is assumed to be context specific and predefined and is typically derived from standards, regulation authorities, customers, etc. In this case, the null hypothesis states that the software component's failure probability is lower than the given predefined upper bound θ_0, whereas the alternative hypothesis states that the software component's failure probability is higher than the given predefined upper bound θ_0.

Furthermore, the authors describe the prior belief in the failure probability ($\pi(\theta)$) of a single software component using two separate uniform probability distributions, one under the null hypothesis and one under the alternative hypothesis. Based on this assumption, the authors show that the number of tests (D) required to obtain an adequate confidence level C_0 so that $P(H_0|D) \geq C_0$, can be significantly reduced compared to the situation where no prior belief regarding the software component is described. By assuming that prior belief in the null hypothesis $P(H_0)$ is 0.01, the predefined upper bound θ_0 is 0.0001, and the confidence level C_0 is 0.99, the authors show that it requires 6831 fault-free tests to reach the confidence level by using Bayesian hypothesis testing compared to 46050 fault-free tests by using classical statistical testing. They also demonstrate that the higher the prior belief in the null hypothesis is, the fewer tests are needed to obtain adequate confidence in the software component.

Although we think that the principles of the Bayesian hypothesis testing approach proposed in Cukic et al. [29] and Smidts et al. [30] are usable, even for compound software, our main concern is related to the use of two separate uniform probability distributions to describe the prior belief in the failure probability of a single software component. This concern is addressed in Kristiansen [31], in which an evaluation of the Bayesian hypothesis testing approach is performed. In this paper, three different prior probability distributions for the failure probability of a software component are evaluated, and their influence on the number of tests required to obtain adequate confidence in a software component is presented. In this evaluation, the first case is based on earlier work done by Cukic et al. [29] and

Smidts et al. [30] and assumes two separate uniform prior probability distributions, one under the null hypothesis and one under the alternative hypothesis. In the second case, the effect of using a flat distribution under the alternative hypothesis is mitigated by allowing an expert to set an upper bound on the failure probability under H_1, i.e. to state a value θ_1 for which the probability of having a failure probability higher than θ_1 is zero. In the third case, the effect of discontinuity in the prior probability distribution is mitigated by using a continuous probability distribution for θ over the entire interval $(0,1)$. A beta distribution is used to accurately reflect prior belief because this distribution is a rich and tractable family that forms a conjugate family to the binomial distribution.

The evaluation in Kristiansen [31] clearly shows that using two separate uniform distributions to describe the failure probability of a software component does not represent a conservative approach at all, even though the use of a uniform probability distribution over the entire interval is usually seen as an ignorance prior. In fact, the number of tests required to obtain adequate confidence in a software component increases significantly when other more realistic distributions for the failure probability of a software component are used. Moreover, it is shown that the total number of tests required by using this approach can both result in fewer and even in more tests compared to classical statistical testing. This means that in the Bayesian hypothesis testing approach, the number of required tests is highly dependent on the choice of prior distribution. It should therefore be emphasized that it is the underlying prior distribution for the failure probability of a software component and underlying assumptions that lead to fewer tests rather than the Bayesian hypothesis testing approach. To choose a prior probability distribution for a software component's failure probability that correctly reflects ones prior belief is therefore of great importance.

3. A COMPONENT-BASED APPROACH FOR ASSESSING RELIABILITY OF COMPOUND SOFTWARE

In Kristiansen [27], we propose a component-based approach for assessing reliability of compound software. In this approach, failure dependencies between software components are addressed explicitly by using Bayesian hypothesis testing [28] on simultaneous failure probabilities. It is assumed that failure probabilities of individual software components are known. The approach consists of five basic steps:

1. Identify the most important component dependencies: based on the structure of the software components in the compound software and information regarding individual software components, identify those dependencies between pairs of software components which are of greatest importance for the calculation of the system reliability [32]. Repeat steps 2-4 for all relevant component dependencies in the system.
2. Define the hypotheses: let $q_{0,ij}$ represent an accepted upper bound for the probability (q_{ij}) that a pair (i,j) of software components fails simultaneously. The upper bound $q_{0,ij}$ is assumed to be context specific and predefined and is typically derived from standards, regulation authorities, customers, etc. Define the following hypotheses:

$$H_0 : a_{ij} \leq q_{ij} \leq q_{0,ij}$$
$$H_1 : q_{0,ij} < q_{ij} \leq b_{ij},$$

where q_{ij} is defined in the interval $[a_{ij}, b_{ij}]$. The interval limits a_{ij} and b_{ij} represent the lower and upper limit for q_{ij}, respectively, and are decided by the restrictions the components' marginal failure probabilities put on the components' simultaneous failure probabilities [32].
3. Describe prior belief regarding probability q_{ij}: establish a prior probability distribution $g(q_{ij})$ for q_{ij} defined in the interval $[a_{ij}, b_{ij}]$ describing the probability that a pair of software components fails simultaneously [33]. This probability distribution is needed for establishing a prior

probability distribution $\pi(q_{ij})$ for q_{ij} defined in the sub-intervals $[a_{ij}, q_{0,ij}]$ and $[q_{0,ij}, b_{ij}]$ and for calculating $P(H_0)$.

4. Update your belief in hypothesis H_0 through testing: based on the prior belief in the null hypothesis $P(H_0)$ from step 3 and a predefined confidence level $C_{0,ij}$, the number of tests required to obtain an adequate upper bound for the probability of simultaneous failure can be found for different numbers of failures encountered during testing. The more failures that occur during testing, the more tests are required to reach $C_{0,ij}$. For further details on when to stop testing see Cukic et al. [29] or Kristiansen et al. [33].

5. Calculate the complete system's failure probability: information regarding failure probabilities of individual software components (which are assumed to be known) and upper bounds for the most important simultaneous failure probabilities (found in step 1-4) can finally be combined to obtain an upper bound for the failure probability of the entire system. This can be performed by various methods, e.g. by discrete event simulation when direct calculation becomes too complicated.

3.1. Assumptions

The component-based approach for assessing reliability of compound software is based on the following assumptions:

1. The states of the software components are associated random variables [35].
2. All data-flow relations between the software components are known.
3. The reliabilities of the individual software components are known [1, 36, 37].
4. The system and its components have only two possible states (functioning and failure) [38].
5. The system has a monotone structure [39].

Furthermore, the research is restricted to on-demand types of situations where the compound software is given an input and execution is considered to be finished when a corresponding output has been produced. In the following, these assumptions are discussed briefly.

3.2. Restrictions on the components' simultaneous reliabilities and failure probabilities

Assumption 1 and 3 put direct restrictions on the components' simultaneous reliabilities and failure probabilities. To show this, let p_i and p_j denote the reliabilities of components i and j in a simple two component system. Furthermore, let p_{ij} denote the simultaneous reliability of components i and j. If it can be assumed that the components' reliabilities do not change due to changes in operational context, the following is true: $p_{ij} \leq min(p_i, p_j)$. This follows directly from the fact that: $p_{ij} = p_{i|j}p_j = p_{j|i}p_i$, where $p_{i|j}$ and $p_{j|i}$ are conditional reliabilities between components i and j.

Furthermore, since we assume that the states of the software components are associated random variables, it follows that [35]: $p_{ij} \geq p_i p_j$. Reasonable constraints on the simultaneous reliability p_{ij} under the given assumptions can therefore be expressed as follows: $p_i p_j \leq p_{ij} \leq min(p_i, p_j)$. In the same way, under the same assumption, it can be shown that: $q_i q_j \leq q_{ij} \leq min(q_i, q_j)$.

How the reliabilities of individual software components put direct restrictions on the components' conditional reliabilities in general systems consisting of two and three components are elaborated in more detail in Kristiansen et al. [32].

3.3. General rules for selecting the most important component dependencies

To identify possible rules for selecting the most important component dependencies, two new concepts were defined in Kristiansen et al. [32]. These concepts contribute to a deeper understanding of how to include component dependencies in reliability modeling, and are given in Definitions 1 and 2.

Definition 1 Data-serial components: two components i and j are said to be data-serial components if either i or j receives data (d), directly or indirectly through other components, from the other.

$$i \xrightarrow{d} j \quad or \quad j \xrightarrow{d} i$$

Definition 2 Data-parallel components: two components i and j are said to be data-parallel components if neither i nor j receives data (d), directly or indirectly through other components, from the other.

$$i \xnrightarrow{d} j \quad and \quad j \xnrightarrow{d} i$$

Based on these definitions, the rules for selecting the most important component dependencies can be summarized as follows [40]:

- Including only partial dependency information may give a substantial improvement in the reliability predictions compared to assuming independence between all software components as long as the most important component dependencies are included.
- It is also clear that dependencies between data-parallel components are far more important than dependencies between data-serial components.

In addition, for a system consisting of both data-parallel and data-serial components, the following applies:

- Including only dependencies between data-serial components may result in a major overestimation of the system's reliability. In some cases, the results are even worse than by assuming independence between all components.
- Including only dependencies between data-parallel components may give predictions close to the system's true reliability as long as the dependency between the most unreliable components is included.
- Including additional dependencies between data-parallel components may further improve the predictions.
- Including additional dependencies between data-serial components may also give better predictions as long as the dependency between the most reliable components is included.

4. CASE

One increasingly popular application of mobile positioning systems is to provide geofencing and backtracking services, for instance for kids or people with dementia. Geofencing is used to alert the person or her family members if the person moves outside of a predefined virtual fence, typically surrounding the neighborhood where the person lives and is familiar. Backtracking systems are typically used for helping the user to find the way back to this familiar area if he/she gets lost. Both these services require reliable positioning of the mobile device. In addition, the backtracker needs a position log, keeping track of the users movements over a period of time. Our example case is a backtracking system.

To ensure effective fault tolerance in the software system, let's assume it is structured as a typical recovery block [41]. Individual system components raise exceptions when they detect errors that their own fault tolerance capabilities are unable to handle. Our example system consists of two independently developed software components capable of establishing and recording the geographical

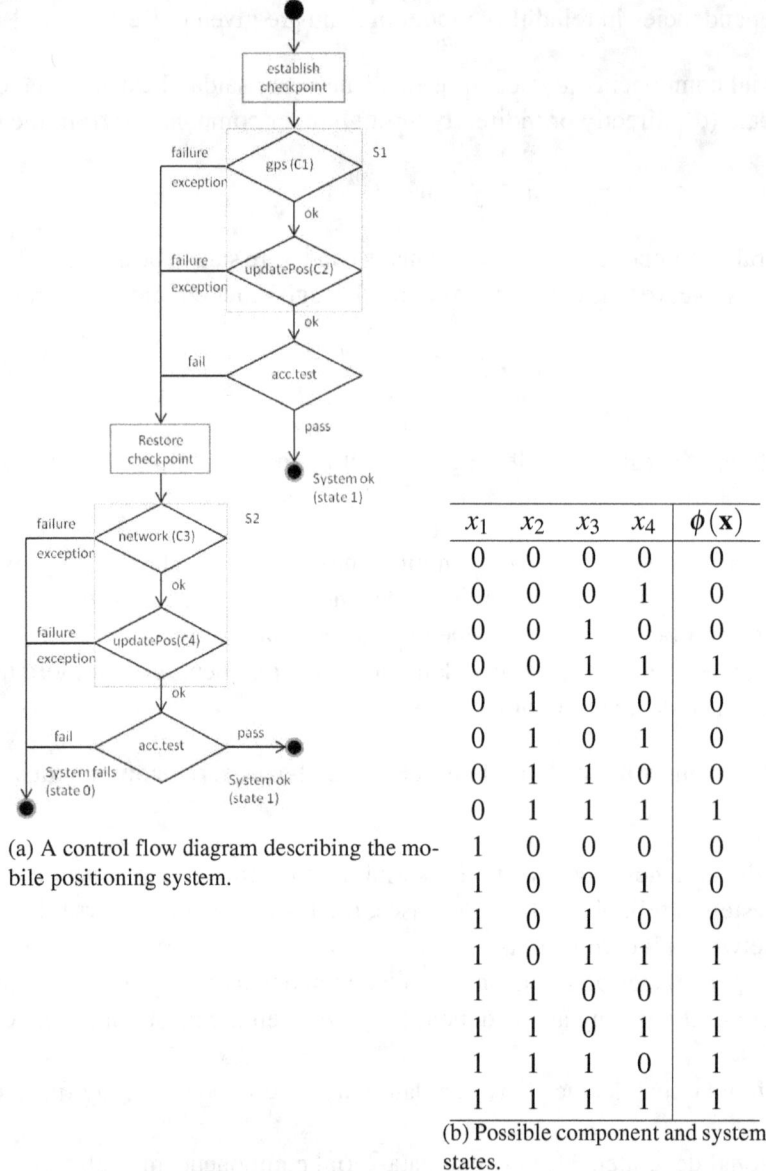

(a) A control flow diagram describing the mobile positioning system.

x_1	x_2	x_3	x_4	$\phi(\mathbf{x})$
0	0	0	0	0
0	0	0	1	0
0	0	1	0	0
0	0	1	1	1
0	1	0	0	0
0	1	0	1	0
0	1	1	0	0
0	1	1	1	1
1	0	0	0	0
1	0	0	1	0
1	0	1	0	0
1	0	1	1	1
1	1	0	0	1
1	1	0	1	1
1	1	1	0	1
1	1	1	1	1

(b) Possible component and system states.

Figure 1: Example case

position of a mobile device. We refer to these two components as super component 1 (S1) and 2 (S2). In addition, a controller is responsible for defining the interactions between the components and performing acceptance tests on the components' output. The system is illustrated in Figure 1(a).

S1 consists of two sub components, C1 and C2. C1 uses the GPS receiver on the mobile phone for establishing the device's geographical position. C2 takes information from C1, transforms it, and updates the position log. The other super component, S2, consists of sub components C3 and C4. C3 uses mobile network and WiFi information to establish the device's geographical position, and C4 updates the position log after transforming the output from C3 to a suitable form. For simplification, it is assumed that the controller, including the acceptance test, is fault free.

S1 and S2 are not run in parallel like in N-version programming. When logging the current device position is requested, the controller first establishes a checkpoint of the system state (including the position log) to permit recovery. Then, C1 is invoked to establish the current device position. On success, the controller passes the result from C1 to C2, which transforms it to a suitable form and adds the current position to the position log. The acceptance test is used by the controller as a final check of the newly logged position. If the acceptance test fails, or if C1 or C2 raises an exception, the controller restores the prior system state, and invokes C3 to provide the position. On C3's success, C4 gets position information from C3, transforms it and adds it to the position log. If the resulting update to the position log fails the acceptance test, or if C3 or C4 raises an exception, the system fails.

Let's assume that the software system and its components have binary states. Let x_i $i = 1, 2, 3, 4$ indicate the state of the ith component at a fixed point in time and let $\mathbf{x} = (x_1, x_2, x_3, x_4)$. Let's further assume that the system state, ϕ, at a fixed point in time is uniquely determined by the states of the components, $\mathbf{x}$, i.e. $\phi = \phi(\mathbf{x})$. For $i = 1, 2, 3, 4$, let $x_i = 0$ if component i fails and 1 if component i functions. The possible component and system states are given in Figure 1(b), which includes the following two system states:

$$\phi(\mathbf{x}) = 0 \quad = \quad \text{system fails}$$
$$\phi(\mathbf{x}) = 1 \quad = \quad \text{system functions}$$

Based on the software system's minimal path sets, the reliability of the software system is given by: $P(\phi(\mathbf{x}) = 1) = p_{12} + p_{34} - p_{1234}$. Let's assume that the marginal reliabilities and failure probabilities for components 1, 2, 3 and 4 are known. This knowledge and the assumption that the states of the software components are associated random variables, put direct restrictions on the components' simultaneous reliabilities and failure probabilities (see Section 3.2 for more details). This is shown in Table 1. Based on the restrictions on the components' simultaneous reliabilities in Table 1, the reliability of the complete system must be within the interval $[0.9988011, 1.0000]$.

Table 1: The components marginal reliabilities and failure probabilities and their restrictions on the components simultaneous reliabilities and failure probabilities.

Marginal reliabilities	Marginal failure probabilities
$p_1 = 0.99$	$q_1 = 0.01$
$p_2 = 0.9999$	$q_2 = 0.0001$
$p_3 = 0.999$	$q_3 = 0.001$
$p_4 = 0.9999$	$q_4 = 0.0001$
Simultaneous reliabilities	Simultaneous failure probabilities
$p_{12} \in [0.989901, 0.99]$	$q_{12} \in [10^{-6}, 10^{-4}]$
$p_{13} \in [0.98901, 0.99]$	$q_{13} \in [10^{-5}, 10^{-3}]$
$p_{14} \in [0.989901, 0.99]$	$q_{14} \in [10^{-6}, 10^{-4}]$
$p_{23} \in [0.9989001, 0.999]$	$q_{23} \in [10^{-7}, 10^{-4}]$
$p_{24} \in [0.99980001, 0.9999]$	$q_{24} \in [10^{-8}, 10^{-4}]$
$p_{34} \in [0.9989001, 0.999]$	$q_{34} \in [10^{-7}, 10^{-4}]$
$p_{123} \in [0.988911099, 0.99]$	$q_{123} \in [10^{-9}, 10^{-4}]$
$p_{124} \in [0.98980201, 0.99]$	$q_{124} \in [10^{-10}, 10^{-4}]$
$p_{134} \in [0.988911099, 0.99]$	$q_{134} \in [10^{-9}, 10^{-4}]$
$p_{234} \in [0.99880021, 0.999]$	$q_{234} \in [10^{-11}, 10^{-4}]$
$p_{1234} \in [0.988812208, 0.99]$	$q_{1234} \in [10^{-13}, 10^{-4}]$

One of the rules for selecting the most important component dependencies in Kristiansen et al. [32], state that including the dependency between the most unreliable data-parallel components gives predictions close to the system's true reliability. In our example, the most unreliable data-parallel components are components 1 and 3. When only including the dependency between components 1 and 3, the reliability of the complete software system becomes: $P(\phi(\mathbf{x}) = 1) = p_1 p_2 + p_3 p_4 - p_4 p_{3|1} p_2 p_1$.

Let's assume that it is required that the reliability of the software system ($P(\phi(\mathbf{x}) = 1)$ is at least 0.9999 with confidence level $C_0 = 0.99$. Based on this requirement and the addition law of probability, a predefined upper bound $q_{0,13}$ for the simultaneous failure probability q_{13} can be calculated, This is shown in Equation 1.

$$
\begin{aligned}
q_{13} &= 1 - p_1 - p_3 + p_{13} \le q_{0,13} \\
&= 1 - p_1 - p_3 + \left(\frac{p_1 p_2 + p_3 p_4 - 0.9999}{p_2 p_4} \right) \\
&= 0.00009891
\end{aligned}
\tag{1}
$$

Based on the upper bound $q_{0,13}$ and the restrictions $[a_{13}, b_{13}]$ on the simultaneous failure probability q_{13} given in Table 1, the following hypotheses can be defined (see step 2 in the component-based approach in Section 3 for details):

$$
\begin{aligned}
H_0 &: 0.00001 \le q_{13} \le 0.00009891 \\
H_1 &: 0.00009891 < q_{13} \le 0.001.
\end{aligned}
\tag{2}
$$

In this case, the null hypothesis states that the simultaneous failure probability q_{13} lies between the lower bound $a_{13} = 0.00001$ and the predefined upper bound $q_{0,13}$, whereas the alternative hypothesis states that the simultaneous failure probability q_{13} lies between the predefined upper bound $q_{0,13}$ and the upper bound $b_{13} = 0.001$.

In the following, the number of fault free tests, n, required to obtain the upper bound $q_{0,13}$ with confidence level $C_{0,13}$ is calculated using:

1. Classical statistical testing, where no prior information about the simultaneous failure probability q_{13} is included.
2. Bayesian hypothesis testing, where only the restrictions imposed on q_{13} by the marginal failure probabilities q_1 and q_3 and the assumption that the states of the software components are associated random variables are taken into account.
3. Bayesian hypothesis testing, where additional prior information about the simultaneous failure probability q_{13} is taken into account.

The number of fault free tests, n, required to obtain the upper bound $q_{0,13}$ at the given predefined confidence level $C_{0,13} = 0.99$, using classical statistical testing, is given in Equation 3 [42].

$$
n = \frac{ln(1 - C_{0,13})}{ln(1 - q_{0,13})} = 46557
\tag{3}
$$

This means that if no prior information in included, 46557 fault free tests need to be carried out to obtain the upper bound $q_{0,13}$ with confidence level $C_{0,13} = 0.99$.

If we only take into account the restrictions imposed on q_{13} by the marginal failure probabilities q_1 and q_3 and the assumption that the states of the software components are associated random variables, the number of fault free tests required to obtain an adequate upper bound for q_{13}, at the given predefined confidence level $C_{0,13} = 0.99$, can be found by solving Equation 4 [33].

$$\frac{\int_{a_{13}}^{q_{0,13}} (1 - q_{13})^n dq_{13}}{\int_{q_{0,13}}^{b_{13}} (1 - q_{13})^n dq_{13}} \geq \frac{C_{0,13}}{(1 - C_{0,13})} \tag{4}$$

In this equation, q_{13} is uniformly distributed (beta distributed with parameters $\alpha = \beta = 1$) over the interval $[a_{13}, b_{13}]$. Taking only these restrictions into account, 51793 fault free tests need to be carried out to obtain the simultaneous failure probability $q_{0,13}$ at confidence level $C_{0,13}$. This means that the number of fault free tests in fact increases when only the restrictions from the marginal failure probabilities are taken into account.

One way to include additional information about the simultaneous failure probability q_{13}, in addition to the restrictions from the marginal probabilities q_1 and q_3, is to identify values for α and β in the beta distribution by visualizing a controlled experiment. In this experiment, n can be considered as the total number of tests and α as the number of simultaneous failures of components 1 and 3. Assuming this prior information, the number of fault free tests required to obtain an adequate upper bound for the simultaneous failure probability q_{13}, at the given predefined confidence level $C_{0,13} = 0.99$ for different choices of α and β, can be found by solving Equation 5 [33].

$$\frac{\int_{a_{13}}^{q_{0,13}} (q_{13} - a_{13})^{\alpha-1} (b_{13} - q_{13})^{\beta-1} (1 - q_{13})^n dq_{13}}{\int_{q_{0,13}}^{b_{13}} (q_{13} - a_{13})^{\alpha-1} (b_{13} - q_{13})^{\beta-1} (1 - q_{13})^n dq_{13}} \geq \frac{C_{0,13}}{(1 - C_{0,13})} \tag{5}$$

In Figures 2(a) and 2(b), the number of tests required are illustrated graphically for different choices of α and β in the beta distribution. From these figures, it is clear that the number of fault free tests, n, is highly sensitive to the choices of α and β. For example $n = 951$ fault free tests are required when $\alpha = 1$ and $\beta = 66$.

In addition to the individual software components' failure probabilities q_1 and q_3, information regarding components' architecture, complexity, programming languages, development processes, etc. might as well be available. This information is also relevant for assessing q_{13}, and will be explored further in future work.

5. SUMMARY, CONCLUSIONS AND FURTHER WORK

In this paper, we have presented a component-based approach for assessing reliability of compound software. This approach applies binary reliability theory to explicitly handle failure dependencies between software components. The approach has been elaborated through several experimental studies [31–33, 40]. To illustrate the approach, we have used a backtracking system typically used for helping people to find their way home if they get lost. This system consists of a recovery block containing two independently developed software components capable of establishing and recording the geographical position of a mobile device.

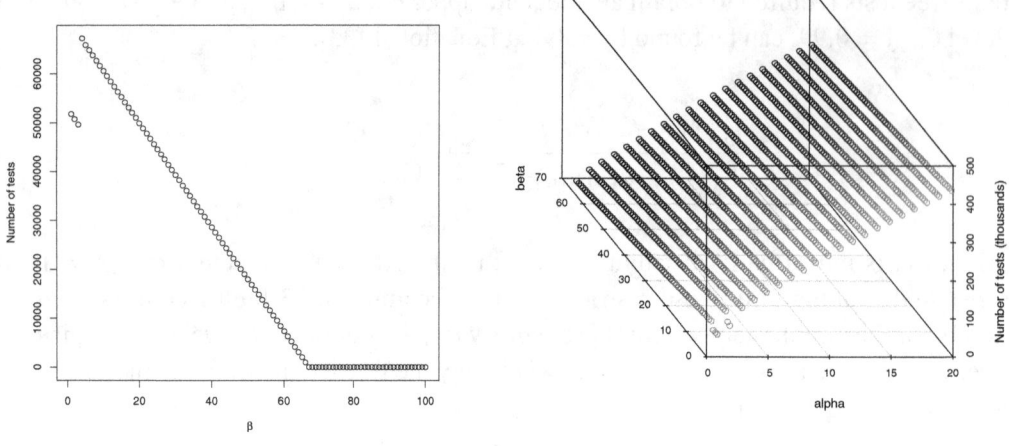

(a) When $\alpha = 1$ and β varies from 1 to 100.   (b) When α varies from 1 to 20 and β varies from 1 to 70.

Figure 2: Number of fault free test required to obtain an adequate upper bound for the simultaneous failure probability q_{13}, at the given predefined confidence level $C_{0,13} = 0.99$.

Using this system, we first illustrate how the components' simultaneous reliabilities and failure probabilities are imposed by a) the components' marginal reliabilities and failure probabilities and b) the assumption that the states of the software components are associated random variables. Secondly, we show how the general rules for selecting the most important component dependencies can be applied on a system consisting of four components. Finally, we calculate the number of fault free tests required to obtain the upper bound for a simultaneous failure probability with a given confidence level. The results show that the number of required tests to obtain an upper bound is highly sensitive to the choices of α and β in the beta distribution. It is noteworthy that the number of tests can both increase and decrease compared to classical statistical testing depending on the choice of these parameters.

Further work includes investigating the main challenges of our component-based approach when we go from binary systems of binary components to multistate systems of binary or multistate components. Especially, we will look at the effect of using only partial dependency information when assessing reliability of multistate systems of binary and multistate components. To find the most important component dependencies in each system state, direct calculation, the Birnbaum measure [38] and Principal Component Analysis [43] will be investigated. A critical question is if these techniques identify the same component dependencies as the most important ones in all system states, or if the most important component dependencies vary between different system states.

References

[1] M. R. Lyu, ed., *Handbook of Software Reliability Engineering*. IEEE Computer Society Press, 1995.

[2] V. Cortellessa and V. Grassi, "A modeling approach to analyze the impact of error propagation on reliability of component-based systems," *Proceedings of the 10th International Conference on Component-based Software Engineering*, pp. 140–156, 2007.

[3] S. S. Gokhale, "Architecture-based software reliability analysis: Overview and limitations," *IEEE Transactions on Dependable and Secure Computing*, vol. 4, no. 1, pp. 32–40, 2007.

[4] S. S. Gokhale and K. S. Trivedi, "Dependency Characterization in Path-Based Approaches to Architecture-Based Software Reliability Prediction," *IEEE Workshop on Application-Specific Software Engineering and*

Technology, pp. 86–90, 1998.

[5] M. Palviainen, A. Evesti, and E. Ovaska, "The reliability estimation, prediction and measuring of component-based software," *Journal of Systems and Software*, vol. 84, no. 6, pp. 1054–1070, 2011.

[6] K. Goseva-Popstojanova and K. S. Trivedi, "Architecture-based approach to reliability assessment of software systems," *Performance Evaluation*, vol. 45, no. 2-3, pp. 179–204, 2001.

[7] D. Hamlet, "Software component composition: a subdomain-based testing-theory foundation," *Software Testing, Verification and Reliability*, vol. 17, no. 4, pp. 243–269, 2007.

[8] D. Hamlet, D. Mason, and D. Woit, "Theory of Software Reliability Based on Components," *International Conference on Software Engineering*, vol. 23, pp. 361–370, 2001.

[9] S. Krishnamurthy and A. Mathur, "On the Estimation of Reliability of a Software System Using Reliabilities of its Components," *Proceedings of the 8th International Symposium on Software Reliability Engineering (ISSRE'97)*, pp. 146–155, 1997.

[10] I. Krka, G. Edwards, L. Cheung, L. Golubchik, and N. Medvidovic, "A comprehensive exploration of challenges in Architecture-Based reliability estimation," *Architecting Dependable Systems VI*, pp. 202–227, 2009.

[11] S. Kuball, J. May, and G. Hughes, "Building a system failure rate estimator by identifying component failure rates," *Proceedings of the 10th International Symposium on Software Reliability Engineering (ISSRE'99)*, pp. 32–41, 1999.

[12] P. Popic, D. Desovski, W. Abdelmoez, and B. Cukic, "Error Propagation in the Reliability Analysis of Component based Systems," *Proceedings of the 16th IEEE International Symposium on Software Reliability (ISSRE'05)*, pp. 53–62, 2005.

[13] R. H. Reussner, H. W. Schmidt, and I. H. Poernomo, "Reliability prediction for component-based software architectures," *Journal of Systems and Software*, vol. 66, no. 3, pp. 241–252, 2003.

[14] H. Singh, V. Cortellessa, B. Cukic, E. Gunel, and V. Bharadwaj, "A Bayesian approach to reliability prediction and assessment of component based systems," *Proceedings of the 12th IEEE International Symposium on Software Reliability Engineering (ISSRE'01)*, pp. 12–19, 2001.

[15] P. T. Trung and H. Q. Thang, "Building the reliability prediction model of component-based software architectures," *Int'l Journal of Information Technology*, vol. 5, no. 1, pp. 18–25, 2009.

[16] M. Vieira and D. Richardson, "The role of dependencies in component-based systems evolution," *Proceedings of the International Workshop on Principles of Software Evolution*, pp. 62–65, 2002.

[17] S. Yacoub, B. Cukic, and H. Ammar, "A Scenario-Based Reliability Analysis Approach for Component-based Software," *IEEE Transactions on Reliability*, vol. 53, no. 4, pp. 465–480, 2004.

[18] D. E. Eckhardt and L. D. Lee, "A theoretical basis for the analysis of redundant software subject to coincident errors," tech. rep., Memo 86369, NASA, 1985.

[19] J. C. Knight and N. G. Leveson, "An experimental evaluation of the assumption of independence in multiversion programming," *IEEE Transactions on Software Engineering*, vol. 12(1), pp. 96–109, 1986.

[20] B. Littlewood and D. R. Miller, "Conceptual Modeling of Coincident Failures in Multiversion Software," *IEEE Transactions on Software Engineering*, vol. 15(12), pp. 1596–1614, 1989.

[21] Y. Dai, M. Xie, K. Poh, and S. Ng, "A model for correlated failures in N-version programming," *IIE Transactions*, vol. 36, no. 12, pp. 1183–1192, 2004.

[22] P. Guo, X. Liu, and Q. Yin, "Methodology for Reliability Evaluation of N-Version Programming Software Fault Tolerance System," *International Conference on Computer Science and Software Engineering*, pp. 654–657, 2008.

[23] B. Littlewood, P. Popov, and L. Strigini, "Modelling software design diversity: a review," *ACM Computing Surveys*, vol. 33, no. 2, pp. 177–208, 2001.

[24] V. F. Nicola and A. Goyal, "Modeling of correlated failures and community error recovery in multiversion software," *IEEE Transactions on Software Engineering*, vol. 16, no. 3, pp. 350–359, 1990.

[25] P. Popov, L. Strigini, J. May, and S. Kuball, "Estimating Bounds on the Reliability of Diverse Systems," *IEEE Transactions on Software Engineering*, vol. 29, no. 4, pp. 345–359, 2003.

[26] L. A. Tomek, J. K. Muppala, and K. S. Trivedi, "Modeling Correlation in Software Recovery Blocks," *IEEE Transactions on Software Engineering*, vol. 19, no. 11, pp. 1071–1086, 1993.

[27] M. Kristiansen, *A component-based approach for assessing reliability of compound software*. PhD thesis, Faculty of Matematics and Natural Sciences, University of Oslo, 2011.

[28] J. O. Berger, *Statistical Decision Theory and Bayesian Analysis*. Springer Verlag, second ed., 1985.

[29] B. Cukic, E. Gunel, H. Singh, and L. Guo, "The Theory of Software Reliability Corroboration," *IEICE Transactions on Information and Systems*, vol. E86-D, no. 10, pp. 2121–2129, 2003.

[30] C. Smidts, B. Cukic, E. Gunel, M. Li, and H. Singh, "Software Reliability Corroboration," *Proceedings of the 27'th Annual NASA Goddard Software Engineering Workshop (SEW-27'02)*, pp. 82–87, 2002.

[31] M. Kristiansen, "Finding Upper Bounds for Software Failure Probabilities - Experiments and Results," *Computer Safety, Reliability and Security (Safecomp 2005)*, pp. 179–193, 2005.

[32] M. Kristiansen, R. Winther, and B. Natvig, "On Component Dependencies in Compound Software," *International Journal of Reliability, Quality and Safety Engineering (IJRQSE)*, vol. 17, no. 5, pp. 465–493, 2010.

[33] M. Kristiansen, R. Winther, and B. Natvig, "A Bayesian Hypothesis Testing Approach for Finding Upper Bounds for Probabilities that Pairs of Software Components Fail Simultaneously," *International Journal of Reliability, Quality and Safety Engineering (IJRQSE)*, vol. 18, no. 3, pp. 209–236, 2011.

[34] M. Kristiansen, R. Winther, and J. E. Simensen, "Identifying the Most Important Component Dependencies in Compound Software," *Risk, Reliability and Safety (ESREL 2009)*, pp. 1333–1340, 2009.

[35] R. E. Barlow and F. Proschan, *Statistical theory of reliability and life testing: probability models*. Holt, Rinehart and Winston, 1975.

[36] B. Littlewood and L. Strigini, "Guidelines for the statistical testing of software," tech. rep., City University, London, 1998.

[37] J. D. Musa, *Software Reliability Engineering*. McGraw-Hill, 1998.

[38] B. Natvig, *Multistate Systems Reliability Theory with Applications*. John Wiley and Sons, Ltd, first ed., 2011.

[39] B. Natvig, *Reliability analysis with technological applications (in Norwegian)*. Department of Mathematics, University of Oslo, 1998.

[40] M. Kristiansen, R. Winther, and B. Natvig, "On Component Dependencies in Compound Software," tech. rep., Department of Mathematics, University of Oslo, 2010.

[41] B. Randell and J. Xu, "The evolution of the recovery block concept," 1995.

[42] J. H. Poore, H. D. Mills, and D. Mutchler, "Planning and Certifying Software System reliability," *IEEE software*, 1993.

[43] R. A. Johnson and D. W. Wichern, *Applied Multivariate Statistical Analysis*. Prentice Hall, New York, 1998.

Automated evolutionary restructuring of workflows to minimise errors via stochastic model checking

Luke Thomas Herbert [a]*, **Zaza Nadja Lee Hansen** [b] **and Peter Jacobsen** [b]

[a]DTU Compute, Lyngby, Denmark
[b]DTU Management, Lyngby, Denmark

Abstract: This paper presents a framework for the automated restructuring of workflows that allows one to minimise the impact of errors on a production workflow. The framework allows for the modelling of workflows by means of a formalised subset of the Business Process Modelling and Notation (BPMN) language, a well-established visual language for modelling workflows in a business context. The framework's modelling language is extended to include the tracking of real-valued quantities associated with the process (such as time, cost, temperature). In addition, this language also allows for an intention preserving stochastic semantics able to model both probabilistic- or non-deterministic branching behaviour. We further extend this formalism to allow for the introduction of error states which allow for both fail-stop behaviour and continued system execution. We explore the practical utility of this approach by means of a case study from the food industry. Through this case study we explore the extent to which the risk of production faults can be reduced and the impact of these can be minimised, primarily through restructuring of the production workflows. This approach is fully automated and only the modelling of the production workflows and the expression of the goals require manual input.

Keywords: Consequence Modeling and Management, Enterprise Risk Management, Industrial Safety and Accident Analysis, Reliability Analysis and Risk Assessment Methods, Safety Assessment Software Tools, Safety Management and Decision Making

* Corresponding author, lthhe@dtu.dk

1. INTRODUCTION

It is vital for all industries, for example the food industry, that workflows are generated which are safe and fulfil specific parameters and criteria like hygiene, cost, efficiency and speed (due to the perishable nature of the products in the food industry). A production workflow is for example cutting, forming or moulding a product or conducting quality control measures. Developing production and business processes is today predominantly an activity in which software tools are used to draw the process maps. The processes are analysed by hand and improved configurations are found by a process of trial and error, often taking too much time to arrive at an optimal practice due to the learning experience involved. There is therefore a need for a more efficient approach.

In this paper we present a framework for the automated restructuring of workflows that allows one to minimise the impact of errors on a production workflow. This framework allows for the modelling of workflows by means of a formalised subset of the Business Process Modelling and Notation (BPMN) language, a well-established visual language for modelling workflows in a business context. The frameworks modelling language is extended to include the tracking of real-valued quantities associated with the process (such as time, cost, temperature). In addition, this language also allows for an intention preserving stochastic semantics able to model both probabilistic- or non-deterministic branching behaviour. We further extend this formalism to allow for the introduction of error states which allow for both fail-stop behaviour and continued system execution. We employ stochastic model checking to efficiently explore the entire statespace of a workflow. The temporal logic PCTL is used to encode properties of interest for model checking (e.g. the probability that the next error-free product will come off the production line in less than 60 seconds).

We present an algorithm that allows for the weighted generation of PCTL queries that may be used to express a desired balance between the occurrence of errors and data quantities associated with the workflow. The expected mean values at points of failure can be calculated in turn, allowing for the expression of queries such as identifying the errors in operations which has the largest impact on production cost and/or time. These queries along with a model of an existing workflow, are used as inputs to an evolutionary algorithm which iteratively, through a process of mutation and cross-over, generates candidate improved workflows. The model checking of a weighted set of queries is used as a fitness function for determining the degree of improvement of candidate workflows. Being an evolutionary algorithm when a candidate workflow shows improvement it is used as the basis for the next round of mutation and cross-over. The software tool SBOAT is presented which implement our approach.

We explore the practical utility of this approach by means of a case study from the food industry. The case company is one of the largest Danish producers of baked goods. Their goal is to reduce the cost of waste in their production workflows. Through this case study we explore the extent to which the risk of production faults can be reduced and the impact of these can be minimised, primarily through restructuring of the production workflows. We discuss both the degree of improvement achieved by use of this evolutionary approach and the limitations of the approach and the additional work needed, compared to other process optimisation techniques, to make use of the approach.

2. RELATED WORK

In general by performing model checking to determine quantitative and qualitative properties of BPMN models, this work draws a comparison with a number of other BPMN analysis techniques outlined below. The selection of analyses discussed is not exhaustive, but covers the main approaches which have been widely referenced in the literature. They can broadly be defined as functional analyses, non-functional analyses and stochastic analyses.

In terms of the analysis of functional qualitative properties a wide range of approaches have been developed for BPMN. These are predominantly focused on the analysis limited sets of functional properties, such as proving the absence of deadlocks. The work of Ouyang et al. [8], [9] is the closest match to the type of translation approach taken in this paper. Here, translation of BPMN models is done directly into the web-services orientated BPEL [12] by means of an algorithm similar to what is presented in this thesis. However, the approach by Ouyang et al. is intended to support simulation through execution of the BPEL services with all the limitations that a simulation approach entails. Limitations are for example that statistical simulation can explore some situations, but cannot observe all behaviours. Safety properties which guarantee that specific behaviour will always, or, can never, occur need to be evaluated under all possible situations which simply cannot be achieved by the simulation method. The other key limitation is that simulations need to be executed for a certain amount of time.

A number of different approaches have been developed to analyse non-functional properties of BPMN models. In particular there has been a focus on determining timing properties of BPMN models. General quantitative analysis has only been identified as being explored by Prandi et. al. [10].

Analysis of BPMN models extended with stochastic properties has seen limited development with only two approaches identified of dealing with general models which exhibit both probabilistic and non-deterministic transitions. Prandi et. al. [10] have identified PRISM as ideally suited to the analysis of stochastic PRISM business processes. This effort involves conversion of PRISM models into a model expressed in the COWS [11], which in turn is converted into a model that can be analysed using PRISM [5]. This approach adds unnecessary complexity in that it is possible to convert the notation of BPMN directly into the PRISM modelling language, and then allow PRISM to impose a semantic interpretation without the additional semantic restrictions of going via COWS. Further, the translations PRISM to COWS and in particular from COWS to PRISM is loosely defined and, in the form described by the authors, not amenable to algorithmic translation. This approach, however, does allow the use of rewards. Consequently, the PRISM model checker is potentially able to perform analysis of both quantitative and stochastic properties of a business process. However, details of how such properties will be included in the original BPMN models is not described.

3. FRAMEWORK

We present a framework which uses an extended version of the BPMN language to automatically restructure of workflows so as to minimise the impact of errors on a production workflow. The software tool which implements this framework is called the Stochastic BPMN Optimisation and Analysis Tool (SBOAT).

3.1. Core BPMN

For the purposes of this paper, only a small subset of BPMN that is sufficient to illustrate the principles of this work will be used. Often called the *core* subset of BPMN, it consists of the eight elements found to be most commonly used in a large survey of real-world BPMN usage [6]; indeed more than 70% of models surveyed consisted only of these elements. The graphical elements of core BPMN are shown in fig. 1 and described below.

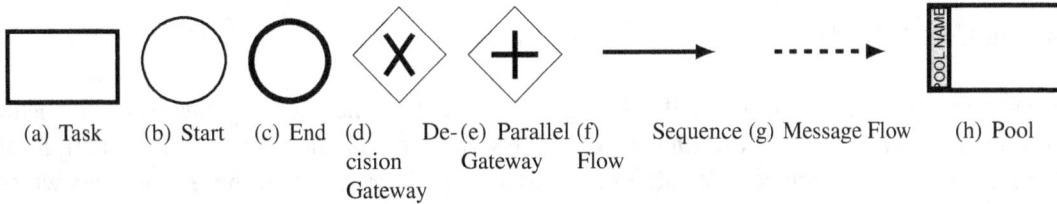

(a) Task   (b) Start   (c) End   (d) De- (e) Parallel (f)   Sequence (g) Message Flow   (h) Pool
                              cision   Gateway   Flow
                              Gateway

Figure 1: Core BPMN elements.

The process of modelling a workflow in BPMN involves composing a number of BPMN elements into a BPD. The intention is that a business process diagram captures the complete workflow of a business process, with separate sub-components of a workflow organised into separate pools. A BPD is formally defined here as simple partitions of a process graph with synchronisation arcs as defined in **??**, with the straightforward addition of a function to capture pools.

Definition 1 (Core BPD). *A core BPD is an extended process graph tuple* $BPD = (\mathbf{N}, \mathcal{F}, \mathbf{P}, \mathsf{pool}, \mathbf{L}, \mathsf{lab})$ *where* $\mathbf{N} \subseteq \mathbf{T} \cup \mathbf{E} \cup \mathbf{G}$, *is a set of nodes composed of the following disjoint sets:*

- *Tasks* $\mathbf{T}$, *are the basic actions performed as part of a business process.*
- *Events* $\mathbf{E} \subseteq \mathbf{E^S} \cup \mathbf{E^E}$, *where the disjoint sets* $\mathbf{E^S}$ *and* $\mathbf{E^E}$ *respectively represent start and end events.*
- *Gateways* $\mathbf{G} \subseteq \mathbf{G^D} \cup \mathbf{G^F} \cup \mathbf{G^M}$, *where the disjoint sets* $\mathbf{G^D}$, $\mathbf{G^F}$ *and* $\mathbf{G^M}$ *respectively represent exclusive decision gateways, parallel fork gateways, and parallel merge gateways.*

$\mathcal{F} \subseteq \mathcal{S} \cup \mathcal{M}$ *is a set of flow relations, where sequence flows* $\mathcal{S} \subseteq \mathbf{N} \times \mathbf{N}$ *relate nodes to each other and message passing* $\mathcal{M} \subseteq \mathbf{T} \times \mathbf{G^M}$ *is a relation between tasks and parallel merge gateways.* $\mathbf{P} \subset \mathfrak{P}(\mathbf{N})$ *is a set of disjoint pools and* $\mathsf{pool} : \mathbf{N} \to \mathbf{P}$ *maps nodes to a pool* $p \in \mathbf{P}$. $\mathbf{L}$ *is a set of unique labels and* $\mathsf{lab} : \mathcal{F} \to \mathbf{L}$ *is a labelling function which assigns labels to flows.*

The definition of a *BPD* given in definition 1 models business processes by using elements of $\mathcal{F}$ to define a directed graph with nodes which are elements of **N**. However, definition 1 allows for graphs which are unconnected, do not have start or end elements, and are free from various other properties which place them outside what is implied to be permitted in standard BPMN models. To ensure that a *BPD* describes a meaningful business process we have developed a set of structural semantics (well-formedness) rules [4] which enforce restrictions on connecting elements, pool boundaries, and message passing.

When performing quantitative analysis of a graph based process language we choose well-formedness conditions such that they impose the minimum semantic interpretation necessary to determine the control flow of a model. In the case of a BPMN *BPD*, it adds no more semantic interpretation than implied by the standard [7]. In the case of BPMN, when we have imposed restrictions they have been made only for simplicity and are discussed at length in previous work [4].

It should be noted that by combining several Core BPMN elements any element of the entire BPMN language can be simulated. Although a few constructs pose more complex issues. Of particular note are inclusive gateways which pose a challenge as the specification of their semantics include a non-trivial and non-local backwards search of the flow graph of the *BPD*. However, these can be addressed through the work of Christiansen, Carbone and Hildebrandt [1] who present a method to translate this construct into other BPMN processes with some restructuring of the overall *BPD*. In general, all elements of the full BPMN language can be incorporated by simply including a preprocessing step in the analysis of BPMN which translates non-core BPMN elements to core BPMN elements.

3.2. Extending Core BPMN

BPMN makes use of external conditions on decision gateways to select the outgoing flow from a decision point. These decisions are modelled by the set **L** and assigned to specific flows by the function lab introduced in definition 1. In practice, decision points in a business process will have outcomes which depend on some inherent property of the task or on outside factors. The idea is that at a decision point an active choice is made, and then that choice results in a number of different possible outcomes.

This behaviour, which is similar to a Markov decision process [13], where we wish to preserve the intention of actors in a process and still enable probabilistic behaviour, can be effectively captured by annotating the possible outcomes of specific decisions with pairs of labels and probabilities (l, p). We employ the following function to ensure meaningful assignment of these intention preserving probabilistic annotations.

Definition 2 (BPD Gateway Flow Probability Function). *Given a BPD, a decision gateway probability function is a partial function $\mathcal{P} : \mathcal{S} \times \mathbf{L} \to [0,1]$ which for a node $g \in \mathbf{G^D}$ and label $l \in \mathbf{L}$, assigns probabilities to all outgoing sequence flows (g,x), such that for a given l:*

$$\sum_{\forall x \in \text{out}(g)} \mathcal{P}((g,x),l) = 1$$

Definition 2 ensures that all decision gateways have an associated probability and that the sum of all probabilities for a given label l is 1.

To enable quantitative analysis of business processes we add numerical data to our models by using the following function which associates positive real numbers with tasks in a *BPD*.

Definition 3 (BPD Task Reward Function). *For a BPD a reward function for a task $t \in \mathbf{T}$ is a partial function $\mathcal{R} : \mathbf{T} \to \mathbb{R}_{\geq 0}$.*

This function captures the notion that certain nodes have some reward or cost associated with the task. The term *reward* is somewhat misleading: there is no practical distinction between costs and rewards, and we can use these annotated values to keep track of whichever quantities may be of interest in a process. We may associate as many reward structures as we wish with a given *BPD*, so that a single task may have multiple different numerical properties which are incremented when the task is performed.

Further, we allow some reward structures to be parametrised, allowing us to model limited resources involved in a process. In this case a reward has an associated upper and lower bound.

Note that the reward structures presented here are simple additions to the BPMN language, or other graph based process languages, and do not alter the implied semantics of the language.

3.3. Probabilistic CTL Property Specification

The main goal with this work is to be able to perform stochastic model checking of BPMN models. We will employ the property specification language called PCTL [5] based on classical continuous stochastic logic [3] extended to probabilistic quantification of described properties. This logic allows us to reason about properties relating to: timing, occurrence and ordering of events, reward values, transient and steady-state probabilities, and best- and worst-case scenarios. An implementation of the PCTL logic is employed by the PRISM model checker in its property specification language and has the following basic elements.

The key constructs in the PRISM property specification language, as it applies to Markov decision processes, are the P and R operators. The P operator refers to the probability of an event occurring, more precisely, the probability that the observed execution of the model satisfies a given specification. The R operator is used to express properties that relate to rewards (more precisely, the expected value of a random variable, associated with particular reward structure) and since a model will often be decorated with multiple reward structures, we augment the R operator with a label. For example, to determine the mean time to exhaust the supply of cake filler we would specify the following property:

$$R_{time} =_? [F^{[0,\infty]} \; filler\_empty]$$

In this property we employ the time-bounded operator F^I, where $I \in [t_1, t_2] \in \mathbb{R}^2$, to describe if a subsequent boolean variable becomes true, it remains true. Further the operator U^I allow for specification of execution paths where one boolean variable holds until another is true, and G^I, which can be seen as the dual of F^I: expresses the fact that a condition remains, rather than becomes, true. Note that these operators can be combined in arithmetic expressions, allowing more complex measures to be expressed.

3.4. SBOAT

SBOAT is a *Stochastic BPMN Optimisation and Analysis Tool*, which allows a user to model, and annotate with rewards and stochastic branching, a business processes as a BPMN *BPD*. Analysis is specified using a PRISM style PCTL query and depending on the nature of the query one or number of results are calculated. At the core of SBOAT is the PRISM model checker [5] which performs analysis of individual models generated by the SBAT. An overview of the design of SBOAT is shown in fig. 2.

Designing a BPMN model and adding annotations to a model is done in the modeller, shown in fig. 3. Within SBOAT we employ an *adjacency list* [2] data structure to store the modelled *BPD* as a directed graph.

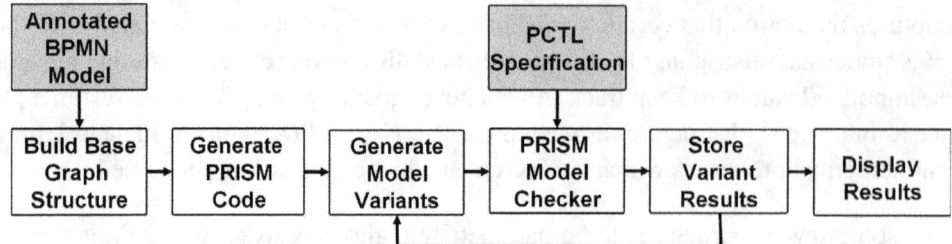

Figure 2: Overall design of SBOAT

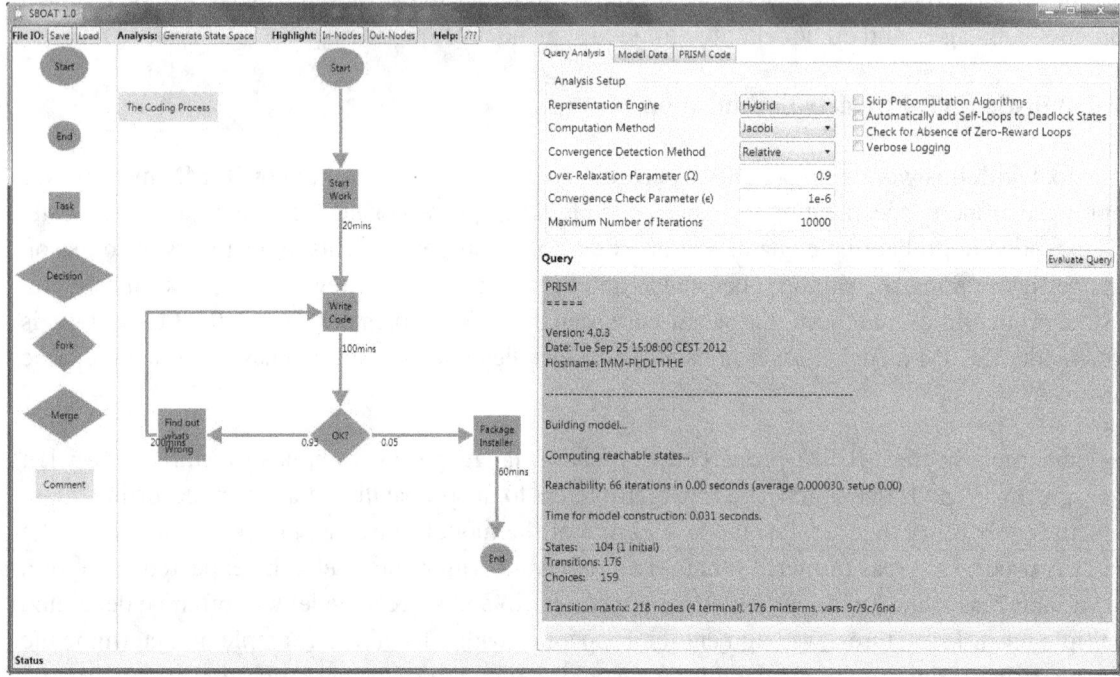

Figure 3: SBAT 1.0 User Interface

This implementation approach allows nodes and edges to be of any data type, where edges implement a simple interface that defines connections to nodes. This flexibility also supports almost any type of reward annotation to be stored in nodes or edges. Further, the data structure is mutable and cloneable making it possible to build a suitable data structure for a wide range of possible graph based process languages. Internally, the data structure keeps a dictionary from nodes to a unordered list of edge elements.

The data structure used is serializable allowing for generation of XML files to store models. In the case of BPMN models, we emit files that comply with the XML based BPMN file format [7], using the annotation features inherent in the BPMN file format to store rewards and stochastic branching probabilities. When importing data files we parse the BPMN XML and at that stage checks for the well-formedness of the model are made as new nodes and edges are added.

3.5. Evolutionary algorithm

The evolutionary algorithm we have developed to optimise business workflow processes employs a genotype-style representation of the optimization problem, variation and selection operators, and a fitness function. However, many details are quite different from typical evolutionary approaches. The algorithm performs optimisation of a BPD by performing modifications directly on the BPD ensuring that the final improved process is also a BPD and requires no special interpretation by end users. The description of the algorithm in this section is intended only to explain the principle of this method and not detail the mathematical proofs behind it.

First, an initial population of BPD variants is generated. Due to the computational expense of performing quantitative model checking of a BPD we filter these variants using well-formedness (structural semantics) criteria and functional requirements, before evaluating their quantitative properties.

The evolution of the initial population takes place for a number of generations determined by limit. For each generation, a population of a given size is generated. This population is produced by selecting pairs from the previous generation, in a fashion that is proportional to their fitness score. This pair is used to generate a new variant BPD using a crossover operator. Mutation is achieved by performing a number of alterations of a BPD dictated by the mutation rates. If a variant proves to be well-formed and meets functional requirements it becomes part of the next generation.

Finally, once the generation limit has been reached, the highest scoring member of the final generation with regard to the optimisation goals, becomes the optimized BPD. Before returning this optimised BPD, any redundant components are removed.

To enable the combination of multiple weighted objectives, and to have sets of optimization goals in which both rewards and event probabilities can be expressed, we employ a set of optimisation goal tuples to define an individual optimisation goal using PCTL formulae. This is because in practice a production process is frequently optimised with regard to multiple quantitative properties. We use a set of optimisation goal tuples for this purpose. For a set of optimization goal tuples, we evaluate the relative improvement of a new production process BPD compared to an existing BPD using an optimisation goals scoring function.

Functional requirements allow the expression of properties which must hold for any future production process BPD derived from a BPD. Like optimization goals, functional requirements will be defined using PCTL formulae, however, in this case we will require that probabilities or reward values within the query are explicitly defined, such that the return value of the query is a Boolean variable. This is to ensure the functional requirements for each individual can be quickly evaluated as either being true or false.

A key step in our algorithm is the selection of members of a current generation used to derive the next generation. Here we employ stochastic sampling with limited replacement. In essence, each member of a current generation is mapped to a contiguous segment of a line, such that each individual's segment is proportional in size to its fitness. A random number is generated and an individual A whose segment spans the random number is selected. The process is repeated to obtain a partner with the restriction that if A is selected a new sample is chosen.

When generating variants we employ the traditional evolutionary algorithm approach of constructing a separate genotype representation upon which to perform modification of a BPD. Our approach allows the genome structure to closely reflect the phoneme structure. Encodings with this property are believed to make the evolutionary algorithm more robust (i.e. reduce the probability of fatal mutations), and also improve the capacity of a system for adaptive evolution.

We employ an adjacency matrix style representation of the underlying graph structure of the BPD for our genotype, where each matrix element is a vector which stores the reward structures associated with the given node of a BPD. The phenotype is simply the BPD that is derived from this matrix representation.

Crossover follows naturally from the structure of the genotype representation. Instead of creating a child by swapping information from two parents based upon one or more points in a linear structure as is commonly done, we use a rectangular section of the matrix structure selected at random. An offspring is then created by using information from inside the rectangle of one parent, and outside the rectangle of the other parent.

Mutation is also defined as a mathematical operator, which is applied to specific elements of the matrix representation of a BPD. This compliments our crossover operator by injecting small local changes to a BPD. We define mutation to allow for considerable variation of a source BPD. This definition allows mutations to have two effects on a BPD:

- **Re-sequencing:** This modification alters the BPD element which defines sequence flows. Specifically, it alters the relation between two nodes in the sequence Low, replacing the destination node with a different node, and reconnecting any excluded nodes to follow after the re sequencing. In effect, randomly reordering the sequencing of a number of tasks in the BPD.
- **Parallelization:** This modification functions by injecting pairs of parallel merge and fork gateways. These can be injected at any point other than at start and end elements, and the nodes between the injected gateways are initially all assigned to one of the parallel paths. Note that when this is combined with the resequencing operator both parallel branches will eventually contain nodes.

4. CASE STUDY

Baked Goods A/S has two production lines. Line 1 develops cakes and pastries and line 2 develops baked goods like sausage rolls and pizzas. The company has an approximate overall waste of 20% for all products during the whole production cycle. The waste is calculated at different places on the production line, but currently it is only possible to get the overall waste percentage of a product. The production process is shown in Figure 2. The dough is prepared, then the product, for example a cake, is prepared and made, then baked or frozen, followed by packaging and then shipment. Dough leftover during production is reused in the preparation of new dough. Quality checks take place during and after production as well as after freezing where defective products are discarded.

Depending on the line and product, the packaging is either automatic or manual and either controlled for number of pieces and weight or just number of pieces.

The processing steps are as follows;

1. Dough mixing. The ingredients for the dough are mixed in vats with a screw-hook mixer. Water and flour are delivered automatically through transport tubes, all other ingredients are manually added including ice.
2. Lamination. If specified the dough is laminated, meaning folded in on itself and rolled out. This takes place continually. Dough is supplied into a dispensing system. The dough is split into two sheets of continues length, and if specified, a layer of fat is supplied in between these layers. The lamination step can be skipped by using the second dispensing machine situated further up the production line.
3. Make-up and filling. Make-up refers to cutting the shapes out from the dough carpet. The product can also be filled with creams etc.
4. Leavening. This is done to induce a faster rising of the product in order to achieve the wanted volume of the product. This is done by raising the temperature and having a high humidity. This will promote the growth of yeast producing gas and the humidity will counteract surface drying allow for the dough to expand without generating stress cracks etc.
5. Baking. A baking step can be used after leavening.
6. Freezing.
7. Packing.

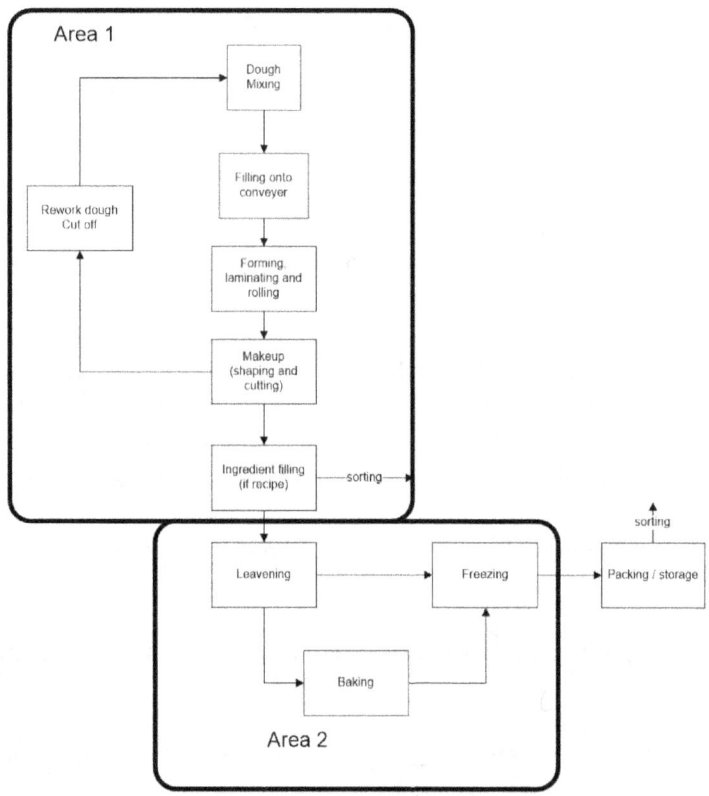

Figure 4: The production process at Baked Goods A/S.

4.1. Analysis of the case study

To illustrate an application of this method we will consider a specific example of a simple production process inspired from the baked goods case study involving bread production. To employ the method, one begins by building a BPD model of an existing production process. fig. 5 is an example of such a process which is annotated with rewards and information about its stochastic behaviour. This naively-designed production process consists of two processes, Conveyor Belt modelling the actions of the machines on a conveyor belt, and Filling Robot which models the actions of robot which fills the dough with cream etc. when needed.

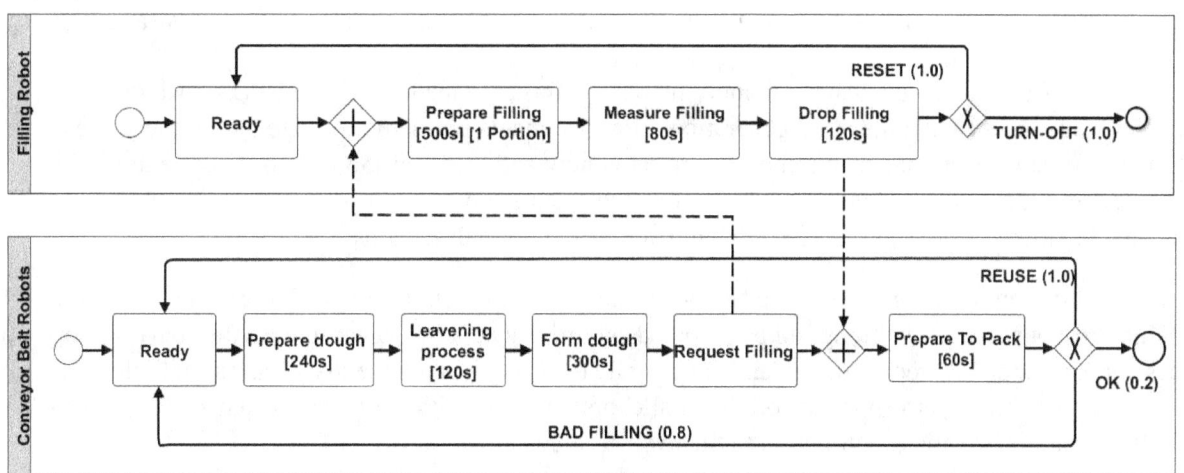

Figure 5: BPD model of a baking workflow process before optimization.

For the production process described in fig. 5 it would be desirable to see an improvement in the time taken for a conveyer belt machine to complete baking a cake. This is in other words the optimisation goal. Further, it would also be desirable that the rate of filling consumption and the consequent probability, given a specific filling stock size, of running out of the filling is kept as low as possible. This is the second optimisation goal.

In addition to the optimisation goals, a number of functional requirements exist for this process (formally expressed using the temporal logic PCTL in SBOAT). These requirements describe the sequences the steps the process have to be in:

1. The baking of the dough should take place before the leaving process of the dough.
2. All dough making, leaving, cut and filling, must take place before a cake is packed:
3. The conveyer belt machine cannot pack the cake before it has received filling.
4. The Filling Robot must ensure that a filling has been prepared and measured before it is sent to the conveyor belt.
5. When the conveyer belt determines that a bad dose of filling has been received it must immediately request a new dose.

Note that the final functional requirement (item 5) is not currently satisfied by the initial BPD shown in fig. 5.

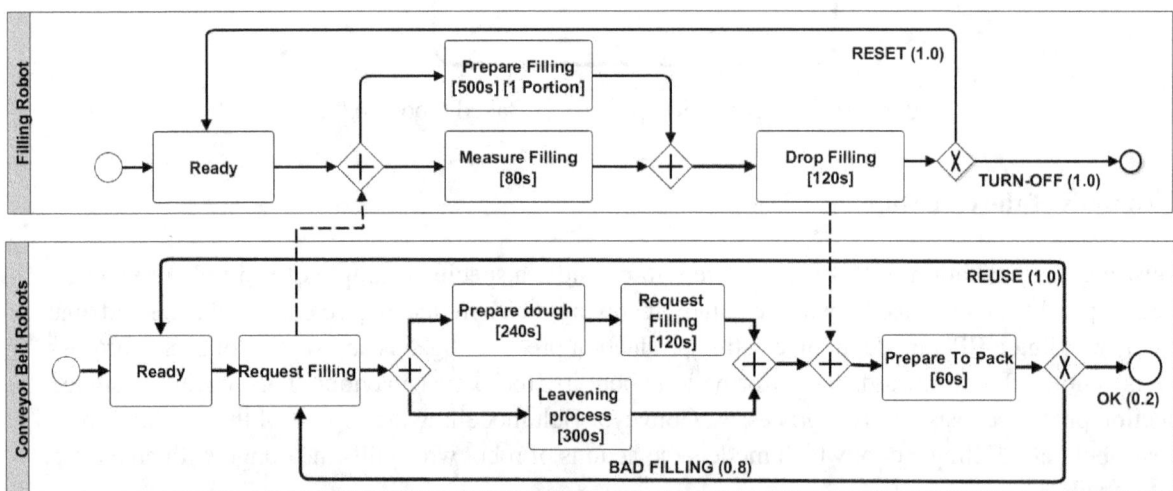

Figure 6: BPD model of a baking system after 28 generations of population size 500 optimisations.

fig. 6 illustrates one possible outcome of applying our optimisation methods to the BPD shown in fig. 5. Specifically, this is the outcome of 28 generational improvements of population size 500 of the process. Note that the new functional requirement (item 5) is now satisfied. In the case of this example the rates for sequencing and parallelizing are set so that the Mutate function ensures that considerably more re-sequencing modifications are performed than parallelization modifications.

In this run of our optimisation method we have identified two opportunities to parallelize actions. Within the Filling Robot process, the filling can be prepared and measured at the same time. In the conveyer belt process, it is possible to prepare dough and cut it while the dough simultaneously leaves. In both cases this saves time, as when performing actions in parallel only the path with the slowest behaviour is counted towards the parallel sections contribution to the reward value.

We have also determined that, in this simplified example, the conveyer belt process always orders filling to finish the cakes but the conveyer belt machines must wait while the Filling Robot performs its operations and then returns the filling. As the filling will inevitably be needed, and only the packing of the cake needs to be done after the dough has been made, it is within the functional requirements for the process, and results in a considerable time saving to order the filling immediately before even preparing the dough. This ensures that there will be no delay imposed on the Conveyer Belt process by the actions of the Filling Robot process. This simplified optimisation example does not violate the functional requirements and results in a significant reduction of the time taken for the execution of the production process.

Existing languages for the modelling of business processes such as BPMN, UML activity diagrams or YAWL lack a formalised semantic basis which would enable formal analysis and subsequent automated scheduling. Further, these languages do not allow for modelling stochastic behaviour or provide mechanisms to effectively track the consumption of resources during execution. These aspects are therefore the key strengths of this optimisation method as no other method, to our knowledge, has all these features. Further, it should be noted that our method by employing the PRISM tool calculates exact values. However, this need for precision also means that a disadvantage of our approach is that it requires detailed knowledge of the workflow processes being optimised. Another disadvantage of our method is that to use the optimisation schedule in practice great computing power is needed which can be both expensive and time-consuming. However, our method allows for automatic optimal scheduling with mathematical precision and within specific parameters which can help organisations limit waste of, for example, energy or material as well as optimise production with regard to parameters such as time, human resources and cost.

5. CONCLUSION

In this paper we have presented a framework for the automated restructuring of workflows that allows one to minimise the impact of errors on a production workflow. We did this by means of a formalised subset of the Business Process Modelling and Notation (BPMN) language. The frameworks modelling language is extended to include the tracking of real-valued quantities associated with the process (such as time, cost, temperature). In addition, this language also allows for an intention preserving stochastic semantics able to model both probabilistic- or non-deterministic branching behaviour. We further extend this formalism to allow for the introduction of error states which allow for both fail-stop behaviour and continued system execution. We employed stochastic model checking to efficiently explore the entire statespace of a workflow. The temporal logic PCTL is used to encode properties of interest for model checking (e.g. the probability that the next error-free product will come off the production line in less than 60 seconds). We presented an algorithm that allows for the weighted generation of PCTL queries that may be used to express a desired balance between the occurrence of errors and data quantities associated with the workflow.

We explored the practical utility of this approach by means of a case study from the food industry. The case showed that the framework could be used to reduce the risk of production faults and that the impact of these can be minimised by restructuring the production workflow.

The key strength of this approach is fully automated and only the modelling of the production workflows and the expression of the goals require manual input.

Further research will focus on refining this framework and the software tool, SBOAT. We hope to release this tool in late 2014 and use it in several case studies. This will allow for a more extensive formalised exploration of the scope and parameters of this method.

ACKNOWLEDGEMENTS

We would like to thank the Inspire project for funding.

REFERENCES

[1] D. R. Christiansen, M. Carbone, and T. Hildebrandt, "Formal semantics and implementation of BPMN 2.0 inclusive gateways", in *Proceedings of the 7th international conference on Web services and formal methods*, M. Bravetti and T. Bultan, Eds., ser. Lecture Notes in Computer Science, vol. 6551, Berlin, Heidelberg: Springer-Verlag, 2011, pp. 146–160, ISBN: 978-3-642-19588-4. DOI: `10.1007/978-3-642-19589-1_10`.

[2] T. H. Cormen, C. E. Leiserson, R. L. Rivest, and C. Stein, *Introduction to Algorithms, Third Edition*, 3rd. The MIT Press, 2009, ISBN: 0262033844.

[3] H. Hansson and B. Jonsson, "A logic for reasoning about time and reliability", *Formal Aspects of Computing*, vol. 6, no. 5, pp. 512–535, 1994, ISSN: 0934-5043. DOI: `10.1007/BF01211866`.

[4] L. Herbert and R. Sharp, "Quantitative analysis of probabilistic BPMN workflows", in *ASME 2012 International Design Engineering Technical Conferences and Computers and Information in Engineering Conference (IDETC/CIE2012))*, ser. ASME Conference Proceedings, (Awaiting Publication), Jul. 2012.

[5] M. Z. Kwiatkowska, G. Norman, and D. Parker, "PRISM 4.0: verification of probabilistic real-time systems", in *Proceedings of the 23rd International Conference on Computer Aided Verification (CAV'11)*, G. Gopalakrishnan and S. Qadeer, Eds., ser. Lecture Notes in Computer Science, vol. 6806, London, UK: Springer-Verlag, 2011, pp. 585–591, ISBN: 978-3-642-22109-5. DOI: `10.1007/978-3-642-22110-1_47`.

[6] M. Z. Muehlen and J. Recker, "How much language is enough? theoretical and practical use of the business process modeling notation", in *Proceedings of the 20th international conference on Advanced Information Systems Engineering*, ser. Conference on Advanced Information Systems Engineering 2008, Berlin, Heidelberg: Springer-Verlag, 2008, pp. 465–479, ISBN: 978-3-540-69533-2. DOI: `10.1007/978-3-540-69534-9_35`.

[7] Object Management Group, "Business process model and notation (BPMN) 2.0", Object Management Group, Needham MA, USA, Standards Document formal/2011-01-03, Jan. 2011. [Online]. Available: `http://www.omg.org/spec/BPMN/2.0/`.

[8] C. Ouyang, M. Dumas, and A. H. M. T. Hofstede, "Pattern-based translation of BPMN process models to BPEL web services", *International Journal of Web Services Research (JWSR)*, vol. 5, no. 1, pp. 42–62, 2007. DOI: `10.1.1.143.3118`.

[9] C. Ouyang, M. Dumas, A. H. M. ter Hofstede, and W. M. P. van der Aalst, "From BPMN process models to BPEL web services", in *Proceedings of IEEE International Conference on Web Services*, Washington, DC, USA: IEEE Computer Society, 2006, pp. 285–292, ISBN: 0-7695-2669-1. DOI: `10.1109/ICWS.2006.67`.

[10] D. Prandi, P. Quaglia, and N. Zannone, "Formal analysis of BPMN via a translation into COWS", in *Proc. of the 10th international conf. on Coordination models and languages*, ser. COORDINATION 2008, Berlin, Heidelberg: Springer-Verlag, 2008, pp. 249–263, ISBN: 3-540-68264-3, 978-3-540-68264-6. DOI: `10.1007/978-3-540-68265-3_16`.

[11] R. Pugliese and F. Tiezzi, "A calculus for orchestration of web services", *Journal of Applied Logic*, vol. 10, no. 1, pp. 2–31, 2012, ISSN: 1570-8683. DOI: `10.1016/j.jal.2011.11.002`.

[12] O. for the Advancement of Structured Information Standards (OASIS), "Web services business process execution language (WS-BPEL) version 2.0", Standards Document WSBPEL-v2.0-OS, Apr. 2007. [Online]. Available: `http://docs.oasis-open.org/wsbpel/2.0/OS/wsbpel-v2.0-OS.html`.

[13] D. J. White, *Markov decision processes*. New Jersey, USA: John Wiley & Sons, 1993, ISBN: 9780471936275.

Enterprise Risk and Opportunity Management for Nonprofit Organizations and Research Institutions

Allan Benjamin[a,1], Homayoon Dezfuli[b,1], Chris Everett[c], Julie Pollitt[d], Dev Sen[c]
[a]Independent Consultant, Albuquerque, NM, USA
[b]Office of Safety & Mission Assurance, NASA Headquarters, Washington, DC, USA
[c]Information Systems Laboratories, Inc., Rockville, MD, USA
[d]Independent Consultant, San Jose, CA, USA

Abstract: Enterprise risk and opportunity management (EROM) concerns the means by which organizations develop and implement their strategic goals through a portfolio of programs, projects, institutional assets, and activities. The overall objective of EROM is to reach an optimal balance between minimizing the potential for loss (risk) while maximizing the potential for gain (opportunity). The focus of this paper is on the development of guiding principles and an overall approach that serves the interests of technically oriented nonprofit organizations and research institutions. These interests tend to place emphasis on performing services and achieving technical gains more than on achieving specific financial goals, which is the province of commercial enterprises. In addition, the objectives of nonprofit organizations may extend to institutional development and maintenance, financial health, legal and reputational protection, education and partnerships, and mandated milestone achievements. This paper discusses the philosophical underpinnings of EROM in the context of nonprofit organizations, the integration of EROM with existing management processes, and the nature of the activities that are performed to implement EROM within this context.

Keywords: Enterprise risk management, opportunity management, strategic goals, desired outcomes, leading indicators.

1. INTRODUCTION

Enterprise risk management (ERM) and enterprise risk and opportunity management (EROM) are synonymous terms used to address the natural desire of an organization to strike a reasonable balance between minimizing the potential for loss (risk) and maximizing the potential for gain (opportunity). These risks and opportunities are addressed within the context of implementing the organization's strategic goals.

General frameworks for EROM have been developed successfully over the past 10-15 years by organizations such as COSO [Committee of Sponsoring Organizations of the Treadway Commission, 2004], and have been encoded within Standards such as ISO-31000 [International Organization for Standardization, 2009]. While these frameworks have undoubtedly provided impetus for the acceptance and practice of EROM, they have tended to focus primarily on monetary gains and losses as would be paramount for organizations whose principal objectives are financial. Furthermore, the frameworks are intentionally presented at a high level wherein the means for implementing them are to be customized by the users. The work described herein is directed instead toward nonprofit organizations such as Government agencies and research organizations, whose principal objectives tend to focus more on performing services or achieving technical gains, most often within frequently changing financial, schedule, and political constraints. The approach presented in this paper along with accompanying ideas for implementation provide a current snapshot of a process that is continuing to evolve and change, and that is scheduled to be documented by NASA in the summer of 2014 [NASA/SP--2014-615, Enterprise Risk and Opportunity Management, Concepts for Implementation within NASA].

---

[1] asbenja@q.com, hdezfuli@nasa.gov

2. PHILOSOPHICAL UNDERPINNINGS

This section discusses the dimensions of opportunity within EROM and the principles for balancing enterprise risk and opportunity, normalizing risk and opportunity management within the enterprise, and analyzing enterprise risks and opportunities over different time scales.

2.1 Dimensions of Opportunity

The term "opportunity" has two definitions in Webster's online dictionary: (1) a favorable juncture of circumstances, and (2) a good chance for advancement or progress. Nonprofit technical organizations are concerned with both types of opportunity. In particular, the first definition applies to events that have a potential to reduce the risk of not meeting one or more desired outcomes; for example, an emerging opportunity for an originating organization to share risks with a partner organization might result in a reduction of risks for the originating organization. The second applies to events that provide an opening to change strategic goals or desired outcomes to align them better with stakeholder expectations; for example, the emergence of a new technology might open up possibilities for the originating organization to achieve strategic benefits that were not previously considered possible.

Significant gains in advancement or progress may involve proactively searching for opportunities, such as putting resources into basic or applied research, with the expectation that on the whole these efforts will bear fruit and speed the rate of progress toward long-term goals. In the words of Francis Bacon ("The Essays," 1612): "A wise man will make more opportunities than he finds."

2.2 Balancing Risk and Opportunity

In order to encourage innovations and because of the drive to address increasingly complex technical challenges with decreased funding, many organizations are having to embark in new directions in which the primary objective is no longer just to minimize risk but rather to balance risk against opportunity. This evolving philosophy was addressed, for example, by the Administrator for NASA in an open letter to all NASA employees dated April 19, 2013:

> "We have to be willing to do daring things. Put another way, *risk intolerance is a guarantee of failure to accomplish anything of significance* [emphasis is the Administrator's]. … While we do this, we must constantly balance our risks and rewards and always, always put the lives and safety of our people first."

The balance between risk and opportunity is a reflection of one's tolerance for risk relative to one's appetite for opportunity. In existing EROM guidebooks, the phrase "risk tolerance" or "risk appetite" is sometimes used to denote the maximum level of risk that a decision maker (DM) is willing to take relative to meeting a strategic goal or desired outcome within given constraints[2]. For each goal or outcome, the DM specifies his or her maximum tolerable level of risk that it will not be accomplished within given constraints on funding, schedule, etc. In this paper we introduce the term "opportunity expectation" to denote the minimum level of opportunity that the DM considers worth pursuing for any strategic goal or desired outcome. In simplest terms, the DM's risk tolerance and opportunity expectation represent his or her balancing point for making decisions regarding responses to emerging risks and opportunities. However, it should be recognized that there is generally a correlation between opportunity expectation and risk tolerance, in that the DM might accept a higher tolerance for risk in conjunction with a higher expectation for opportunity.

---

[2] By the term "strategic goals," we mean the planned objectives that an organization strives to satisfy in keeping with its mission. The term "desired outcomes" refers to the specific accomplishments or milestones that must be achieved in order to satisfy the strategic goals.

For example, suppose a project at a research organization dedicated to fundamental physics has an objective to determine whether or not a postulated subatomic particle exists, and suppose the project has an associated funding constraint. In this case, there is a risk that the money will be spent without a resolution to the question of whether or not the particle exists. If it can be proved that the particle does exist, on the other hand, there is an opportunity for significantly advancing mankind's understanding of the laws of nature. Suppose the DM is willing to accept a likelihood of 20% that there will be no resolution if there is at least a 5% likelihood that the particle will be found to exist. In addition, the DM may state that he/she will accept a 40% likelihood of no resolution if there is a 10% minimum likelihood of proving existence. In terms of that particular risk and that particular opportunity, the opportunity expectation is 5% if the risk tolerance is 20%, and 10% if the risk tolerance is 40%.

In addition to balancing risks and opportunities in a generic sense, organizations must frequently manage risks and opportunities in a diversified manner. The organization may have higher standards (lower tolerance for risk) relative to preserving its core capabilities and human lives and safety, while at the same time having looser standards (tolerating higher risk) relative to accepting the possibility of losing hardware in the pursuit of pioneering or capability-expanding activities that create new opportunities to more effectively advance the organization's mission. This considered diversification of risk tolerance is essential for progress and success over the long term. It creates areas where the organization learns rapidly, in part through acceptable numbers of setbacks, as well as promoting areas where the gains made through high risk activities are consolidated and institutionalized into a more capable organization.

There is a well-known tendency for such balances to be made based on psychological factors that are not always in the interest of making the optimum decision. It was originally pointed out in the so-called Ellsberg paradox (Quarterly Journal of Economics, 1961) and subsequently in many treatises concerning risk aversion, that when people are confronted with two choices where the balance between opportunity for success and risk of loss is neutral or even moderately favorable to the opportunity, they will tend to choose the path with lower risk. Use of EROM in a structured approach helps to ensure that strategic decisions are made more objectively.

2.3 Normalization of Risk and Opportunity Management within the Enterprise

The EROM process should lead to a common approach for conducting risk and opportunity management throughout the organization, wherein such tasks as identifying, analyzing, ranking, and responding to risks and opportunities are performed in a consistent manner across the various entities that comprise the enterprise. Such commonality of approach provides several advantages:

- It makes it much easier to understand how risks cut across the various entities within the enterprise.

- It simplifies and improves the accuracy of the roll up of risks and opportunities from entity levels to the enterprise level, thereby providing increased confidence in the strategic decisions that the organization must make.

- It improves upon the ability of the organization to respond in an agile and timely manner to risks and opportunities that require immediate attention.

At the same time, the EROM framework should promote creative approaches to solve diverse problems and does not insist on a prescriptive approach wherein one size fits all.

2.4 Risks and Opportunities over Difference Time Scales

The EROM process treats the organization's goals and desired outcomes within various time scales. For example, the goals and desired outcomes might be expressed in terms of the following time intervals: (1) less than one year, (2) one to five years, (3) five to ten years, and (4) longer than ten years. The EROM approach strives for the successful attainment of the desired outcomes within each of these time frames.

3. INTEGRATION WITH EXISTING MANAGEMENT PROCESSES

Although the need for EROM in nonprofit technical organizations may be driven by a need to provide innovative technical solutions, it is also necessary and desirable to implement EROM within the current management framework of the organization. While the detailed structure of each organization differs, most nonprofit technical organizations are organized under the top level into the three main functions shown in Figure 1: enterprise strategic management, institutional management, and project management. The authority for strategic management sets the overall strategic goals and desired outcomes for the enterprise; develops a high-level plan for implementation, including the definition of major programs and projects and specification of institutional support requirements; evaluates performance in terms of the degree to which its strategic goals and desired outcomes are being realized; and makes major course correction or course resetting decisions when conditions warrant it. The authority for institutional management provides the same goal setting and execution oversight with respect to the institutional capabilities of the enterprise, including the sufficiency of the workforce, availability of facilities, and integrity of procurement and quality control practices. The authority for project management does the same for the projects that the enterprise funds to achieve its strategic goals and desired outcomes. Communication among these three functions occurs through a process of reporting and informing as shown in the figure.

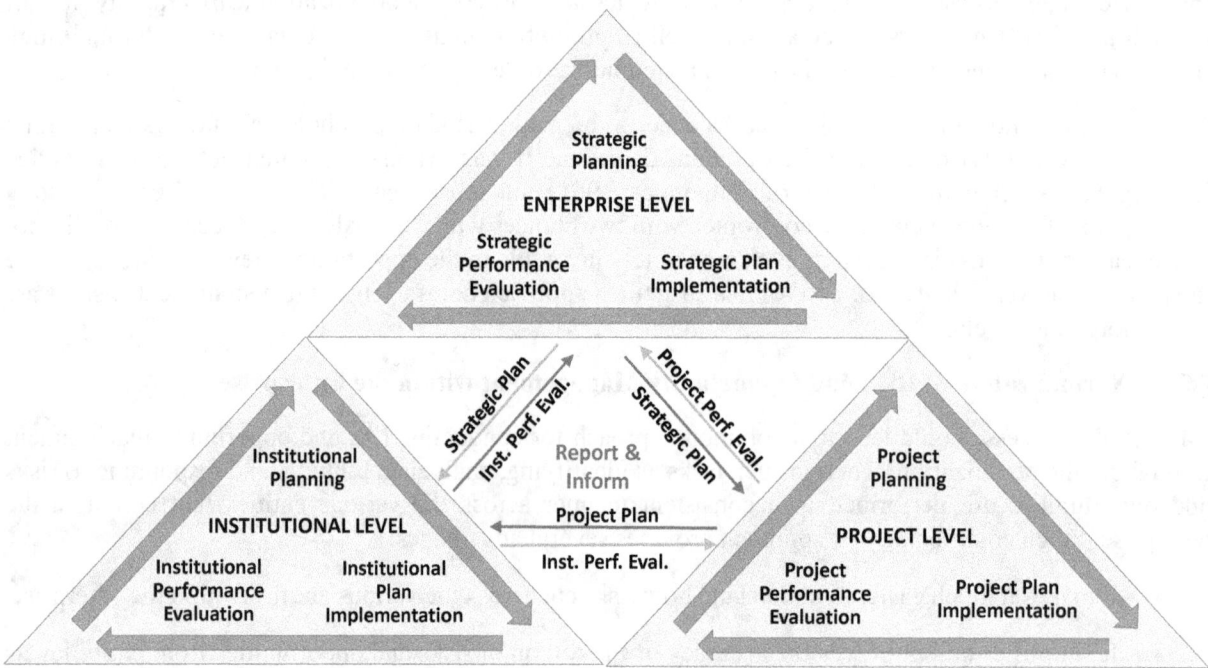

Figure 1. Generic Illustration of Organizational Functions and the Interfaces between Enterprise, Project, and Institutional Levels.

At the enterprise level, the processes of strategic planning, strategic plan implementation, and strategic performance evaluation are guided by information obtained from both external and internal sources, as shown in Figure 2. The needed information includes knowledge and understanding of the constraints that are imposed by Government and other sources, as well as recognition of the problems that occur during the execution of the strategic plan, the opportunities that present themselves, the risks from potential adverse events that have not yet occurred, and the leading indicators that portend possible problems, opportunities, and risks.

ENTERPRISE-LEVEL RISK AND OPPORTUNITY MANAGEMENT

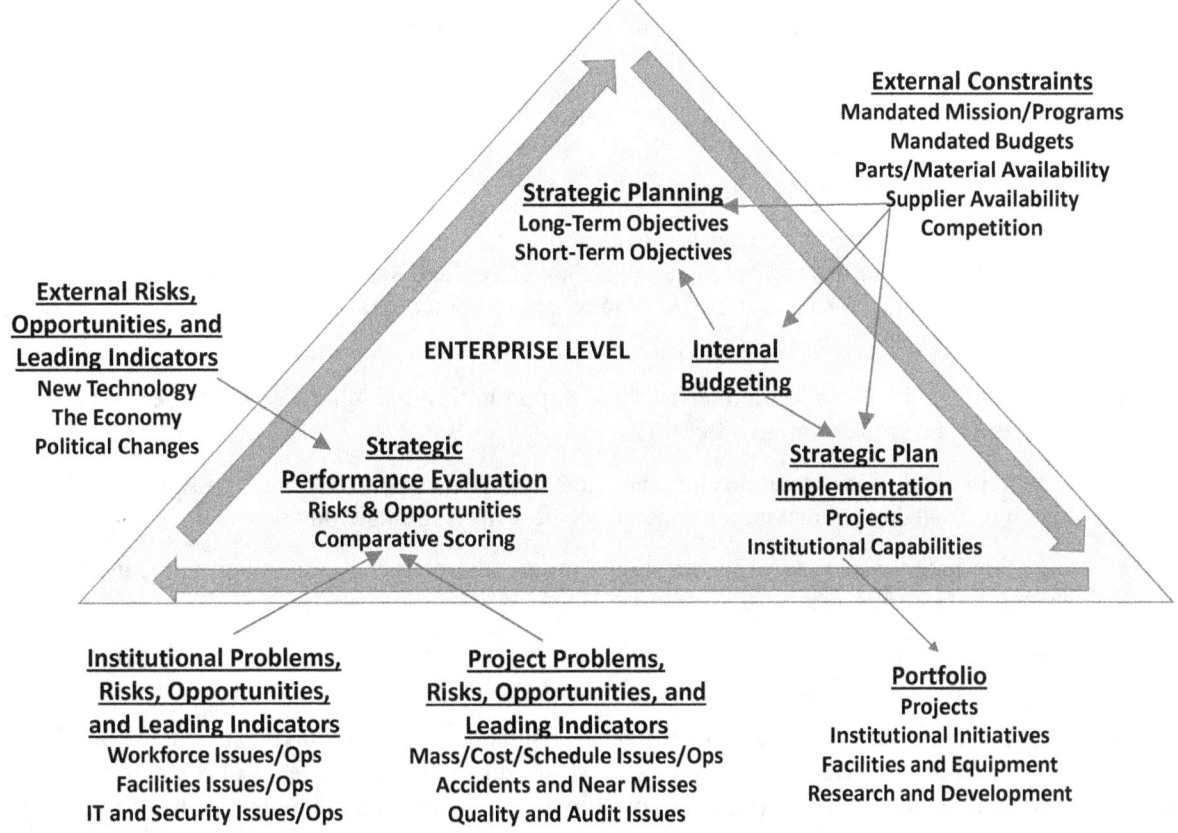

Figure 2. Generic Illustration of the Transfer of Information Into, Out of, and Within the Enterprise Level.

A well-conceived EROM approach should support each of these functions in the following areas:

- Planning within the strategic management function in setting strategic direction, goals, architecture, and policies; establishing metrics against which to measure strategic performance; projecting future performance; establishing mission and budget priorities; establishing enterprise-level performance requirements; and approving major new initiatives

- Planning within the institutional management and project management function and management of risks and opportunities for new institutional initiatives

- Data gathering for high-level planning and review meetings; acquisition strategy, procurement, and planning meetings; and performance review meetings conducted under each of these functions

- Identification of risk and opportunities to be pursued based on findings from these high-level planning and review meetings together with results from other sources, such as reports from external agencies and independent review councils

4. EXAMPLES OF EROM ANALYSIS ACTIVITIES

This section discusses some of the analysis activities involved in developing enterprise-level risk and opportunity taxonomies, identifying corresponding leading indicators, composing enterprise risk and opportunity statements, correlating strategic success likelihoods with leading indicator values, rating present indicators and success likelihoods, and incorporating potential opportunity into strategic planning.

4.1   Developing Enterprise-Level Risk and Opportunity Taxonomies

A taxonomy is a tree structure of classifications that begins with a single, all-encompassing classification at the root of the tree, and partitions this classification into a number of sub-classifications at the nodes below the root. This process is repeated iteratively at each of the nodes, proceeding from the general to the specific, until a desired level of category specificity is reached.

Taxonomies can be used to group enterprise risks and opportunities into categories that reflect, first, the types of strategic goals and desired outcomes that they affect, and second, the types of events that could create risk and opportunity for each strategic goal or desired outcome. Risk and opportunity taxonomies provide the following benefits:

- They assist in the identification of risks and opportunities that otherwise might be missed (e.g., by facilitating the brainstorming process)

- They help identify leading indicators that can be used to rank the likelihood (at least qualitatively) that a postulated event that either threatens or benefits a strategic outcome will occur

- They facilitate the process of identifying planning alternatives to effectively mitigate the risks or exploit the opportunities

- They assist in properly allocating resources among the entities or organizational units of the enterprise (e.g., to mitigate a risk or exploit an opportunity)

Figure 3 illustrates an example three-level enterprise risk and opportunity taxonomy that is applicable to nonprofit technical organizations. For each categorical unit in the bottom level of the taxonomy, it also provides an example individual risk (R) or opportunity (O). As noted on the figure, each bottom-level sub-category can be further decomposed into one or more strategic goals or desired outcomes that apply to that categorical unit. For example, new technology pursuits pertaining to mission performance is comprised of different individual technology pursuits, each of which represents a strategic goal or desired outcome of the enterprise. Thus, the taxonomy in Figure 3 may be construed as having an unseen bottom level corresponding to the strategic goals and desired outcomes.

4.2   Identifying Corresponding Leading Indicators

The degree to which each of the organization's strategic goals and desired outcomes is being satisfied can be inferred, at least qualitatively, by tracking a set of quantitative and/or qualitative surrogate measures referred to here and elsewhere as leading indicators. Table 1 suggests a set of potential leading indicators for each of the lowest-level categorical units in Figure 3 and identifies whether each indicator emanates from an internal source or an external source. For strategic goals and desired outcomes relating to new technology development for mission performance, for example, the leading indicators may include results from assessments of the technical state-of-the-art both within the organization and in other organizations, the number and type of patents obtained within the organization, and the rate of progress in advancing the associated technology readiness levels (TRLs). The manner in which the status of the leading indicators can be used to qualitatively infer the likelihood of success for each strategic goal or desired outcome will be discussed further in Section 4.4.

4.3   Composing Enterprise Risk and Opportunity Statements

Enterprise-level risk and opportunity statements are concise descriptions of credible scenarios that could potentially affect the organization's ability to meet its strategic goals and desired outcomes, either positively or negatively. In its simplest form, a risk or opportunity statement contains three basic parts: a Condition, a Departure, and a Strategic Result.

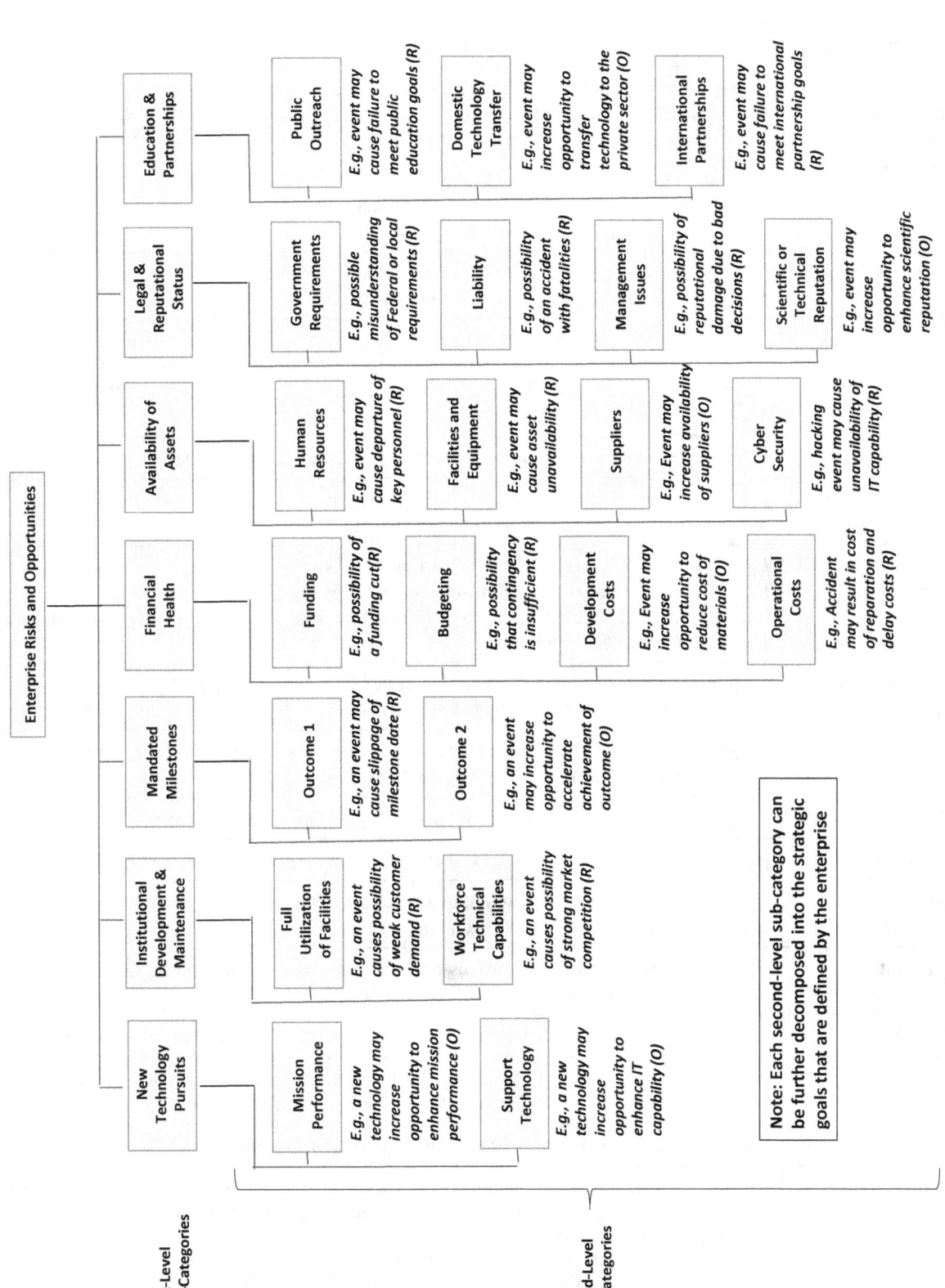

Figure 3. Example Taxonomy for Enterprise Risks and Opportunities

Probabilistic Safety Assessment and Management PSAM 12, June 2014, Honolulu, Hawaii

Table 1. Example Leading Indicators for Each Taxonomy Unit

Taxonomy Unit(s)	Example Leading Indicators
New technology pursuits: mission performance and support technology	*Internal*: Initiation of and results from internal state-of-the-art assessments; number and types of patents obtained; rate of progress in technology readiness level (TRL). *External*: Technology trends in areas pertinent to the organization's missions and support capabilities.
Utilization of facilities	*Internal*: Scheduling backlog. *External*: Market demand for facility capabilities; capabilities of competitive facilities.
Workforce technical capabilities	*Internal*: Educational and experience backgrounds; technical training courses taken and passed; number and type of technical papers published.
Mandated milestones	*Internal*: Schedule compared to other programs/projects in the organization; number of missed intermediate milestones and slippage amount; number of unresolved action items and uncorrected problems. *External*: Mandated changes in prioritization of the organization's outcomes.
Funding sufficiency	*External*: Economic indicators; Congressional makeup; changes in national priorities.
Budgeting sufficiency	*Internal*: Contingency compared to other programs/projects; rate of spending compared to other programs/projects; unresolved assignment of roles and responsibilities
Development and operational costs	*Internal*: Monthly cost reports; scores on self-assessments and audits; earned-value reports; precursor, anomaly, and mishap reports. *External*: Price trends; threats of foreign conflicts or political changes (affecting rare material costs, e.g.); supplier financial problems; Government shutdown.
Availability of human resources	*Internal*: Age of workforce; workplace morale (from surveys, e.g.). *External*: Changes in competitive labor market; demographic changes.
Availability of facilities & equipment	*Internal*: Number of unplanned maintenance actions; age of equipment. *External*: Terrorism trends; changes in OSHA regulations.
Supplier availability	*External*: Market factors (demand versus supply); supplier financial or legal problems.
Cyber security	*Internal*: Number of unaddressed vulnerabilities. *External*: Hacking trends; new viruses.
Government requirements	*Internal*: Quality of ethics program; quality of record keeping (e.g., for OSHA requirements). *External*: New regulations.
Legal liability	*Internal*: Increased use of hazardous or toxic materials; accident precursors. *External*: Trends in Court decisions regarding liability.
Management issues	*Internal*: Findings of independent reviews; resolution of internal dissenting opinions.
Scientific or technical reputation	*Internal*: Number of technical papers published; number of patents granted. *External*: Number of citations in technical papers; number of nominations or awards received.
Public outreach	*Internal*: Missed milestones; low enrollment in educational programs.
Technology transfer	*Internal*: Missed milestones; number of technology transfer agreements. *External*: Lack of interest or progress from potential commercial partners; trends regarding the sharing of sensitive information and materials.
International partnerships	*Internal*: Missed milestones; *External*: Lack of interest or progress from potential international partners; new regulations regarding sensitive information; competition from a foreign country.

Probabilistic Safety Assessment and Management PSAM 12, June 2014, Honolulu, Hawaii

- The *Condition* is a single phrase that describes current key fact-based situation in terms of the status of the appropriate leading indicators

- The *Departure* is an event or set of events that could potentially occur which, if they do occur, produce a departure from the baseline assumptions behind the implementation plan for the organization's strategic goals and desired outcomes.

- The *Strategic Result* is a single phrase that describes the foreseeable positive or negative impact(s) on the ability of the organization to meet one or more of its strategic goals or desired outcomes or to further its mission by enabling a new strategic goal or desired outcome.

It is important to the EROM process that risk and opportunity statements be composed without regard to potential modifications of the implementation plan, such as risk mitigation or opportunity exploitation options that may suggest themselves. The statements do not presume anything that is not in the current baseline implementation plan, other than the Condition, which has its basis in fact.

Table 2 illustrates some example risk and opportunity statements that may apply to some of the categorical units that were identified in the taxonomy of Figure 3. The Conditions in these statements refer to leading indicators that were identified in Table 1.

A more detailed description of risk statements and narrative that accompany them may be found in the NASA Risk Management Handbook [NASA/SP-2011-3422, 2011].

Table 2. Example Risk and Opportunity Statements for Selected Taxonomy Units

Taxonomy Unit(s)	Example Risk or Opportunity Statement
New technology pursuits: mission performance	*Opportunity:* Given that [*CONDITION*] the rate of progress in technology readiness for technology X is 20% faster than anticipated, there is a possibility that [*DEPARTURE*] the technology will be ready in time for Program Y resulting in [*STRATEGIC RESULT*] an ability to exceed the performance requirements associated with Strategic Goal Z.
Funding sufficiency	*Risk:* Given that [*CONDITION*] economic indicators suggest the possibility of a recession, there is a possibility that [*DEPARTURE*] overall funding for the organization will be cut substantially resulting in [*STRATEGIC RESULT*] the need to scale back on Strategic Goal X.
Government requirements and reputational issues	*Risk:* Given that [*CONDITION*] an audit of ethics training has indicated shortcomings in the contents and attendance of the training, there is a possibility that [*DEPARTURE*] there will be a serious ethical infraction resulting in [*STRATEGIC RESULT*] Government sanctions and reputational damage to the organization.
Public outreach	*Risk:* Given that [*CONDITION*] the schedule for establishing participatory engagement activities with the public has slipped 6 months, there is a possibility that [*DEPARTURE*] the quality of such engagement will be less than desired resulting in [*STRATEGIC RESULT*] the public education initiative being out of compliance with the strategic plan.

4.4 Correlating Strategic Success Likelihoods with Leading Indicators (An Example Approach)

As mentioned earlier, the likelihood of successfully satisfying a strategic goal or achieving a desired outcome is estimated from the status of the leading indicators. Leading indicators may be considered to be quantitative or qualitative measures that define the present condition from which risks and opportunities may emanate. They are also surrogate measures with respect to the success likelihood of each strategic goal or desired outcome. Because they are surrogates, they do not necessarily comprise the complete set of factors that affect the success likelihood. At best, therefore, the success likelihood for each strategic goal or desired outcome can only be estimated within a range of uncertainty.

Superimposed on this estimate of success likelihood is the decision maker's risk tolerance. Generally speaking, the DM specifies his/her risk tolerance as a minimum likelihood of failure that he/she is willing to accept for the desired outcome in question at the present point in time. Conversely, the DM's minimum success expectation is by definition the complement of his/her risk tolerance. Thus, if the DM has a risk tolerance of 0.20 for the likelihood of not meeting the strategic goal or desired outcome, he/she has a minimum success expectation of 0.80.

Suppose, for example, that based on the current status of the leading indicators, the success likelihood for a desired outcome is estimated to be between 0.50 and 0.70. Since the DM's expectation is 0.80, we would say that the risk is intolerable. If, however the estimated success likelihood were between 0.90 and 0.95, we would consider the risk to be tolerable. If it were between 0.70 and 0.90, straddling the minimum success expectation, we would consider the risk to be marginal.

Once the DM has specified a risk tolerance or minimum success likelihood expectation for each strategic goal or desired outcome, it is incumbent on the analyst to determine how these expectations map to threshold values of the leading indicators. The following discussion presents an example approach. To start with, the analyst assumes that each leading indicator except the one being examined is at a representative best-estimate value. Then, he/she specifies two thresholds for the leading indicator being examined. This is done for each strategic goal or desired outcome to which the leading indicator applies. The two thresholds correspond to the lower bound and upper bound of the uncertainty range in the correlation of the strategic goal/desired outcome to the leading indicator.

For example, referring to Table 3, the DM has specified that he/she requires an 80% minimum tolerable likelihood of success for Strategic Goal A. The analyst determines that if other leading indicators are at their representative values, the DM's expectation could conceivably be satisfied if the technology readiness level, or TRL, for a particular technology development is at a value of 7, and that it should definitively be satisfied if the TRL is at 8. He/she therefore specifies two thresholds for the technology readiness level, which is designated as leading indicator 1 in Table 3: an "optimistic" threshold of 7 and a "pessimistic" threshold of 8. For interpretation, the risk of not satisfying the DM's success expectation is tolerable if the actual TRL is 8 or higher, is intolerable if it is at 7 or lower, and is marginal if between 7 and 8.

Table 3. Example Development of Leading Indicator Threshold Values Consistent Corresponding to the Decision Maker's Minimum Tolerable Likelihoods of Success

Strategic Goal or Desired Outcome	Minimum Tolerable Likelihood of Success	Leading Indicator Levels	Values for Leading Indicator 1 (Technology Readiness)	Values for Leading Indicator 2 (Public Support)	Values for Leading Indicator 3 (Material Cost)	Etc.
A	80%	Optimistic Threshold	7	N/A	$20M	
A	80%	Pessimistic Threshold	8	N/A	$25M	
B	75%	Optimistic Threshold	N/A	High	N/A	
B	75%	Pessimistic Threshold	N/A	Moderate	N/A	
C	90%	Optimistic Threshold	8	N/A	N/A	
C	90%	Pessimistic Threshold	9	N/A	N/A	
Etc.						

It should be noted that although this example has started from the presumption that all the leading indicators are composed of individual metrics, it is quite possible that some may represent composite metrics. For example, if high confidence in the success of a strategic goal or desired outcome required that *both* metric X and metric Y satisfy threshold values, the leading indicator could be defined as a composite of metric X and metric Y.

4.5 Rating Present Leading Indicators and Future Success Likelihoods (An Example Approach)

Once the correlations between strategic success likelihoods and leading indicator values have been established, it is relatively straightforward to rate both of them in terms of the DM's minimum success likelihood expectations. As shown in Table 4, the present status of each leading indicator is first rated as green, yellow, or red for each goal/outcome based on how its value relates to the optimistic and pessimistic threshold values. The future success likelihood for the goal/outcome as a whole is then rated as green (acceptable), yellow (marginal), or red (unacceptable) based on the ratings for the leading indicators that apply to it. Generally, the future success likelihood rating for the goal/outcome will be the same as the rating for the worst-case leading indicator. There will sometimes, however, be ameliorating factors that would cause the rating for the future success likelihood to be better than the rating for the worst-case leading indicator. An example leading to this result is cited in Note 1 in Table 4.

Table 4. Example Development of Ratings for the Present Leading Indicators and the Future Success Likelihoods of the Strategic Goals and Desired Outcomes

Strategic Goal or Desired Outcome	Minimum Tolerable Likelihood of Success	Leading Indicator Levels	Values for Leading Indicator 1 (Technology Readiness)	Values for Leading Indicator 2 (Public Support)	Values for Leading Indicator 3 (Material Cost)	Etc.	Overall Rating of Future Success Likelihood
A	80%	Optimistic Threshold	7	N/A	$20M		Marginal (Note 1)
		Pessimistic Threshold	8	N/A	$25M		
		Present Value	8.5	N/A	$30M		
B	75%	Optimistic Threshold	N/A	High	N/A		Acceptable
		Pessimistic Threshold	N/A	Moderate	N/A		
		Present Value	N/A	Very High	N/A		
C	90%	Optimistic Threshold	8	N/A	N/A		Marginal
		Pessimistic Threshold	9	N/A	N/A		
		Present Value	8.5	N/A	N/A		
Etc.							

* Note 1: The overall future success likelihood is rated higher than the rating for Leading Indicator 3 because material cost for goal/outcome A is considered to be less important than other leading indicators, and because it may be compensated by savings in other areas.

4.6 Incorporating Potential Opportunity into Strategic Planning

Potential opportunity occurs when a new condition arises and promotes the possibility of an opportunity coming to play at a later time. For example, suppose that a new technology has been under research for a long time and suddenly the possibility of an unexpected breakthrough is announced. The reality of the breakthrough cannot be verified until it is corroborated by another set of experiments. If the breakthrough is indeed real, the possibility will open for achieving a new strategic goal that was not previously considered possible. Before instituting any changes to the current plan, however, it is necessary to see whether the additional testing results in a positive result.

Let us examine how Table 4 would be influenced by the emergence of this new potential opportunity. Space limitations prevent a detailed exposition of this example, but the principal steps would be as follows. First, a new strategic goal D would be added after goal/outcome C to indicate that a new initiative may be possible, and a new leading indicator #4 would be added after indicator #3 to provide a measure of the status of corroboration of the breakthrough. As was the case for the other leading indicators, the present value of the new leading indicator would be rated green, yellow, or red for goal D and for any other goal/outcome to which it applies. These leading indicator ratings would again be based on threshold values that the analyst derives from the DM's minimum success likelihood expectations. Using the approach described in the preceding section, the future success likelihood of goal D would then be rated green, yellow, or red based on the ratings of all the leading indicators that affect it, and the ratings for other goals/outcomes that are affected by the new leading indicator would be modified as appropriate to reflect the influence of the new indicator.

This process provides a means for judging whether the new potential opportunity eventually will be worth pursuing. A contingency plan would perhaps be developed at this point so that if the potential opportunity were to actualize, it could be exploited in a timely fashion.

5. CONCLUSIONS

Enterprise risk and opportunity management (EROM) addresses the natural desire of an organization to strike a balance between minimizing the potential for loss (risk) while maximizing the potential for gain (opportunity). These risks and opportunities are addressed within the context of implementing the organization's strategic goals and focus on the achievement of broad outcomes. Whereas previous frameworks for EROM have tended to emphasize monetary gains and losses, this paper has discussed an approach and presented application examples that pertain to nonprofit organizations whose principal objectives are to perform technical services or achieve technical gains, often within frequently changing financial, schedule, and political constraints. The EROM approach assists the planning activities of the strategic/executive function of the enterprise in developing strategic direction, goals, architecture, and policies, establishing metrics against which to measure strategic performance, projecting future performance, setting mission and budget priorities, deriving enterprise-level performance requirements, and selecting potential new initiatives.

Taxonomies are used to group enterprise risks and opportunities into categories that assist in the identification of leading indicators, facilitate the identification of planning alternatives, and assist in properly allocating resources among the entities or organizational units of the enterprise. The degree to which each of the organization's strategic goals and desired outcomes is likely to be satisfied is inferred by tracking the leading indicators relative to a set of threshold values. A process is suggested wherein the success likelihood for each strategic goal or desired outcome can be judged to be acceptable or not based on the decision maker's expectations. The process includes a means for examining potential new opportunities as they arise to estimate their future likelihood of success, so that strategic planning decisions can be risk and opportunity informed.

Programmatic Assessment of RG-MOX Utilization Following Participation in the DOE Surplus Plutonium Disposition Program

David H. Johnson[a*], Andrew A. Dykes[a], Andrew G. Sowder[b], and
Albert J. Machiels[b]
[a] ABSG Consulting, Irvine, CA, USA
[b] Electric Power Research Institute (EPRI), Charlotte, NC, USA

Abstract: EPRI is building a suite of tools for assessing nuclear fuel cycle options based on a platform of software, simplified relationships, and explicit decision-making and evaluation guidelines. This paper summarizes an example of an assessment from a utility perspective regarding continuing MOX utilization with commercial reactor-grade mixed-oxide fuel (RG-MOX) following successful utilization participation in the DOE Surplus Plutonium Disposition Program. This assessment reflects potential opportunities and problems based on topic familiarity and the perspective embedded in the scenario definition, as follows: (1) economic considerations will represent a primary driver for utilities operating in the U.S. commercial environment, and (2) back-end management issues must be flagged due to the number and magnitude of constraints in used-fuel management at U.S. nuclear plants for both wet and dry storage (and the important interface between them). While economic considerations are seen as the primary utility decision drivers with respect to RG-MOX use under the stylized conditions defined here, this assessment also showed that technical waste management issues could be showstoppers if not adequately resolved.

Keywords: PgRA, LWR, MOX Fuel, Advanced Nuclear Fuel Cycles.

1. INTRODUCTION

To address challenges and gaps in nuclear fuel cycle option assessment and to support research, develop and demonstration (RD&D) programs oriented toward commercial deployment, EPRI is seeking to develop and maintain an independent analysis and programmatic risk assessment (PgRA) capability by building a suite of assessment tools based on a platform of software, simplified relationships, and explicit decision-making and evaluation guidelines. The assessment tools support a decision analysis framework. The framework is intended to support and facilitate:

- Clear delineation of the issues associated with a fuel cycle option and the requisite activities needed to achieve a nuclear fuel objective.
- References to source material (e.g., reports, peer-reviewed manuscripts, and expert knowledge) to provide clear pedigree for inputs.
- Assignment of a readiness or "favorability" Figure of Merit (FOM) of an option and its uncertainty, reflecting the state of knowledge upon which the assessment is based.
- Ability to record the reasoning for each evaluation such that the overall basis for a decision path can be accessed and assembled in a summary report format.
- Identification of actions needed to address gaps in the state of knowledge, needed research, additional infrastructure, and regulatory and licensing requirements.

The framework provides a method for assembling and structuring available and relevant information for transparent, auditable assessments. It provides a structured, phased approach to evaluating or comparing nuclear fuel cycle options based on the level of detail desired and the amount of information available (Figure 1). The strategic assessment (Level 1) evaluates the alignment of the "what is being proposed and why" with strategic objectives whose satisfaction is of primary

* djohnson@absconsulting.com

importance. For its nuclear fuel cycle options assessment work, EPRI has selected five criteria: Resource Utilization, Proliferation Resistance and Security, Waste Management, Fuel Cycle Safety, and Economic Competitiveness. EPRI Report 1025208 [1] defines and describes application of the decision framework in detail and provides a more complete listing of references in a preliminary evidence database format.

Figure 1: Basic Layered Structure of EPRI Decision Analysis Framework

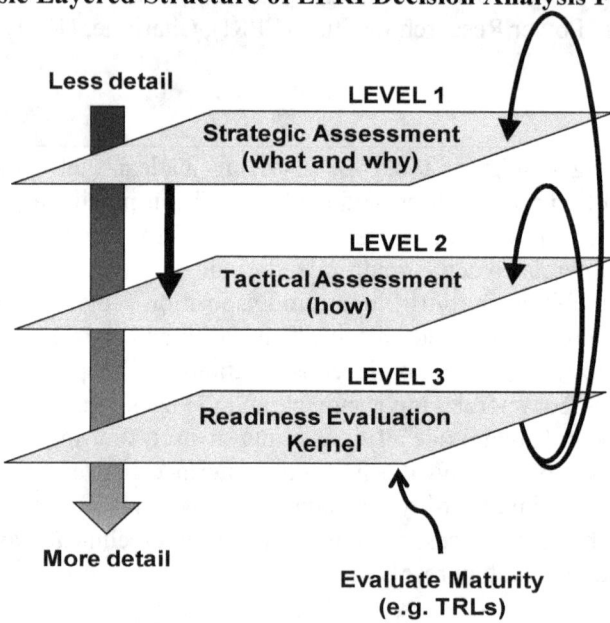

At the initial stages of such an assessment, the metrics can be quite broad. For this type of assessment, qualitative figures of merit provide a reasonable categorization of assessment results (i.e., favorability) and uncertainty; i.e., confidence. A simple three-color scheme of red, yellow, or green is used to display the evaluation results in summary fashion. In general, the colors convey common interpretations of the suitability and/or difficulties of proceeding with the proposed program. In the context of this assessment, the specific meanings of the color scales are summarized in Table 1.

Table 1: Assessment Metrics and Guidelines for Their Use

Figure of Merit	Framework Element		
	Level 1	Level 2	All Levels
	Alignment with Strategic Objectives	Favorability with Respect to Overcoming Barriers and Challenges	Confidence in Assessment
ALERT	Option is either not aligned with strategic criteria or represents a critical condition, requiring resolution, clarification, or further evaluation.	Option difficult to implement. Actions required to address barriers.	Low level of confidence in basis for assessment. Actions required to increase confidence.
CAUTION	Option possibly not aligned with strategic criteria. Evaluations during Level 2 assessments should identify barriers and their significance.	Option challenging to implement. Actions should be considered to reduce barriers.	Some uncertainty in basis for assessment. Possible conflicting evidence. Actions should be considered to increase confidence.
SUITABLE	Option aligns with strategic criteria. No additional evaluation is needed.	Option can be implemented.	High level of confidence in basis for assessment. No additional evidence required.

This paper summarizes an example of an assessment from a utility perspective regarding continuing MOX utilization with commercial reactor-grade mixed-oxide fuel (RG-MOX) following successful utilization participation in the DOE Surplus Plutonium Disposition Program. It was selected because existing nuclear plants participating in this program will have gained the capability and experience for the utilization of MOX fuels. Therefore the programmatic issues involved in extending the use of MOX via the procurement of RG-MOX presented an opportunity to exercise and test the framework on a reasonable well defined question. The following sections summarize the assessment accomplished in EPRI Report 1025208 [1].

2. STRATEGIC ASSESSMENT

The first stage in applying the EPRI Decision Analysis Framework involves the assessment of alignment of the RG-MOX use scenario against the five strategic criteria (introduced above) that EPRI has adopted for its nuclear fuel cycle assessment work. Ratings may be based primary on documented evidence. The EPRI framework is supported by an evidence database, and documents and citations are linked to each evaluation. In addition, expert judgments and opinions incorporated into the assessment can and should be documented in the evidence database for transparency and future review. The strategic assessment not only evaluates the alignment of with strategic criteria, but also provides insight into which criteria could present significant barriers to implementation of the program.

The EPRI team judged that the successful utilization of WG-MOX over the duration of the DOE Surplus Plutonium Disposition Program provided adequate confidence that the strategic objectives of Resource Utilization, Proliferation Resistance and Security, and Fuel Cycle Safety can be met with RG-MOX. The option does not conflict with the sustainability criterion. At any time a utility has an option to return to use of UOX and there is clear evidence that there is an ample supply of natural uranium and manufacturing capacity for UOX fuel. Proliferation Resistance and Security and Safety are not impacted by implementation of the option at the utility and plant level. These issues are assumed to have been evaluated and found acceptable in preparation for and subsequent utility participation in the DOE SPD program, and any external, higher-level concerns, such as non-proliferation policy considerations, lie outside the scope of this assessment. No additional issues and requirements have been identified with the continued use of MOX. Accordingly, these three criteria were screened out from further evaluation. Two strategic issues,

However, Table 2 shows the strategic assessments of Economic Competitiveness and Waste Management revealed considerable uncertainty (red confidence findings) and significant concern regarding problems associated with the long term use of RG-MOX fuel (yellow and red alignment/favorability findings) to warrant a tactical assessment of technical and programmatic issues in these two areas.

The next section summarizes those results and discusses how the evidence evaluated by the EPRI team that conducted the assessment contributed to these results.

Table 2: Results from Level 1 Strategic Assessment Indicating Non-Alignment of Fuel Cycle Option with Criteria

Criterion	Assessment FOM	Confidence FOM	Basis	Result
Economic Competitiveness	ALERT	LOW	RG-MOX fuel cost relative to UOX expected to be the primary commercial decision driver. Substantial cost savings for plant modifications and upgrades will be realized from WG-MOX program participation. However, costs associated with continued MOX use and storage require evaluation. Discounted or incentivized RG-MOX fuel purchases could offset these other costs.	Level 2 assessment on costs for fuel procurement, waste management, O&M, etc.
Waste Management	CAUTION	MEDIUM	Greater decay heat and higher neutron dose rates for used RG-MOX (vs. UOX and WG-MOX) may have significant impacts for onsite wet and dry storage with continued MOX use beyond limited WG program unless arrangements for early offsite transport are provided.	Level 2 assessment on wet and dry storage challenges and barriers

3. TACTICAL AND PROGRAMMATIC ASSESSMENTS

The more detailed assessments of Economics and Waste Management issues are summarized in the following two subsections. Space limitations prevent the inclusion of all assessment here.. Findings indicating significant concerns regarding favorability (yellow or red) or low confidence (red) for technical Waste Management issues are summarized in Reference [2]. The scope and approach used to accomplish all assessments are documented fully in Reference [1].

3.1. Assessment of Economic Competitiveness

The economic competitiveness of continuing MOX utilization with RG-MOX fuel assemblies involves offsetting the potential increased cost of operating the plant with MOX fuel against the savings that can be achieved through a smaller cost of MOX fuel assemblies or incentives that produce savings in other areas. As these benefits are not yet known and purely speculative, the tactical assessment focuses on the increased costs that would accompany the continued utilization of RG-MOX fuel after a limited period of WG-MOX use under the DOE SPD program. At this preliminary stage it focuses on identifying and characterizing the additional requirements associated with MOX operations and identifies potential follow-on action items to pursue.

For this initial assessment example, the following simple breakdown of nuclear electricity generation, or total cost of electricity, is adopted. It consists of

1. Capital Costs
2. Fuel Procurement (i.e., front-end services)
3. Operation and Maintenance (O&M)
4. Cost of Waste Management (i.e., back-end services)

The first cost category area (i.e., capital expenditures for construction and major equipment upgrades) is excluded from this assessment. The capital expenditures necessary to accommodate WG-MOX fuel cycles are treated as sunk costs, and those modifications are assumed to be sufficient to permit the plant to continue operations with RG-MOX fuel. The other three cost categories, fuel procurement, O&M, and waste management, are examined further in the Level 2 tactical assessment.

3.1.1. Fuel Procurement

Fuel procurement addresses the potential differences in costs for fuel and associated services to support operation on partial (nominally 35%) RG-MOX cores versus the reference case of a return to 100% UOX. This includes the following considerations and activities:

- Fuel procurement cost of both RG-MOX and UOX fuel assemblies needed for the fuel cycles in which RG-MOX will influence the core load.
- Additional core design engineering and analysis needed to support partial MOX core loadings and operation.
- Changes to new fuel acceptance and inspection procedures to protect personnel from the increased radiation field emitted from assemblies containing recycled Pu.
- MOX parity with UOX in terms of fuel burnup and core design and management. Disparities between maximum burnups licensed for UOX fuel and MOX limits core design flexibility and fuel utilization, resulting in additional cost burdens in terms of fuel procurement and heterogeneous core management.

Table 3, reproducing Table 5-5 of [1] summarizes an initial assessment of the costs associated with the purchase and utilization of RG-MOX fuel, and Table A-3 of Appendix A of [1] contains citations contains the relevant citations cited in the table. In order to be cost competitive, a reduction in the cost of RG-MOX fuel procurement relative to UOX will likely be required to offset any additional costs that arise from all the requirements that must be implemented to operate the plant.

3.1.2. Operations and Maintenance

The operations and maintenance category accounts for the cost impact of additional requirements imposed on plant personnel to support plant operations with RG-MOX fuel. As used WG-MOX will already be present in the spent fuel pool (SFP) and onsite independent spent fuel storage installation (ISFSI), only the differential costs associated with the use of RG-MOX fuel are considered. The issues identified in this assessment include:

- O&M of systems required exclusively for the safe operation of the core containing RG-MOX.
- Changes to radiation protection programs in terms of additional personnel, equipment, training, and monitoring to meet radiation protection objectives for plant personnel.
- Additional procedures and activities to ensure regulatory compliance for plant operations with RG-MOX.
- Material degradation and aging management activities needed to monitor and mitigate potential impacts of RG-MOX use on the integrity and performance of systems, structures and components within the plant. These issues will need to be reflected in the aging management plan and may require mitigation.

Table 4, reproducing Table 5-6 of [1] summarizes an initial assessment of the additional costs associated with operations and maintenance with RG-MOX fuel, and Table A-5 of Appendix A of Reference [1] contains the relevant citations cited in the table. Use of RG-MOX is expected to incur costs associated with WG-MOX operations plus any additional costs, but the anticipated experience with WG-MOX provides high confidence that they can be controlled.

3.1.3. Costs Associated with Waste (Used Fuel) Management

Used Fuel Management addresses the additional costs of safely handling and storing the additional RG-MOX that will be discharged from the reactor. Issues selected for consideration include:

- Criticality control, which includes the costs of measures to ensure that criticality margins are maintained during all aspects of used fuel handling and storage.
- Heat load management, which addresses the costs associated with ensuring that used fuel remains within its thermal limits under all storage conditions.
- Radiation protection, which encompasses the cost of worker protection associated with the increased radiation from used MOX fuel assemblies.
- Dry storage, includes both the incremental cost of dry storage cask and canisters (DSC) capable of storing RG-MOX, loading and transport operations, in addition to ISFSI activities needed to meet the first three requirements.

Table 5, reproducing Table 5-7 of [1], summarizes the issues that could drive cost penalties associated with the management of used RG-MOX. Table A-4 of Appendix A of [1] contains the relevant citations cited in the table. Although there may be some increase in cost associated with reviews of criticality margins and radiation protection support, the economic issue flagged as being of greatest concern (red) relates to the additional costs incurred due to the impacts of greater cooling times for RG-MOX on the ability to offload fuel to dry storage in a manner compatible with the used fuel management requirements, such as maintaining full core reserve.

Table 3: Assessment of Fuel Cost Issues Associated with Use of RG-MOX Following Completion of the U.S. DOE SPD Program

Issue	Assessment FOM	Confidence FOM	Basis	References	Follow On Actions
Fuel Procurement Cost EC_FUEL01	ALERT	LOW	Fuel costs represent the primary economic (and likely overall) driver. Procurement of RG-MOX will need to be discounted relative to UOX or otherwise incentivized to offset other additional costs and "hassle". Future price of uranium and front-end services are highly uncertain, although natural U supplies remain adequate for next 50–100 years. Fuel costs are ultimately subject of proprietary commercial arrangements; external predictions of market characteristics are therefore speculative.	OECD-IAEA Redbook 2009 - 2	Monitor UOX cost projections.
Core Design Analysis EC_FUEL02	SUITABLE	HIGH	MOX fuel introduces heterogeneities that must be addressed in designing and controlling the reactor. Experience with WG-MOX will demonstrate that this can be done, but the effort and cost will likely be greater than that required for a full UOX core.	EPRI 1018896 - 31 ORNL TM-13421 - 1	None.
New Fuel Acceptance EC_FUEL03	SUITABLE	HIGH	Fresh RG-MOX assemblies exhibit higher dose rates relative to fresh UOX fuel, but these increases are relatively modest. Radiation protection measures will increase costs associated of the acceptance and inspection. However, once a suitable area and process is established to accomplish the required inspections, the impact should be minor and definable compared to other costs.	EPRI 1018896 - 34	Estimate cost of plant modifications (if any) and procedures needed to meet ALARA requirements.
MOX Parity EC_FUEL04	CAUTION	HIGH	Generally lower burnup limits for MOX do not align with industry trend toward higher fuel burnup. French plants operating on MOX were pursuing MOX/UOX equivalency, but the applicability to U.S. operations is doubtful. The Caution ranking reflects the challenge of achieving MOX parity, especially with respect to U.S. licensing.	EPRI 1018896 - 10 EPRI 1018896 - 49 IAEA TRS415-38 IAEA TRS415-39	Track progress with MOX utilization and it equivalency for fuel cycles expected to be used in the U.S.

Table 4: Assessment of Operations and Maintenance Costs Associated with Use of RG-MOX

Issue	Assessment FOM	Confidence FOM	Basis	References	Follow On Actions
O&M of systems needed only for MOX EC_OM01	CAUTION	HIGH	The principal additional O&M burden from MOX operation appears to be the use of enriched boric acid systems for reactor control and shutdown margin. Such systems incur additions costs due to required O&M resources, active monitoring, and routine replenishment of ^{10}B but may also produces benefits that could reduce O&M burdens.	EPRI 1003124 (multiple citations)	**Evaluate enriched boric acid systems further with appropriate readiness (Level 3) assessment** (See Table 5-8)
Radiation protection EC_OM02	SUITABLE	HIGH	Additional radiation protection actions may be required for the handling of fresh and used RG-MOX fuel assemblies. Accordingly, additional costs will be associated with radiation protection and fuel handling; however, these costs are definable and likely to be manageable.	ORNL TM-13421-3 EPRI 1018896-34 EPRI 1025206-2	Estimate costs of maintaining the occupational dose for RG-MOX operation.
Procedures and licensing EC_OM03	SUITABLE	HIGH	Except for the additional enhanced decay heat and radiation from used RG-MOX verses WG-MOX, it is anticipated any licensing requirements will bounded by licensing for WG-MOX use or their implementation will be relatively straight forward.	EPRI 1021048-6	Estimate costs of any additional analyses and reviews to support licensing and regulatory compliance.
Material degradation monitoring and aging management EC_OM04	CAUTION	HIGH	Increased embrittlement of reactor pressure vessel due to hardened neutron flux is a concern for economic life of plant but is considered to be manageable; e.g., via interior placement of MOX assemblies in core. Such considerations will have been taken into account for finite WG-MOX operation period, but continued operation on RG-MOX may warrant review.	ML993620025 – 3 (NRC, 1999)	Review core design and plant aging management plan and update if necessary to account for any changes for RG-MOX operation

Table 5: Assessment of Used Fuel Management Cost Issues Associated with Use of RG-MOX

Issue	Assessment FOM	Confidence FOM	Basis	References	Follow On Actions
SFP criticality control EC_WM01	SUITABLE	HIGH	Criticality calculations and precautions necessary to maintain sub-critical limits should not be significantly different from UOX. There may be additional restrictions and requirements for absorber materials, but these should be manageable using existing technologies and practices. Any differences should be bounded by the WG-MOX experience.	EPRI 1018896-38	None at this time.
SFP thermal management EC_WM02	CAUTION	HIGH	Continued use of MOX fuel will require the SFP cooling systems to support a greater heat removal capacity over a longer duration than if operations returned to exclusive UOX use. The ability to transfer used fuel to dry storage may be impacted by greater inventories of RG-MOX fuel that remains hotter longer relative to used UOX and WG-MOX.	EPRI 1018896-35 EPRI 1021048-14 EPRI 1025206-13 ORNL/TM-2011/290-4	Determine the maximum heat load the existing system can safely accommodate without additional capital expenditures.
Radiation protection for fuel receipt and back-end management EC_WM03	SUITABLE	HIGH	Used RG-MOX yields higher neutron fields than UOX. This may require additional radiation protection measures and the costs associated with them. These costs, however, are well-understood and should be relatively minor.	EPRI 1018896-38 EPRI 1025206-15 to 18	None at this time.
Dry storage system procurement and management EC_WM04	ALERT	MEDIUM	Dry storage systems can be designed for higher heat loads from MOX, and existing systems can be adapted for use with MOX. These options will incur greater costs than for UOX due to design modifications for increased heat removal or reductions in loading capacities. While dry storage of MOX is a mature technology internationally, the potential disruption to a U.S. utility's used fuel management practices and changes in DSC designs and unit costs indicate the need for adequate review and preparation.	EPRI 1025206-12 EPRI 1021048-3	DSC and ISFSI design modifications should be considered as part of integrated back-end used fuel management approach.

3.2. Assessment of Technical and Programmatic Waste (Used Fuel) Management Issues

The Level 2 Waste Management assessment addresses the safety, O&M and licensing impact of introducing used RG-MOX fuel assemblies into the back-end infrastructure of an operating plant. Two distinct phases of waste management follow the progression of handling used fuel at the plant: wet and dry storage.

Wet storage addresses the period from discharge from the reactor vessel through the total time of storage in the spent fuel pool. Dry storage begins with closure and drying following transfer of fuel to a dry storage canister (or cask) and prior to transfer to an onsite ISFSI. These two phases define appropriate distinct stages for consideration and evaluation. Only on-site used fuel management is considered in the present example.

3.2.1. Tactical Assessment of Wet Storage

Key issues for wet storage consideration include:

- What are the implications for spent fuel pool management strategy and planning?
- What are the implications to on-going operation and maintenance activities, procedures, training and costs?
- Are any new criticality issues introduced?
- Are existing cooling mechanisms adequate (spent fuel pool cooling for wet storage/air cooling for dry storage)?
- Are additional shielding and radiation protection required?
- Are additional accident analyses required?
- What are the implications for licensing and regulatory compliance?

Table 5-3 of [1] summarizes the overall assessment of wet storage issues, and Table A-1 of Appendix A contains the relevant citations. Due to space limitations it will not be repeated in this paper. The primary concern for safety should be addressed under the regulatory and operation envelope for WG-MOX use, but needs to be reviewed and revised as necessary for application to RG-MOX as well. The ability of the SFP cooling system to accommodate the heat load of additional MOX fuel assemblies in the SFP will also be a concern. It may become necessary to transfer MOX fuel assemblies with higher heat loads to dry storage to maintain the ability to off-load fuel from the core in order to maintain plant operation. Failure to maintain adequate reserve capacity in the pool could lead to a costly, prolonged reactor shutdown

3.2.2. Tactical Assessment of Dry Storage

Key issues for dry storage consideration include:

- Impact of the heat dissipation capability of DSC systems on the storage capacity of MOX assemblies.
- Satisfaction of criticality limits must be demonstrated for all credible conditions, including during loading of canister and cask systems in the pool and under hypothetical transportation accident scenarios.
- Shielding and radiation protection limits within and at the boundary of the ISFSI.
- Ability of DSCs to meet criteria for transportation when off-site storage or disposal becomes available.

Table 5-4 of [1] summarizes the overall assessment of wet storage issues, and Table A-2 of Appendix A contains the relevant citations. Due to space limitations it will not be repeated in this paper. The primary concern is the availability of DSCs that are designed and licensed for the higher long term heat loads associated with used MOX fuel assemblies. As the used WG-MOX fuel should be the first used MOX fuel assemblies transferred to dry storage, it is anticipated that suitable DSCs will be qualified and licensed for both WG and RG-MOX when needed.

4. CONCLUSIONS

Application of the EPRI decision-support framework to a stylized, limited scenario illustrated the feasibility of the tool's use for broader, more ambitious assessments such as the transition to a closed fuel cycle employing fast spectrum reactor technology.

At one level, application of the decision framework provided the opportunity to synthesize existing static information sources (technical reports and peer-reviewed literature) and more ephemeral information sources (expert knowledge and industry experience) into a structured, concise, and reproducible format. The exercise itself illustrated the demands associated with conducting even a rudimentary assessment in a comprehensive, documented, evidence-based manner.

From EPRI's perspective, the most promising RD&D paths are those that leverage existing and proven technologies, infrastructures, and institutions to pursue as a core approach an evolutionary approach to revolutionary end states. One question that guides EPRI's development of assessment tools and expertise is how society might transition from the established LWR-based once-through nuclear fuel cycle to one reliant on unproven but promising fast neutron spectrum reactor technology (for increased natural resource utilization) while keeping the electricity generation safe, affordable, and reliable. EPRI envisions directing the application of the decision-support framework and other assessment tools to this larger technology implementation challenge in future reports. Development of the decision framework and associated assessment tools is ongoing. EPRI welcomes feedback in the form of comments, suggestions, and potential applications of interest.

References

[1] "*Program on Technology Innovation: EPRI Framework for Assessment of Nuclear Fuel Cycle Options – Framework Description and Example Application for Evaluating Use of Reactor-Grade Mixed Oxide Fuel in U.S. Light Water Reactors,*" EPRI, Palo Alto, CA, 1025208 (2013).
[2] A. Dykes, D. Johnson, A. Sowder, and A. Machiels, "*Evaluating Feasibility of Reactor Grade Mixed Oxide Fuel Use in U.S. Reactors: Application of EPRI's Decision Analysis Framework,*" Proceedings of the 14th International High-Level Radioactive Waste Management Conference, Albuquerque, NM, April 28–May2, 2013, Manuscript No. 6927, American Nuclear Society, (2013).

A Jointly Optimization of Production, Delivery and Maintenance planning for multi-warehouse/muli-delivery problem

Hajej Zied, Turki Sadok, and Rezg Nidhal
LGIPM-University of Lorraine, Metz, France

Abstract: This paper develops à jointly optimization problem in order to establish an optimal production, delivery and maintenance strategy for a manufacturing system subjected to a random failure. The problem consists on several warehouses allow to satisfy random demands during a finite horizon, under service level. In order to assure an economical objective, we have determined the optimal production/maintenance plan and the economically delivery quantities plan considering the delivery time for each warehouse. The aim of the proposed approach is to show a jointly production/maintenance/delivery optimization, with a constrained stochastic production-delivery-maintenance planning problem under hypotheses of service level, delivery time for each warehouse and failure rate, which minimizes the total production, inventory, delivery and maintenance costs. A numerical example confirms the analytical results

Keywords: Delivery time, failure rate, random demand, service level, multi-warehouse.

1. INTRODUCTION

Throughout the last two decades, the determination of an optimal maintenance and production plan which minimizes the total cost including production, inventory and maintenance is one of the first actions of a hierarchical decision making process. This plan allows to finding the optimal production plan and maintenance strategy required by the company to manufacture their products which satisfy a random demand over future periods. [6] developed the linear decision rule and which has considered an important contribution for production planning decisions. This analytical rule determine the optimal solution for aggregate inventory, production and workforce levels by minimizing the quadratic cost functions subject to inventory and workforce balance equations. However, this work considers a production/inventory problem and not maintenance. [2] is among the first authors who studied the problem of integrated production and maintenance strategies. He studied the role of buffer stocks in increasing the system productivity. Also, [1,3,11] studied the strategy of a safety stock building to meet demands during periods of production interruption due to maintenance actions. A joint optimization of strategic stock and age type maintenance policy is proposed by [5]. In reality, the failure rate increases with time and the use of the equipment. Besides, in the literature, this phenomenon it's rarely considered, except [7] who discussed the conditions of optimality of the hedging point policy for production systems in which the failure rate of machines depends on the production rate. Indeed, most of the researchers focus on a perfect manufacturing system and a perfect service level, and do not present the effect of the service level and the proportion of defective items on relevant performance measures and costs. Recently, [8, 9] studied a randomly failing manufacturing system which has to satisfy a random demand during a finite horizon given a required service level. More recently, [13] determined simultaneously the economical production planning, optimal maintenance strategy and the optimal delivery plan taken into account the delivery time, machine failures , random demand and withdrawal right to minimize the sum of inventory, maintenance, production and transportation costs. The authors studied the impact of delivery time and withdrawal right on optimal production/maintenance planning and transported quantity. Indeed, delivery time and transported quantity are important characteristics of manufacturing systems. Thus many manufacturers are working to reduce transportation delays such as the delivery time, which is the period of time that the part takes between a manufacturing store and a purchase warehouse (customer), and which usually has great impact on performance measures. studied the impact of delivery time on the optimal buffer level take into account to the machine failures and random demands, [4] presented a model for supply planning under lead time uncertainty and proposed a method to determine the optimal value of the planned lead time under lead time uncertainty. [10] developed a model for supporting investment strategies about inventory and preventive maintenance in an imperfect production system take into account the delivery time to the customer. In recent years, another important characteristic of transport is considered in the manufacturing system study such as transported quantity, which is the quantity of parts to be transported between the manufacturing store and the purchase warehouse [12]. However, this above works studied systems with single warehouse. In this paper we study a more complicate system consists on several warehouses allow to satisfy random demands. The originality of this work is that determined the optimal production/maintenance plan and the economically delivery quantities plan considering the delivery time for each warehouse. Indeed, in order to respect the service level, each warehouse should contain enough parts for satisfying customer demands. Thus, an optimal planning of the transported quantity between the manufacturing store and purchase warehouses should be determined based on the relationship with the production/maintenance planning and the service level.

The objective of this paper is to determine simultaneously the economical production planning, optimal maintenance strategy and the optimal delivery plan for different warehouses taken into account the delivery time, random demand to minimize the total cost of inventory, maintenance, production and transportation costs.

This remainder of this paper is organized as follows: Section 2 presents and formulates a general stochastic production, delivery and maintenance problem with the different production and maintenance policies. Section 3 develops the analytical expression of production and maintenance policies considering the influence of the delivery time on the production, delivery and maintenance plans. A simple numerical example is presented in section 4. Finally, the conclusion is given in Section 5.

2. PRODUCTION/MAINTENANCE/DELIVERY PROBLEM

2.1. Notation

The following parameters are used in the mathematical formulation of the model:

τ_i: delivery time for warehouse S_i
L: number of warehouse
Δt: length of a production period
H : number of production periods in the planning horizon
$H.\Delta t$: length of the finite planning horizon
$u(k)$: production rate of machine M during period k ($k=0, 1,..., H-1$)
$U=\{u(0), u(1), ..., u(H-1)\}$
$Q_i(k)$: delivery rate during period k ($k=0, 1,..., H-1$) for each warehouse
$Q_i=\{Q_i(0), Q_i(1), ..., Q_i(H-1)\}$
$\hat{d}_i(k)$: average demand during period k ($k=0, 1,..., H$) for each customer
$V_{di(k)}$: variance of demand during period k ($k=0, 1,..., H$) for each customer
$S(k)$: inventory level of S at the end of period k ($k=0, 1,..., H$)
$S_i(k)$: inventory level of S_i ($i:0...L$) at the end of period k ($k=0, 1,..., H$) for each warehouse
C_p: unit production cost of machine M
Cs: inventory holding cost of one product unit during one period at the first store S.
Cs_i: inventory holding cost of one product unit during one period at the warehouse S_i ($i:0...L$)
Cl: delivery cost
Q_v: delivery vehicle capacity
C_M: total maintenance cost
C_{pm}: preventive maintenance action cost

C_{cm}: corrective maintenance action cost
mu: monetary unit
U_{max}: maximal production rate of machine M
U_{min}: minimal production rate of machine M
θ_i: probability index related to each customer i satisfaction and expressing the service level.

2.2. Problem description

In this section, a joint optimization of production, delivery and maintenance planning problem is presented. In this proposed model, we assumed that the manufacturing system consists a single machine M which produces one type of product, a principal manufacturing store S (where the manufactured products are stored) and multi purchase warehouse (S_0, S_1,....S_L) (where the customer receives his demand (products)). Each warehouse aims to satisfy a several random demand under a given service level θ_i, over a finite horizon H. Hence the delivery time considered between the principal store S and each warehouse S_i (i=0,...,L), denoted by τ_i (i=0,...,L) (see figure1). In other words, if we suppose that the products leave S at period k, they will arrive at the period $k+\tau$ to the warehouse. Also, the products preparing for delivery is characterized by a cost called order preparation cost.

The machine M is subject to a random failure. The degradation degree of machine is influenced by the production rates, consequently, the failure rate $\lambda(t)$ increases with time and production rate $u(t)$.
Our objective lies in establishing the best production, delivery and maintenance strategy. To do this conjugated production, delivery and maintenance optimization, we minimize the sum of the inventory costs at the different stores, the manufacturing and delivery costs along with the costs associated with the maintenance strategy.

Figure 1: Problem description

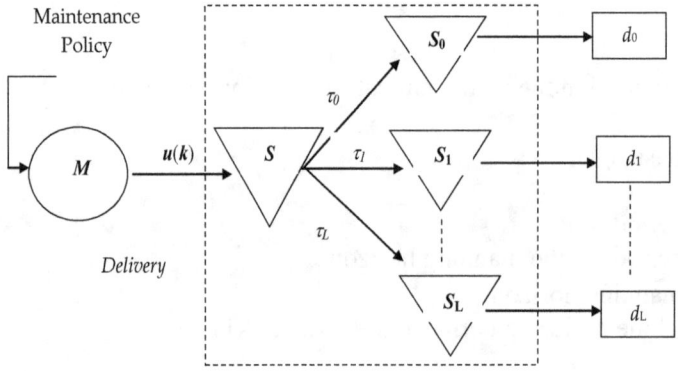

2.3. Problem formulation

Aiming at organizing and optimizing the production system, our objective is to minimize the expected costs related to production, inventory, delivery and maintenance over the finite time horizon. It's assumed that the horizon is portioned equally into H periods with a length equal to Δt. Moreover, we assume that the fluctuation of the demands is a normal process with mean and variance given respectively by $\hat{d}_i$ and V_{d_i} and the demands are satisfied at the end of each period.
The stochastic problem as follows:

$$\underset{(U,Q,N)}{Min}\left(\left\{\sum_{k=0}^{H-1}f_k\left(S(k),S_0(k),....,S_L(k),u(k),Q_0(k),Q_1(k),.....,Q_L(k)\right)\right\}+\left\{\Gamma_M(U,N)\right\}\right) \qquad (1)$$

Where
$f_k(.)$ denotes the expected production, inventory and delivery costs of each production period. $\Gamma_M(.)$ the total cost of preventive and corrective maintenance actions.

Subject to
The inventory balance equation level of the principal store S is given by the following equation:

$$S(k+1)=S(k)+u(k)-\sum_{i=1}^{L}Q_i(k) \qquad (2)$$

(k=0, 1,..., H-1)

The inventory level of each warehouse inventory S_i at the period $k+1$ is given by the following equation:

$$S_i(k+1)=S_i(k)+Q_i(k-\tau_i)-d_i(k) \qquad (3)$$

(k=0, 1,..., H-1) and (i=1,..., L)

The inventory level of each warehouse S_i at the period $k+1$ equals to the inventory level of S_i at period k plus the rate of products that arrives to S_i (i.e. $Q_i(k-\tau)$) minus the customer demand d_i at period k.

The service level requirement constraint for each warehouse at each period k is expressed by the following constraint.

$$\text{Prob}\lfloor S_i(k+1)\geq 0\rfloor\geq\theta_i$$
(k=0, 1,..., H-1) and (i=1,..., L) $\qquad (4)$

The following constraint defines an upper and lower bounds on the production level during each period k.

$$U_{min}\leq u(k)\leq U_{max} \qquad (5)$$

2.4. Production/Delivery/Maintenance policies

In this section, we represent a constrained production/delivery/Maintenance problem under service level, delivery time, and random demand using a HMMS model.

Inspired from the HMMS model, we had the idea to make emphasis on production rate, inventory levels and delivery rates in order to make the best and optimal production planning and maintenance strategy. Also, in our work, we make some changes on the model keeping its linear quadratic form. Furthermore, we take into account some constraints on the decision variables to make our approach more realistic and to ensure its applicability in real industrial cases.

2.2.1. Production/Delivery Policy:

The idea of the proposed model is the use of a quadratic cost function allows penalizing both excess and shortage in the inventory level.
The expected cost including production and holding costs for the period k is given by:

$$f_k(S(k), S_{i(t:0...L)}(k), u(k), Q(k)) = f_{u(k)}(u(k)) + f_{s(k)}(S(k), S_0(k), ..., S_L(k)) + f_{Q(k)}(Q(k)) \qquad (6)$$

Where the expected production cost for period k

$$f_{u(k)}(u(k)) = C_p \times E\{u(k)^2\} \qquad (7)$$

The expected holding costs of period k

$$f_{s(k)}(S(k), S_0(k), ..., S_L(k)) = C_s \times \left(E\{S(k)^2\}\right) + \sum_{i=1}^{L} C_{s_i} \times \left(E\{S_i(k)^2\}\right) \qquad (8)$$

The expected transported cost for period k

$$f_{Q(k)}(Q_i(k)) = C_l \times \left(\sum_{i=0}^{i=L} E\left\{\left(\frac{Q_i(k)}{Q_v}\right)^2\right\}\right) \qquad (9)$$

Note that $E\{\}$ denotes the mathematical expectation operator.

The total expected cost of production, inventory and delivery over the finite horizon $H.\Delta t$ can then be expressed as follows:

$$f(u) = C_s \times \left(E\{S(H)^2\}\right) + \sum_{i=0}^{L} C_{s_i} \times \left(E\{S_i(H)^2\}\right) + \sum_{k=1}^{H-1} \left[\begin{array}{l} C_p \times \left(E\{u(k)^2\}\right) + C_l \times \left(\sum_{i=0}^{i=L} E\left\{\left(\frac{Q_i(k)}{Q_v}\right)^2\right\}\right) \\ +C_s \times \left(E\{S(k)^2\}\right) + \sum_{i=0}^{L} C_{s_i} \times \left(E\{S_i(k)^2\}\right) \end{array} \right] \text{ with } k \in \{0,1,...,H-1\} \quad (10)$$

2.2.2. Maintenance Policy

In this subsection, the resolution of maintenance planning problem consists in minimizing costs related to preventive and corrective maintenance. Other, to make correct and suitable decisions, it's important to determine the best scheduling of carrying out maintenance. The maintenance strategy considered in this work is a preventive maintenance with minimal repair. Preventive actions should be scheduled over the finite horizon H which divided equally into N parts of duration T. We suppose that performing a preventive action corresponds to times $k.T$ ($k=1,2,....N$) consists in replacing some critical parts restoring the production unit to an as good as new condition. However, between successive preventive interventions, breakdowns may happen, minimal repair is performed. It is assumed that the repair and overhaul durations are negligible.

So, it is necessary to keep in mind and to consider that the status of machine depend on their production and transported plan variation. Otherwise, it is worth considering the influence of production on the degradation of machine and consequently in maintenance planning as it was studied in [8].

In this section, considering the joint optimization strategy, we aim at determining the optimal maintenance strategy which allows the firm minimizing its maintenance costs. Such a strategy is characterized by the optimal number N^* of preventive maintenance actions and the most adequate time between them noted T^*.

$$C_M\left(U\left(u(1), u(1), ...u(H-1)\right), N\right) = C_{pm} \times (N-1) + C_{cm} \times \varphi_M(U, N) \qquad (11)$$

Where $\varphi(U, N)$ is the average number of failures as a function of the production plan defined by the vector U and the number of preventive maintenance actions N.

3. ANALYTICAL STUDY

In this section, we would like to show an analytical approach that can be used to solve the above stochastic problem and that by transforming it to a deterministic equivalent problem. Some of these approaches are based on the certainty equivalent principle.

3.1. Deterministic Equivalent problem

An approach that transforms the stochastic problem into a deterministic equivalent is necessary. This deterministic problem maintains the main properties of the original problem.

Before proceeding, the following notation is introduced:

Mean variables:

$$E\{S(k)\} = \hat{S}(k), E\{S_i(k)\} = \hat{S}_i(k) \ i:0....L, E\{u(k)\} = u(k), E\{Q_i(k)\} = Q_i(k) \ i:0....L$$

Variance variables: $V_{u(k)} = V_{Q_i(k)} = 0; i:1....L$. (Note that this reflects the fact that the control variables $u(k)$ and $Q(k)$ are deterministic).

- The production, delivery and inventory costs simplified as:

Lemma1:

$$f\left(S(k), S_{i\{r0...L\}}(k), u(k), Q_{i\{r0...L\}}(k)\right) = \sum_{k=0}^{H}\left(C_s \cdot \hat{S}(k)^2 + \sum_{i=0}^{L} C_{s_i} \cdot \hat{S}_i(k)^2\right) + \sum_{k=1}^{H-1}\left(C_p \times u(k)^2 + C_l \times \sum_{i=0}^{L}\left(\frac{Q_i(k)}{Q_v}\right)^2\right) + [H+1] \cdot \frac{H}{2} \cdot \sum_{i=0}^{L} C_{s_i} \cdot \sigma_{d_i}^2$$

(12)

- The inventory balance equation (2) can be reformulated as:

$$\hat{S}(k+1) = \hat{S}(k) + u(k) - \sum_{i=0}^{L} Q_i(k) \quad k = 0,1,......,H-1$$

Likewise, the inventory balance equation (3) can be reformulated as:

$$\hat{S}_i(k+1) = \hat{S}_i(k) + Q_i(k-\tau_i) - \hat{d}_i(k) \quad k = 0,1,......,H-1; \text{ and } (i=1,..., L)$$

- The service level constraint:

In this part, we introduce the service level constraint. To continue transforming the problem into a deterministic equivalent, the equation describing the service rate for the principal store and each warehouse can be transformed as follows [8]:

Lemma 2

For $k=0,1,..,H-1$ and for $i=0,....,L$ we have:

$$\mathrm{Prob}\left(S_i(k+1) \geq 0\right) \geq \theta_i \Rightarrow \left(Q_i(k-\tau_i) \geq V_{d_i(k)} \times \varphi^{-1}(\theta_i) - S_i(k) + \hat{d}_i(k)\right) \quad k = 0,1,....,H-1 \text{ and } i = 0,1,....,L$$

(13)

φ : Cumulative Gaussian distribution function with mean $\hat{d}(k)$ and finite variance $V_{d(k)}$.

φ^{-1} : Inverse distribution function

3.2. Maintenance Cost

Motivated especially from the work [9], in our present work, we take into account the degradation of machine while forecasting the maintenance actions. Thus, the maintenance strategy depends strongly on production planning which is on accordance with the principle of joint production and maintenance planning.

We recall that the optimal maintenance strategy characterized by the optimal number N^* of preventive maintenance actions and the time between them T^*, as given by Eq. (14).

$$T^* = \frac{H}{N^*} \tag{14}$$

Recall that analytic expression of the total maintenance cost is as follows, with $N \in \{1, 2, 3 .....\}$.

$$C_M(U, N) = (N-1) \cdot C_{pm} + C_{cm} \cdot \varphi_M(U, N) \tag{15}$$

Where $\varphi_M(U,N)$ corresponds to the expected number of failures that occur during the horizon H, considering the production rate in each production period Δt.

Furthermore, we assume that the equipment degradation is linear, we assume that $\lambda(t)$ represents the linear failure rate function at production period k is expressed as following :

$$\lambda_k(t) = \lambda_{k-1}(\Delta t) + \frac{u(k)}{U_{max}} \cdot \lambda_n(t) \quad \forall t \in [0, \Delta t] \tag{16}$$

$\lambda_n(t)$: failure rate for nominal conditions which is equivalent to the failure rate with maximal production.

The average failure number over the horizon $H.\Delta t$ is:

$$\varphi_M(U,T) = \sum_{j=0}^{N-1} \left[\sum_{i=In(j \times \frac{T}{\Delta t})+1}^{In\left((j+1) \times \frac{T}{\Delta t}\right)\Delta t} \int_0^{In\left((j+1) \times \frac{T}{\Delta t}\right)+1)\times \Delta t} \lambda_i(t) + \int_0^{(j+1)\times T - In\left((j+1) \times \frac{T}{\Delta t}\right)\times \Delta t} \lambda_{In\left((j+1) \times \frac{T}{\Delta t}\right)+1}(t)dt \right. $$
$$\left. + \int_{(j+1)\times T} \frac{\left(\left(In\left((j+1) \times \frac{T}{\Delta t}\right)+1\right)\right)}{U_{max}} \times \lambda_n(t)dt \right] \tag{17}$$

We now replace $T=H/N$:

$$\varphi_M\left(U, H/N\right) = \sum_{j=0}^{N-1} \left\{ \begin{array}{l} \left(In\left((j+1)\times\dfrac{H}{N\cdot\Delta t}\right) - In\left(j\times\dfrac{H}{N\cdot\Delta t}\right)\right)\times\Delta t\times\lambda_0\left(t_0\right) + \\[2em] \dfrac{\lambda_0\left(\Delta t\right)\times\Delta t}{U_{max}}\times\displaystyle\sum_{i=In\left(j\times\frac{T}{\Delta t}\right)+1}^{In\left((j+1)\times\frac{H}{N\cdot\Delta t}\right)}\sum_{l=1}^{i-1}u(l)\,dt \;+\; \dfrac{1}{U_{max}}\cdot\displaystyle\sum_{i=In\left(j\times\frac{H}{N\cdot\Delta t}\right)+1}^{In\left((j+1)\times\frac{H}{N\cdot\Delta t}\right)}\int_0^{\Delta t}u(i)\cdot\lambda_0(t)\,dt + \\[2em] \displaystyle\sum_{l=1}^{In\left(\frac{(j+1)\times H}{N\cdot\Delta t}\right)}\dfrac{u(l)}{U_{max}}\cdot\lambda_0\left(\Delta t\right)\cdot\left((j+1)\times\dfrac{H}{N}-In\left(\dfrac{(j+1)\times H}{N\cdot\Delta t}\right)\times\Delta t\right) \\[2em] +\displaystyle\int_0^{(j+1)\times\frac{H}{N}-In\left(\frac{(j+1)\times H}{N\cdot\Delta t}\right)\times\Delta t}\dfrac{u\left(In\left(\dfrac{(j+1)\times H}{N\cdot\Delta t}\right)+1\right)}{U_{max}}\cdot\lambda_0(t)\,dt \\[2em] +\dfrac{u\left(In\left(\dfrac{(j+1)\cdot H}{N\cdot\Delta t}\right)+1\right)}{U_{max}}\times\displaystyle\int_{In\left((j+1)\cdot\frac{H}{N\cdot\Delta t}\right)\times\Delta t}^{(j+1)\frac{H}{N}}\lambda_0(t)\,dt \end{array}\right.$$

$$(18)$$

4. NUMERICAL EXAMPLE

Let us consider a system that produces one type of products to meet the delivery for $L=2$ warehouses that will satisfy random demands below. Basing in the models described in previous sections, we will establish the optimal production plan and the optimal maintenance strategy minimizing the total cost over a finite planning horizon: $H=18$ periods each of period length $\Delta t=1$. We supposed that the standard deviation of each demand of product is the same for all periods and for each demand $\sigma_{di(\{i:1,2\})}=1.2$ and the initial inventory level we assume that $S(0)=0$.

- Lower and upper boundaries of production capacities: $U_{min}=0$ and $U_{max}=44$.
  $C_p =2\ mu$, $Cs =0.2\ mu/k$, $Cs_i=0.2\ mu/k$ $\{i:1,\ldots L=2\}$, $Cl=12$, $Q_v=6$, $S_i(0)=40$ with $\{i:1,2\}$
- We assume that customer satisfaction degree, associated with each warehouse stock constraint, is equal to 90% ($\theta_i=0.9$ $(i=1, 2)$).

Concerning the system reliability, we suppose that the failure time of machine M has a degradation law characterized by a Weibull distribution. To calculate the failure rate and average number of failures functions given by equations 16 and 17, we assume that the nominal degradation follows a Weibull distribution $W(\alpha,\beta)$ with scale and shape parameters are respectively $\beta=100$ and $\alpha=2$ given by :

$$\lambda(t) = \frac{\alpha}{\beta}\cdot\left(\frac{t}{\beta}\right)^{\alpha-1}$$

The cost associated with a corrective and preventive maintenance action are respectively $C_{cm} =3000$ mu $C_{pm}=500$ mu (monetary unit).

The average demand is presented in table 1 below:

Table 1: Average demands for customer of warehouse 1

$d_1(0)$	$d_1(1)$	$d_1(2)$	$d_1(3)$	$d_1(4)$	$d_1(5)$
15	17	15	15	15	14
$d_1(6)$	$d_1(7)$	$d_1(8)$	$d_1(9)$	$d_1(10)$	$d_1(11)$
16	14	16	15	15	15
$d_1(12)$	$d_1(13)$	$d_1(14)$	$d_1(15)$	$d_1(16)$	$d_1(17)$
15	15	15	13	15	15

Table 2: Average demands for customer of warehouse 2

$d_2(0)$	$d_2(1)$	$d_2(2)$	$d_2(3)$	$d_2(4)$	$d_2(5)$
16	13	15	15	14	16
$d_2(6)$	$d_2(7)$	$d_2(8)$	$d_2(9)$	$d_2(10)$	$d_2(11)$
16	16	14	15	15	14
$d_2(12)$	$d_2(13)$	$d_2(14)$	$d_2(15)$	$d_2(16)$	$d_2(17)$
15	16	14	16	14	14

The optimal delivery time for each warehouse is presented in figure 2. Other hand, The economically production and delivery plans for each warehouse 1 and 2 are presented respectively in table 3, table 4 and table 5 and the optimal maintenance scheduling and figure 3.

Table 3: Optimal Production Plan

$u^*(1)$	$u^*(2)$	$u^*(3)$	$u^*(4)$	$u^*(5)$	$u^*(6)$
44	29	25	32	32	42
$u^*(7)$	$u^*(8)$	$u^*(9)$	$u^*(10)$	$u^*(11)$	$u^*(12)$
41	42	36	23	14	44
$u^*(13)$	$u^*(14)$	$u^*(15)$	$u^*(16)$	$u^*(17)$	$u^*(18)$
30	18	19	36	27	25

Table 4: Optimal delivery plan for warehouse 1 for $\tau^*_1=3$

$Q_1^*(1)$	$Q_1^*(2)$	$Q_1^*(3)$	$Q_1^*(4)$	$Q_1^*(5)$	$Q_1^*(6)$
25	25	25	25	22	9
$Q_1^*(7)$	$Q_1^*(8)$	$Q_1^*(9)$	$Q_1^*(10)$	$Q_1^*(11)$	$Q_1^*(12)$
16	21	7	20	23	22
$Q_1^*(13)$	$Q_1^*(14)$	$Q_1^*(15)$			
11	16	12			

Table 5: Optimal delivery plan for warehouse 2 for $\tau^*_2=3$

$Q_2^*(1)$	$Q_2^*(2)$	$Q_2^*(3)$	$Q_2^*(4)$	$Q_2^*(5)$	$Q_2^*(6)$
16	12	15	16	25	17
$Q_2^*(7)$	$Q_2^*(8)$	$Q_2^*(9)$	$Q_2^*(10)$	$Q_2^*(11)$	$Q_2^*(12)$
10	25	7	8	20	13
$Q_2^*(13)$	$Q_2^*(14)$	$Q_2^*(15)$			
23	12	25			

Figure 2: Total cost in function of the delivery time τ_1 andτ_2

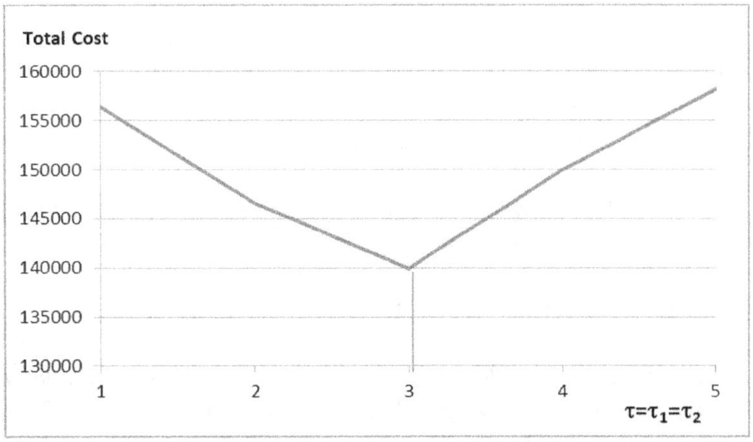

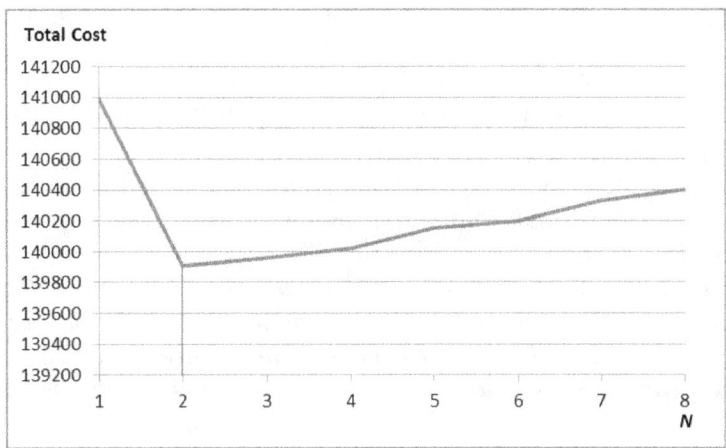

Figure 3: Total cost in function of preventive maintenance actions number N

The above tables 3-5 illustrate the optimal production and delivery plans of the minimum total cost for different values of the delivery times τ_1 (for warehouse 1), τ_2 (for warehouse 2) and the number N of PM actions to be performed.

In what follows, we interest to find the values of τ_1, τ_2 and N which corresponds to the lowest total cost value. Figure 2, Figure 3 show the total cost (production, inventory, delivery and maintenance) in function of the delivery times τ_1 and τ_2 and number of preventive maintenance actions N. we can see that the lowest total cost value corresponds to $\tau^*=3$ for the first and second warehouse and $N^*=2$. Thus, the optimal delivery time denoted by $\tau_1^*=3$ for the first warehouse and $\tau_2^*=3$ for the second warehouse and the optimal number of preventive maintenance actions denoted by $N^*=2$. Therefore, over the finite horizon H of 18 months, Two preventive maintenance actions should be done, i.e. for every period equals to $T^*=H/N^*= 9$ *tu* a preventive maintenance action should be done.

5. CONCLUSION

This paper dealt an approach involving in optimizing a constrained stochastic production, delivery and maintenance planning problem for several warehouse considering a several delivery time, a several random demand, a several service level and a randomly failing production system. The contribution of this study is that it formulates and solves the related stochastic production/delivery/maintenance problem under a service level. An optimization has been performed obtaining an optimal production, delivery plans as well as the corresponding preventive maintenance intervals taking into account the influence of production rates on the machine degradation and the delivery time on the problem optimization.

References

[1]   E.H. AGHEZZAF, M.A.JAMALI, D.AIT-KADI, '*'An integrated production and preventive maintenance planning model*'', European journal of operational research, 181, 676-685, (2007).

[2]   Buzacott, J.A. "*Automatic transfer lines with buffer stocks*". International Journal of Production Research, 5 (3), 183. (1967).

[3]   Chelbi, A. and Ait-Kadi, D. "*Analysis of a production/inventory system with randomly failing production unit submitted to regular preventive maintenance*". European Journal of Operational Research, 156 (3), 712-718 (2004).

[4]   Dolgui, A. and Ould-Louly, M.A. "A model for supply planning under lead time uncertainty", International Journal on Production Economics, Vol. 78, pp.145–152, (2002).

[5]   Cheung, K. L, Hausmann, W. H.,"*Joint optimization of preventive maintenance and safety stock in an unreliable production environment*"; Naval Research Logistics, 44, 257-272, (1997).

[6]   C.C. Holt, F. Modigliani, J.F.Muth, and H.A.Simon, "*Planning Production, Inventory and Work Force*", Prentice-Hall, NJ, (1960).

[7]   Hu, J., Vakili, P. and Yu, G. "*Optimality of hedging point policies in the production control of failure prone manufacturing systems*". Automatic Control, IEEE Transactions, 39 (9), 1875-1880. (1994).

[8]   Hajej, Z., S. Dellagi and N. Rezg.,"An optimal production/maintenance planning under stochastic random demand,service level and failure rate". *IEEE International Conference on Automation Science and Engineering*, pp. 292 –297, (2009).

[9]   Hajej Z., Dellagi S., Rezg N., "*Optimal integrated maintenance/production policy for randomly failing systems with variable failure rate*" International Journal of Production Research, 49, 19, 5695-5712,(2011).

[10]   Lee, H.. "*A cost/benefit model for investments in inventory and preventive maintenance in an imperfect production system*". Computers & Industrial Engineering, 48, 55-68, (2011).

[11]   Rezg, N., Xie X., Mati Y., "*Joint optimisation of preventive maintenance and inventory control in a production line using simulation*", International Journal of Production Research, 44, 2029-2046. (2004).

[12]   Richard Y.K. Fung · Tsiushuang Chen, "*multi agent supply chain planning and coordination architecture*", Int J Adv Manuf Technol. 25: 811–819, (2011).

[13]   Turki, S.; Z. Hajej, and N. Rezg. "*Impact of delivery time on Optimal Production/Delivery/Maintenance Planning*", 8th IEEE International Conference on Automation Science and Engineering, 335 – 340, (2012).

Investigating the Role of Statistical Models in Water Distribution Asset Management: A Semi-structured Interview Approach

Vikram M. Rao*, and Royce A. Francis
The George Washington University, Washington DC, USA

Abstract: A robust asset management plan needs to be in place for water utilities to effectively manage their distribution systems. Of concern to utilities are broken pipes, which can lead to bacteria entering the water system and causing illness to consumers. Typically, water utilities allocate a portion of funds every year for renewal of pipes and valves. However, pipe renewal is largely based on replacing current broken pipes, and long- term asset management planning to replace pipes is not a priority for water utilities. Water utilities are beginning to use probabilistic break models and other statistical tools to predict pipe failures. These models incorporate variables such as pipe length, diameter, age, and material. These models are emerging in the water industry; however, their direct impact on long term asset planning remains to be seen. In addition, the effectiveness of these models is questionable, as there is currently little research done to evaluate the ability of these models to assist in asset management planning. This paper discusses the role of probabilistic pipe break models in structuring long-term asset management decisions. We determine that there are many factors that are needed to contribute to the feasibility of statistical models in a water asset management program, including data availability, funds, and shared information.

Keywords: Water Distribution, Asset Management, Pipe Renewal and Replacement, Mental Models, Structured Interviews.

1. INTRODUCTION

Water utilities are facing challenges regarding their water distribution systems and maintenance of their assets such as pipes, valves, and other elements of their water network. One of the major issues is repairing and replacing aging assets that have been placed underground for fifty or more years. All utilities are facing the need to replace assets in order to avoid breaks and disruptions in water flow, and provide consistently high quality water to their customers. As pipes age, it becomes more imperative to find effective solutions to plan repair and rehabilitation schedules, and to make effective use of limited funds and resources.

One of the ways in which utilities can approach the issue of managing assets is to develop a comprehensive asset management plan. EPA and AWWA (American Water Works Association) provide guidelines for developing an asset management plan, including providing replacement and rehabilitation guidelines, providing decision making strategies for capital improvements, and outlining risk assessment and characterization methods[1],[2],[3]. However many utilities, constrained by limited funds derived from revenues or allocated from city or local governance, are not able to create effective management plans. The strategies they employ consist of replacing current broken pipes, with little attention to long term asset management planning. To support such planning, a framework which promotes effective decision making is needed.

A practice that has been introduced recently to support asset management decision making are statistical models that serve to predict pipe break rates[4]. These models are designed to take various inputs including pipe age, diameter, material, as well as covariates such as soil conditions, weather, and geographical location. The output of the model provides information on which pipes are most likely to break, through probability distributions or discrete values, which allows the utility to create effective prioritization strategies. Many utilities are already adopting models and are beginning to

---

* To whom correspondence should be addressed: vrao81@gwu.edu

implement them as part of their asset management programs. There are many statistical models available, and a review of recent literature shows many models that have been developed and compared with historical break data obtained from a local utility[5],[6],[7],[8],[9]. The accuracy of the models is good, and continuously improving. Table 1 below shows the accuracy of several models developed in the literature.

Table 1: Accuracy of Current Statistical Models in Literature

Type of Model	Accuracy Measure	Result	Conclusion
Time Linear Model[8]	Akaike Information Criterion (AIC)	14,267	Time Exponential Model is the best predictor of non-zero pipe breaks [8]
Time Exponential Model[8]	AIC	1066	
Poisson GLM[8]	AIC	14932	
I-WARP (Individual Water mAin Renewal Planner)[10]	Pipe-dimension coefficient of determination (pR^2), and time-dimension coefficient of determination (tR^2)	$pR^2 = 0.43$ $tR^2 = 0.61$	Rather successful at estimating the total number of breaks/year, not as successful at estimating breaks/pipe[10]
Weibull/Exponential Model[5]	R^2	$R^2 = 0.39$	Explained by simplicity of the model and random processes. Future research will consider additional risk factors[5]

However, there is a need to understand the role of models and the value they bring to an organization. This is an area of interest because utilities are constrained by lack of funds and limited capital improvement budgets. Considering that implementing models can take a lot of time and energy, it is worthwhile to explore the role of models and their value towards long term asset management planning. There is also a need to understand the perceived accuracy of the models and whether they are worth the time and monetary cost of implementing them. This information can prove valuable to utilities and decision makers who seek to adapt models in the hopes of mitigating risks associated with pipe breaks such as adverse health effects and diminished water quality. The tradeoffs associated with adopting models and using them need to be understood in order to provide decision makers with the best information on how to direct their asset management plans.

The objective of this article is to investigate the role of statistical methods in drinking water distribution asset management. We use semi-structured interviews as part of a mental models approach to understand the issues facing water asset managers. These include the role of statistical models in asset management, their value to an organization, and their perceived accuracy. We highlight the issues that are obtained from our interviews to show what can be improved with models, and how utilities can best use them as part of their asset management plan. The value of our research lies in the need to evaluate pipe break models as to their effectiveness and usefulness in long term asset management planning. We believe our results will help models become more effective tools for utilities to employ, leading to more prevalent use among water utility firms.

2. BACKGROUND

Pipe break models have been developed and commented on in recent literature. Clark[6] utilized a condition assessment model to analyze risk of pipe breaks. Their model was a Cox Proportional

Hazards model which was incorporated into a survival model. The authors used a frailty term, which refers to random, unidentified factors which may cause failure (such as soil) as part of the model. Data came from a utility in Laramie, WY.

Debon[7] used several models including the Cox proportional hazard model, the accelerated lifetime model, and generalized linear models to compare and contrast risk factors such as age, size, and pipe diameter. The goal was to examine which variables contributed most to the incidence of pipe breakage. The authors used receiver operating characteristics curves to study the accuracy of each of the models. Their findings showed that the characteristics that contribute to pipe breaks are short pipe length, large pipe diameters, and low pressure.

Yamijala[8] carried out an assessment of several statistical models of pipe break failure. The goal was to compare how accurate various models were at predicting pipe breaks. The models included time linear ordinary least squares regression, time exponential ordinary least squares regression, generalized linear models, and logistic GLM. Data came from a water utility in Texas. The authors fit the various models and compared actual versus predicted breaks, discovering that Poisson GLM is a better predictor of breaks (zero breaks and non-zero breaks) than the time linear model. But the Logistic GLM is a better fit for the data, since it can handle a large percentage of zeroes in the data set (referring to pipes that do not break).

Kleiner[9] developed a cost effective distribution network renewal plan using a dynamic programming approach, which sought to minimize the total discounted costs associated with rehabilitation of pipes. A procedure known as Multistage Procedure for Rehabilitation Analysis of Water Distribution Systems was used to solve. Kleiner[10] also developed a new study using a computer model called Individual Water mAin Renewal Planner, which is used to examine historical pipe breakage patterns using a non-homogenous Poisson process to model individual breaks. The model uses dynamic factors such as climate and pressure change as part of its analysis.

These models cited in literature are innovative models that can provide a means for predicting breaks and other events based on historical information. They can be of benefit to water distribution asset managers if they are implemented as part of a formal asset management plan. So far, many of the models used by water utilities are off-the-shelf software packages or basic statistical models such as Weibull analysis.

3. METHODS

Our methodology in this study is to perform semi-structured interviews with water utility firms, specifically including asset managers, to understand the role of statistical models in long term asset planning. The goal is to develop understanding of the role of models and identify the strengths and weaknesses of models, the perceived benefits of using models, and ways to improve models to make them a more effective tool for utilities.

Semi-structured interviews as part of a mental models framework in the context of risk communication are described in Johnson-Laird and Morgan[11],[12]. The goal is to improve risk communication through a five step process, including the creation of an expert model, conducting mental models interviews to elicit beliefs about the hazard, followed by structured initial interviews, and drafting and evaluating risk communication measures.

In mental models interviews, the goal is to examine a person's mental models, or psychological representations of situations. Researchers have used a mental models approach to determine a person's system of beliefs regarding a topic at hand. The mental models framework consists of a semi-structured interview process designed to give the respondent the most flexibility in providing a response, while limiting bias that the interviewer may inadvertently provide. Mental models interviews have been used to gather information on subjects' beliefs about a number of topics. One such example, done by Magzan[13], is to use mental models to explore leadership effectiveness in business environments.

The first step in developing a mental models framework is to create an expert model, or influence diagram, that summarizes the current expert knowledge. Based on our literature review, we propose an influence diagram, shown in Figure 1 below, which captures the current state of expert knowledge regarding models in water distribution asset management. This influence diagram serves as a basis for developing our interview questions.

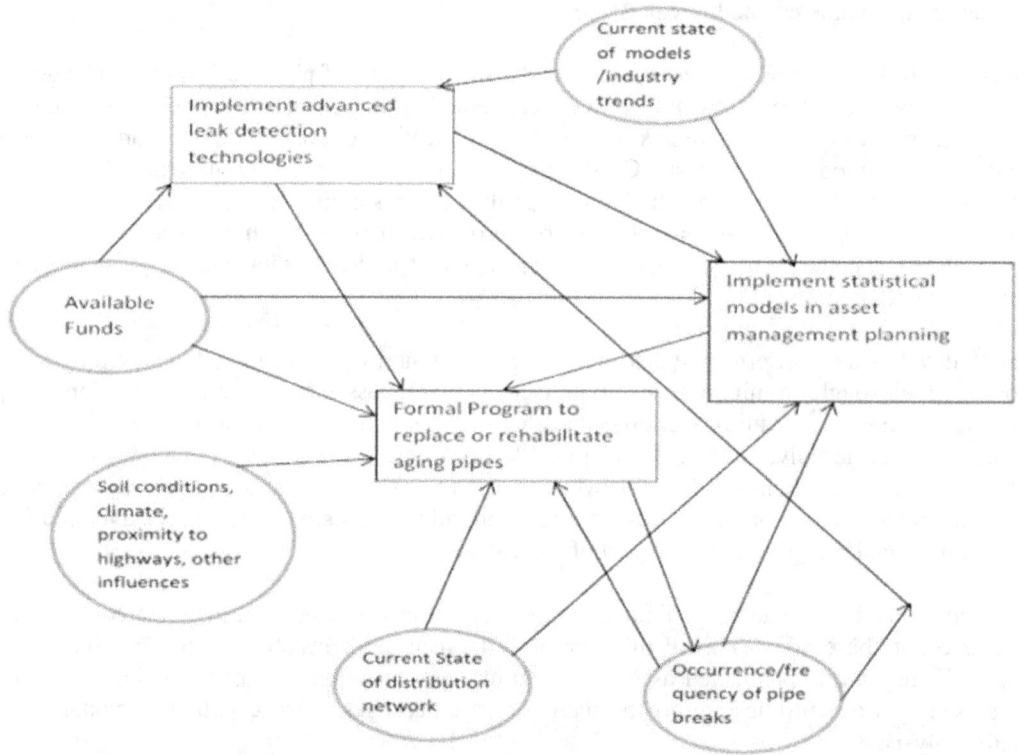

Figure 1: Influence Diagram for Water Asset Management

Our approach was to contact various water utility firms located across the United States, each serving a medium to large sized city. We began by holding informal discussions with asset program managers or other individuals deemed knowledgeable with asset management programs and the use of models. We chose six participants based on their degree of agreeability to the study and knowledge and experience about the subject matter. We provide a list of the utilities and qualifications of the various individuals in Table 2 below.

Table 2: Description of Participants in our Study

	Title of Individual	Years of Experience at Utility
Utility A	Asset Management Coordinator	5
Utility B	Senior Engineer	9
Utility C	Asset Management Program Manager	4
Utility D	Manager, Water Infrastructure Planning	10
Utility E	Asset Strategy Manager	15
Utility F	Managing Engineer, Infrastructure	4.5

We found that the utilities were agreeable to participate and were receptive to our inquiries. We provided information about our study including our data collection, which consisted of audio recordings of our conversation. We ensured that the confidentiality of the data would not be compromised, and that the utilities would not be mentioned by name in our report.

Our mental models interviews were developed upon completion of the informal interviews, which provided background information to help develop our formal interview questions. Following the guidelines provided by Johnson-Laird and Morgan, we kept the questions as open-ended as possible, to exclude biases or intervention on the part of the interviewer. The questions were worded to allow the respondents to answer as freely as they wished. Our interview checklist consisted of several question topics and subsequent follow up questions. We began with an initial, open-ended prompt, to begin the discussion, and as the respondents brought up various topics, we followed up with the topic questions. In this way the respondents dictated the flow of the conversation. We have included a copy of the interview checklist in Appendix A.

We archived the data recordings and transcribed the conversations into documents. These transcriptions were as verbatim as possible, though we removed words such as 'uh' and 'um' which we decided were irrelevant and inapplicable to our study. The rest of the transcription captured the recording as closely as possible, and we verified by comparing the audio recording with our written transcription. We did not conduct member checks, however we felt our transcriptions were as accurate as possible and did not compromise the integrity of the data. We ensured rigor by verifying that our questions captured the spectrum of issues pertaining to statistical models with our participants. All participants either admitted they had no additional information, or added information that was already mentioned and deemed a relevant theme, so we considered this as no new information.

Our goal was to understand themes that pertain to facilitating or inhibiting the use of models in water asset management. These themes were derived from coding the interviews using Atlas software which allows for analysis of qualitative information. We found codes that represented the salient points of the respondents' answers, including areas for concern or areas in need of improvement. We validated these codes by counting the frequency of occurrence across all interviews, providing support that these were indeed themes that were regarded as important by the majority of respondents.

4. RESULTS

4.1 Codes

Our analysis began by compiling a list of codes and their relative frequency. A Pareto chart showing the codes, their relative frequency, and the cumulative distribution is shown in Figure 2.

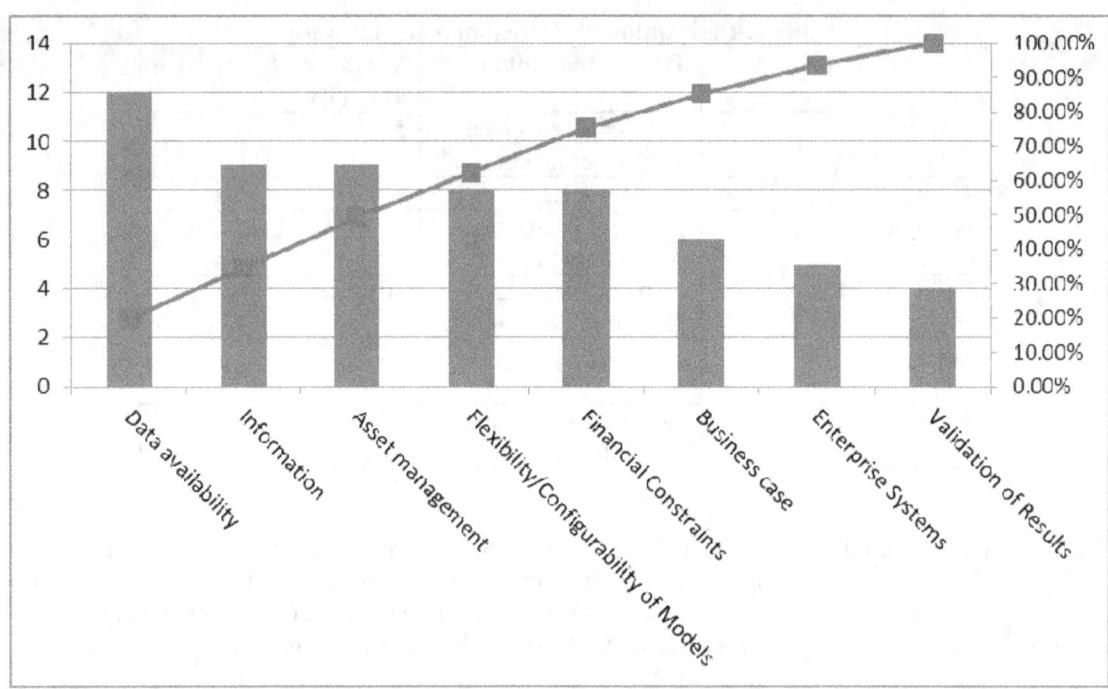

Figure 2: Pareto Chart for Descriptive Codes

We describe the codes in detail. First, data availability refers to the idea that the effectiveness and accuracy of the models is dependent on the availability of the data. The model can only work if the right data is available and readily accessible. Numerous respondents claim that since this is the input to the model, a lack of proper data can hinder the accuracy of the models, or make models unsuitable for use by a utility.

Next is information, which refers to information about the various models, how they work, what can they inform utility managers, and what data is required. Asset managers need to know this information before they can plan to use models in their asset programs. A common theme among the respondents is that there is not enough information about models and how to use them. One participant recommended a project be underway to review models to see what they require and what they produce. This would serve to inform asset managers how to use models and what it takes to use them.

Asset management refers to the understanding that an asset management program framed using statistical models as a basis for determining replacement prioritization is recognized as necessary. Most participants we surveyed recognized that models are becoming more useful to asset management. Predicting a schedule of prioritization is especially beneficial since utilities are dealing with limited capital improvement budgets, and planning using models can help determine the most effective way to use available funds.

Flexibility and configurability of models refers to the idea that utilities want to see models that are user friendly and can be customized to meet the specific needs of the utility in question. Water utilities want to be able to configure models to handle various inputs, and be able to adapt to changing geographical or spatial conditions Too many packages are canned and do not allow for user configurability. Since each utility has different sets of pipes, different data collection methods, and different needs for models, each model needs to be tailored for a specific utility, or have the ability to undergo customization to meet specific needs.

Financial constraints refers to working under limited financial resources. Utilities are working with restricted capital expenditure budgets which limit how much they can spend on models and training. Some utilities understand that empirical testing of pipe, using methods such as acoustic leak detection, can be very expensive, and are seeking other means of assisting asset planning. Models can be a cheaper alternative to physical testing, and all the utilities surveyed agreed that a balanced approach to asset management, using both modelling and physical testing, is ideal.

Similar to financial constraints, there is an idea that a business case needs to be in place to justify the expense of models. Part of this involves communicating the need for models to executives or a management team who may not be able to understand the technical details of models and may not understand their value.

Another theme is enterprise systems. This refers to a trend towards developing enterprise systems-based asset management, which encompasses business processes and information flow across multiple systems, including engineering, accounting, and operations and maintenance. An enterprise system based asset program would allow all departments to have information readily available and shared within the system's network, allowing for more collaboration and data flow. Modelling would be easy to share with other departments, resulting in prompt response from financial departments towards predicted capital allocation budget requests and operations and maintenance response towards predicted pipe failures. Models need to be able to fit in the context of enterprise systems, in order to play a role in future asset management.

Lastly is validation of results. Most of the participants cited want to see validation of the statistical results by comparing them with historical data or pipe condition assessment (physical pipe inspection). The utilities surveyed had varying responses to this theme. One stated satisfaction with the results, another saw a need to improve the accuracy of the models, but was generally satisfied, and another claimed there was no satisfaction with models and accuracy needed to improve greatly.

4.2 Questions on Current State of Models and Future

Many of our interview questions were specific towards the use of models, how accurate they were, and what benefit was derived from them. However, we did have a cohort of questions that were directed towards the current feelings towards models as well as their future. We feel these questions are among the most important in our interview list, and we have selected three questions and presented them here. The responses serve to highlight the issues that need to be addressed before models become more prevalent among water utility firms.

- **Question 1:** Are water utilities quick to embrace models?

  Answers: "*I think yes and no. It depends on where you are in the utility. From a planning context, people have a very high regard, an early adopter phase to modeling. From an operational context, the value of modeling is not seen as much. However that is changing, as technology is becoming more commonplace.*"

  "*No, generally utilities are comfortable doing what they have been doing. Moving beyond that is hard. Especially with a model that takes understanding that people might not have. It's less likely that it will be embraced. There's recognition that there is value there but it's hard to know what to do.*"

  "*I want to say no. In our utility, we do use them. We are a very large utility, and have therefore the need and resources, based on the number of assets, we can apply statistical models, rather than rely on observation. With smaller utilities, with a smaller number of assets or a smaller geographical region, there is a thought that people know their assets well, better than a computer model. This is an old school mentality.*"

Based on these responses, it is evident that there is still some hesitation on the part of utilities to embrace models. Many users recognize the value of models, but there are some who are used to practicing old ways of managing assets, and are not comfortable embracing new technologies or new methods.

- **Question 2:** How do you feel about the future of models in asset management planning?

  Answers: *"Well I think there's a place for them. But they can't live as standalone entities. The biggest challenge is that the public utilities lag behind the private utilities by a decade or more, in terms of their IT systems or internal systems and the enterprise connectivity of them. None of our data is accessible with other departments. We are working to improve that with the master data management system, so everyone is working with the same source data. Everyone's data goes into it, so there is a system of record, so you can see what is happening."*

  "The models are going to be critical for developing asset management plans for the future. The systems are getting more and more complex, the operations and maintenance of the systems and rehabilitation are getting more and more expensive, the funds are going to be limited, even in the future so the models are going to play a critical role in the prioritization of your asset management program."

  "I think there's a great future for it. It is the way of the future. Modeling can give you answers to questions that there is no substitute for."

These responses are more optimistic about the future of models than the first question above. There is recognition that models will play a role in future asset management. Since water systems are complex, and utilities have limited funds to address capital improvements, facilitating prioritization of pipe replacement with models will become more important. Universally, the future for models is favorable, and there are beliefs that more utilities will embrace them to assist in managing complex systems.

- **Question 3:** How easy or difficult is it to incorporate models into your current asset management approach?

  Answers: *"It's very difficult, or easy and difficult. The difficult part is that you have to have data, and you may not have the data, or the data is somewhere and you don't know where it is. And it costs money, and these days for us you have to plan ahead to get a budget allocation to do a project of that type. That's the difficult part, access to data and money."*

  "It's a challenge for us because we have so many different assets, it's not a challenge where we use statistical analysis. It is a challenge where we bring data together, our challenge is we are drowning in data. We need to bring it together, in one area, so we can apply analysis. Not a challenge to do modeling, it's more of management of information."

  "It's a lot of work, because the more thorough and sophisticated the model is, the complexity is exponential. An additional factor requires connections to other systems, you just can't do it out of the blue. So it can get complicated quickly."

We see from these responses that there are issues with data collection and storage, and that the model is only as useful as the data that is provided. Models need the right data available as well as systems in place to allow data sharing to ensure they can provide meaningful results. Management of information is just as important as the model itself. This is a reason why enterprise data systems are valuable – they allow sharing of information with teams and departments. Shared information allows departments to collaborate and meet needs in a more effective manner.

5. DISCUSSION AND CONCLUSIONS

Based on the information we have collected, it is clear that utilities encounter difficulties introducing models into their asset management plans. The reasons for this include lack of available data, lack of user-friendliness and configurability of models, difficulties tying models into current asset management plans, lack of information about models and their benefits, and lack of funds to pursue modeling.

There are several ways to ameliorate these issues. A first step would be to provide guidelines to water utilities as to how models are used and what benefits they can provide. Utilities can also be informed about what data they need to have collected before they can effectively use models. Introducing such information can help clear up misconceptions and assuage fears that utilities may have, encouraging more adoption of models. A second step would be to encourage the use of enterprise systems to facilitate data sharing and information flow. Models would be more valuable in such a context because their predictions and outputs would be seen by many departments and can be addressed in a timelier manner. The benefit would be swifter action to address issues such as pipe breaks and loss of water flow.

However, despite the issues stated above, all the utilities surveyed remain optimistic about the future of models, citing the need for developing more effective asset management plans. There is recognition that models can provide a useful tool in conjunction with condition assessment to allow utilities to better manage their assets.

We can take the information that has been synthesized from this survey and create a network diagram, as shown in Figure 3 below.

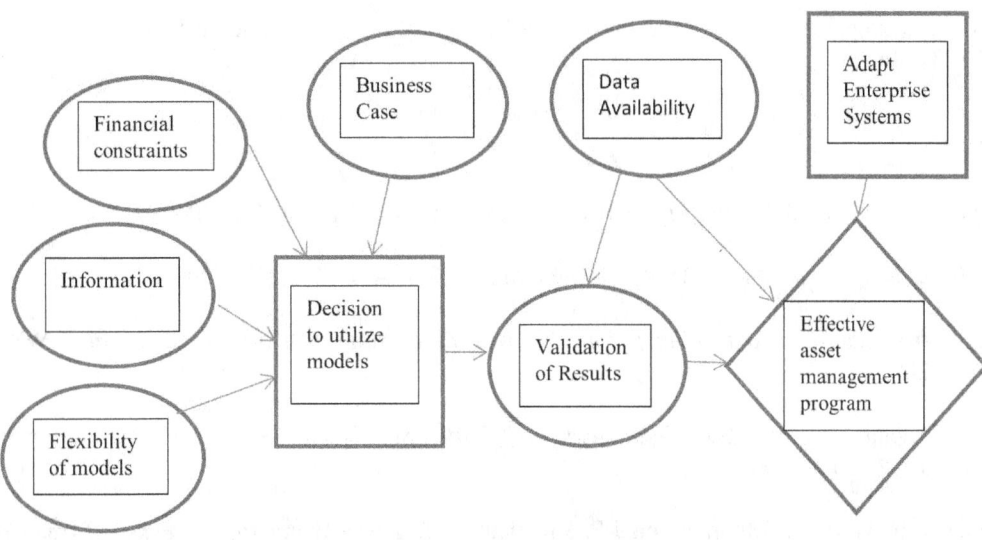

Figure 3: Network Diagram for Models in Water Distribution Asset Management: Influences That Contribute to an Effective Asset Program .

This diagram shows the various influences that play a role in the decision to adopt models and their impact on developing a more comprehensive and effective asset management program. We derive this diagram based on our knowledge of the major issues obtained from our interviews. We have labelled several of the nodes such as financial constraints, information, and data availability as unknowns since in many cases they are not readily quantifiable and information is not readily available to the decision maker. This diagram shows the major factors that contribute to the success of statistical models in asset management planning, and what information decision makers need to be aware of before they choose to implement models. When comparing Figure 3 to our influence diagram (Figure 1), we see that we have gained more information about the types of influences that are directly affecting models and their use in water asset management, including data, enterprise systems adaptation, and flexibility of models. These influences were not apparent from our literature review, which explains the value and contribution of our semi-structured interviews.

In conclusion, we have shown that our mental models interview approach can help illuminate answers to questions regarding the role and outlook of statistical models in asset management planning. We have seen a number of requirements that need to be satisfied for models to be viable. These include robust and available data, more information about the model requirements and needed inputs, and user-configurability and ease of use of the model. We also note that utilities are constrained by limited financial resources and many times a business case must be made to justify the use of models. Utilities also want to see increased accuracy of models when compared with physical testing results. This serves as validation that the model is working properly. Utilities also want to see models become part of advanced enterprise systems that connect multiple parts of the utility firm to allow sharing of data and information. Despite these issues, we see that all of our study participants agree that the future of models is optimistic. Models will allow utility firms to make more informed decisions regarding rehabilitation and repair of pipes, and will be a valuable part of future long term asset planning.

Our next stage of the study would be to use a Q-sort methodology to elicit opinions and determine shared ways of thinking among water utility firms regarding the use of models. Current existing knowledge of the Q sort methodology is widespread, and the methodology is widely used in areas relating to humanities, social sciences, and psychology. However, to the authors' knowledge, no current Q sort study has been performed on water industry experts. We believe this will shed more information about the role and future of models, and how they can better assist utility firms.

Acknowledgments

The authors are supported for work on this project by a grant from the National Science Foundation Project # 1031046.

References

[1] Stubbart, John M. *AWWA water operator field guide.* AWWA, Denver, CO. 2004.

[2] *Rehabilitation of Water Mains: AWWA Manual M28.* AWWA, Denver, CO. 2001.

[3] *Water Infrastructure at a turning point: the road to sustainable asset management.* AWWA, Denver, CO. 2006.

[4] Alegre, Helena. *Strategic Asset Management of Water Supply and Wastewater Infrastructures.* IWA Publishing, London, 2009.

[5] Pelletier, Genevieve, A. Mailhot, and J.P. Villeneuve. *Modeling Water Pipe Breaks – Three Case Studies,* Journal of Water Resources Planning and Management. Vol. 129, No. 2, pp. 115-123, (March 2003).

[6] Clark, Robert et al. *Condition assessment modeling for distribution systems using shared frailty analysis,* Journal AWWA Vol. 102, No. 7, pp. 81-91, (2010).

[7] Debón, A. Et al. *Comparing risk of failure models in water supply networks using ROC curves,* Reliability Engineering and System Safety, vol. 95, no. 1, pp. 43–48, (Sep. 2009).

[8] Yamijala S., S. D. Guikema, and K. Brumbelow, *Statistical models for the analysis of water distribution system pipe break data,* Reliability Engineering and System Safety, vol. 94, no. 2, pp. 282–293, (Feb. 2009).

[9] Kleiner Y., B. Adams, and J. Rogers, *Water distribution network renewal planning,* Journal of Computing in Civil Engineering, vol. 15, no. 1, pp. 15–26, (2001).

[10] Kleiner, Y and B. Rajani, *I-WARP: Individual Water Main Renewal Planner*, Drink. Water Eng. Sci., vol. 3, no. 1, pp. 71–77, (2010).

[11] Johnson Laird, P.N. *Mental Models: towards a cognitive science of language, inference, and consciousness.* Harvard University Press, 1983, Cambridge MA.

[12] Morgan, M. Granger, et al. *Risk Communication: A mental models approach.* Cambridge University Press, 2002, Cambridge, UK.

[13] Magzan, Masa. *Mental Models for Leadership Effectiveness: Building Future Different Than the Past.* Journal of Engineering Management and Competiveness. Vol.2, No. 2, pp. 57-63, (2012).

[14] Saldana, Johnny. *The Coding Manual for Qualitative Researchers.* 2nd Ed. Sage Publications, 2013, Los Angeles.

[15] Seale, Clive and D. Silverman. *Ensuring Rigor in Qualitative Research.* European Journal of Public Health. Vol.7, pp. 379-384, (1997).

Appendix A: Our Interview Checklist

The following is our interview checklist for our semi-structured interview approach. We check off each topic as it is mentioned by the participant, and proceed with the set of sub-questions below each section, one by one.

The interview begins with an initial prompt: "What I'd like to ask you is just to talk about statistical models and their role in water asset management. That is, just tell me what you know about models."

__I__ **Benefits/Costs**
___ Can you tell me what are the benefits of implementing statistical models?
___ Do you feel that models are worth the cost of implementing them?
___ Do you feel you are able to make better decisions regarding your asset management plan when using models?

__I__ **Information**
___ How did you find out about statistical models and their use in asset management?
___ Did you hear about models from industry trade journals or conferences?
___ Do you feel there is enough information about models that utility firms are well-informed about them?

__I__ **Asset Management**
___ What role do statistical models play in asset management planning?
___ Do you feel a balanced approach between models and empirical testing of pipes is ideal?
___ Are water utilities quick to embrace models?
___ How do you feel about the future of models in asset management planning?

__I__ **Implementation/Choosing Models**
___ How easy or difficult is it to incorporate models into your current asset management approach?
___ How do you choose a model and evaluate it?
___ What do you feel about the accuracy of current models?
___ Are you satisfied with the performance of models that you have used?
___ What can be done to improve models so that they can better assist utility firms?

Modeling of pollutant dispersion in street canyon by means of CFD

Davide Meschini[a], Valentina Busini *[a], Sjoerd W. van Ratingen[b], Renato Rota[a]

[a] Department of Chemistry, Materials and Chemical Engineering "G. Natta", Politecnico di Milano, Italy

[b] TNO, Utrecht, Netherlands

Abstract: Nowadays, pollution from traffic remains one of the major sources for contamination in urban areas and it is widely known that substances emitted by vehicles represent a serious hazard to human health; some traffic-related pollutants, such as NO, NO_x and CO are responsible for both acute and chronic effects on human health. This is often the case near busy traffic axis in city centers or street canyons. Purpose of this work is to validate the CFD model predictions against the field measurements of pollutants dispersion in an actual urban environment: Göttinger Strasse, Hanover, Germany. In the location, the population exposure to traffic-related pollution is expected to be high. Steady-state simulations have been performed for 18 different wind directions, with an increment of 20°, in order to cover the whole wind rose. A grid and a Schmidt number sensitivity analysis have been carried out in order to determine both the most suitable resolution of the computational geometry and the most suitable parameter to model the turbulence conditions in the street canyon. All CFD simulations have been performed for neutral atmospheric conditions and have been carried out with the CFD code FLUENT 12.1.

Keywords: CFD, Street Canyon, Atmospheric Turbulence, RANS, Schmidt Number

1. INTRODUCTION

In urban environments, and especially in those areas where population and traffic density are relatively high, human exposure to hazardous substances is expected to be significantly high. This is often the case close to busy traffic axis in city centers or street canyons. The term street canyon refers to a relatively narrow street with buildings lined up along both sides, where prediction of pollutant flow dispersion is particularly challenging because of flow recirculation induced by its configuration.

It is therefore difficult to predict pollutant dispersion with certain accuracy due to the complex interaction between atmospheric flow and flow around buildings. Computational Fluid Dynamics (CFD) techniques are widely utilized to study the wind field and pollutant transport near and around building, because they offer some advantages compared to other methods: they are less expensive than field or tunnel tests and provide results for every point in space. However, this is often accompanied by the difficulty in choosing universal modeling parameters such as the grid resolution and the iterative convergence, in order to obtain reliable results. Moreover, when CFD codes are used to model urban air quality, a proper validation is necessary and the validity of the results is often limited to street geometries and dispersion conditions similar to those for which the validation has been carried out. A difficulty that has been encountered is the fact that field measurements are usually made at a few selected points, which should be representative for the case under examination: for example, it is important to check the presence of concentrations gradients that could mislead the analysis.

The aim of this work is to validate the predictions of a CFD model against the field measurements of pollutants dispersion in an actual urban environment: Göttinger Strasse, Hanover, Germany. In the location, the population exposure to traffic-related pollution is expected to be high. Steady-state simulations were performed for 18 different wind directions, with an increment of 20°, in order to cover the whole wind rose. A grid and a Schmidt number sensitivity analysis were carried out in order to determine both the most suitable resolution of the computational geometry and the most suitable parameter to model the turbulence conditions in the street canyon. All CFD simulations were performed for neutral atmospheric conditions and carried out with the CFD code FLUENT 12.1 [1].

2. MATERIAL AND METHODS

*valentina.busini@polimi.it

Typically, there are four boundary conditions required for gas dispersion modeling: inlet, outlet, top and bottom of the computational domain. The analysis was conducted for 18 different wind directions, considering, each time, a change of flow direction of twenty degrees. It has to be considered that the real urban configuration has a 17°-rotation with respect to North. Thus, velocity inlet boundary conditions were applied at two of the four lateral boundaries, while pressure outlet conditions were applied at the remaining two. For wind directions equals to 0° and 180° (i.e. when the wind blows exactly from North or South), symmetry boundary conditions were applied to the lateral boundaries, where no interaction between the borders of the domain and the flow of pollutants is foreseen. At the inlet of the domain the mean wind speed is represented by a logarithmic law instead of a power law in order to generate the initial velocity profile, in accordance with the M-O profile for neutral stratification which states that the mean velocity profile in the lower part of the atmosphere may be adequately described by:

$$u = \frac{u^*}{k}\ln\left(\frac{z}{z_0}\right)$$

where u^* is the friction velocity, k is the von Karman constant (equal to 0.41) and z_0 is the aerodynamic roughness (equivalent to the height at which the wind speed is zero) and assumes different values if related to the buildings or to the terrain: in the first case the value is set at $z_0 = 0.01$m; in the latter case this changes to $z_0 = 0.05$m. The logarithmic law is strictly valid only in neutral conditions (i.e. when the temperature profile in the surface layer is closed to adiabatic).

The parameters k_s (sand-grain roughness height) was determined from a relationship developed by Blocken and co-workers [2], as:

$$k_s = \frac{9.793 \cdot z_0}{C_s}$$

where C_s is a roughness constant, set equal to 9.5.

In this work a standard turbulence k-ε model was used [1], which solves the Reynolds-averaged Navier-Stokes equations using turbulence terms derived from separate transport equations for the turbulent kinetic energy k and the turbulent dissipation rate ε. This turbulence closure model was chosen because of its relative simplicity and reliability. The CFD code produces a steady-state solution for the mean flow components and the turbulence variables k and ε. Second-order discretization schemes were used to increase the accuracy of the numerical solution.

3. RESULTS AND DISCUSSION

3.1. Geometrical Configuration

Göttinger Strasse is SSE-NNW oriented (17°-rotation towards West with respect to North), as shown in Figure 1. It is a busy inner-city street, 25 m wide and has 4 lanes, with a traffic volume of about 30,000 vehicles per day and a high share of trucks (16%). On both sides of the lanes there are complexes of buildings (most of which are joined) with an average height of about 20 m. Hence, the aspect ratio of this canyon is slightly less than one. On the tallest building (32 m) there is a 2 m high mast and on this one, in turn, there is a 10 m high measuring mast. At the top of the mast, wind speed, wind direction and the background concentration of pollutants have been registered. The road-measuring point is situated before the NLÖ-building, at a location about 1.5 m high and 1m away from the road edge; also at this point, the wind direction and hourly averaged values of NO, NO_2, CO and Benzene are monitored [3]. The measuring point is located not only in close proximity to the road, but also in a highly structured area of the building of NLÖ; hence, for some wind directions the complex geometry can have a significant impact on the measured concentrations. As for the speed of the cars passing by the monitoring point, no data were available; nevertheless, increase or decrease of the allowed urban speed can be excluded, since the street does not present any slope. This means that neither the cars heading northwards, nor the cars heading southwards will pass the monitor under heavy acceleration or in a deceleration phase.

Figure 1: Orientation of the Göttinger Strasse street canyon and height of surrounding buildings.

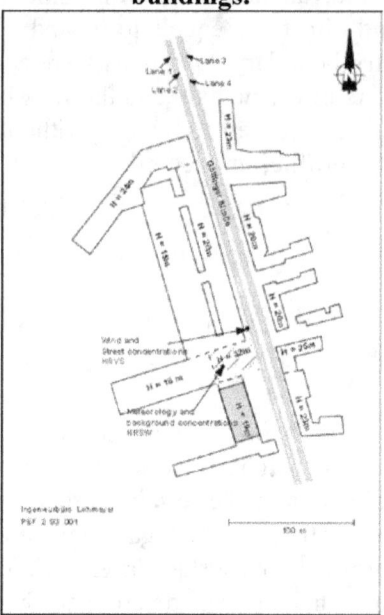

3.2. Evaluation of field data

Testing and validation of a CFD code relies heavily on the availability of experimental data, which have to be compared to the output of the CFD simulations. The State Environmental Agency of Lower Saxony operates a permanent monitoring station in this four-lane street canyon. Detailed field data measurements, and specifications of the case, were obtained from the database of the DGXIITMR TRAPOS. From the provided time series for:

- the number of vehicles on each of the four lanes, divided into small vehicles and trucks;
- the wind speed and wind direction at the station at 42m above the street level;
- the concentrations of NO, NO_2, CO and Benzene at the station and on the road.

For the comparison with the model predictions the mean dimensionless concentration c^* [4] was used:

$$c^* = \frac{C \cdot U_{42} \cdot H}{Q/L}$$

where C is the measured concentration referred to the measuring point minus the background load measured on the roof (it has to be underlined that in a city environment with numerous sources and large local concentration differences it is not easy to determine a meaningful background: in case of the Göttinger Strasse the background concentration is measured on the top of the highest building about 32 m above street level), U_{42} is the wind speed at a height of 42 m (where the monitoring mast is located), H is the characteristic height of the buildings (20 m), L = 180 m is the length of the line sources, Q is the source strength in [mg/s] and Q/L is the specific source strength.

To get a value of source strength, Q, the starting point is to estimate the traffic volume throughout the monitoring period: the analysis of time series generates an average daily traffic volume of about 30,000 vehicles, with a truck share of about 16%, as shown in Table 1.

Table 1: Hourly traffic load in Göttinger Strasse.

Traffic Lane	Vehicles	Trucks	% trucks
1	352	56	17
2	220	32	14
3	324	58	17
4	350	50	15

For the calculation of the emission strength of the line sources, emission factors were used; they represent the relationship between the amount of pollution produced and the amount of raw material processed or burned and depend on the following factors:

- the time-varying number of vehicle;
- the truck share;
- the average speed;
- the percentage of vehicles with cold engine.

To obtain an estimate as reliable as possible, continuous traffic measurements, which also separate vehicles into two categories, were used, thus being able to determine the total emission and the emission factors for each category. It has to be noted that, currently, there are several sets of emission factors in use [3]; Table 2 contains the values used for this validation case.

Table 2: Emission factor both for cars (from the German acronym PKW: Personenkraftwagen) and trucks (from the German acronym LKW: Lastkraftwagen).

Benzene		NO_x		CO	
PKW	LKW	PKW	LKW	PKW	LKW
0.075	0.057	1.03	9.6	15.6	4.1

The emission is then calculated for each lane, considering this equation:

$$Q = P \cdot ePKW + L \cdot eLKW$$

Where Q is the emission of one lane in [mg/s], P is the number of vehicles per lane per second, L is the number of trucks per lane per second, ePKW is the specific emission factor of a vehicle in [mg/m·PKW], and eLKW is the specific emission factor of a truck in [mg/m·LKW].

Plotting all the available data together on a single graph some qualitative trend can be identified, as shown in Figure 2.

Figure 2: Normalized hourly mean concentration values as a function of wind direction measured over a period of one year at the street monitoring station Göttinger Strasse in Hanover.

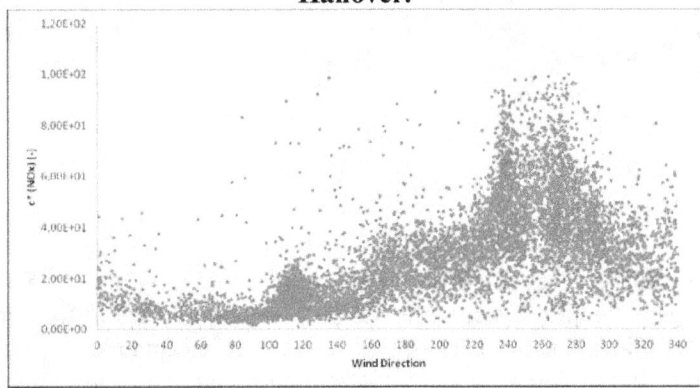

However, to obtain a meaningful parameter, the measurements that fell within the same interval of wind direction (step of 20°) and grouped together to calculate a mean value, in order to get an average of the emitted pollutant (i.e., NO_x) in that direction. This same procedure was also applied to the wind speed at 42 m. Figure 3 shows the computed averaged dimensionless concentration c* for the complete range of wind directions:

Figure 3: Dimensionless concentration from field data.

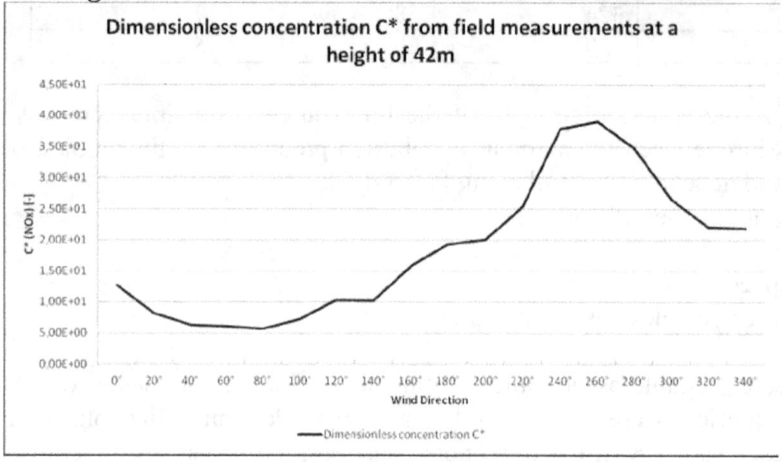

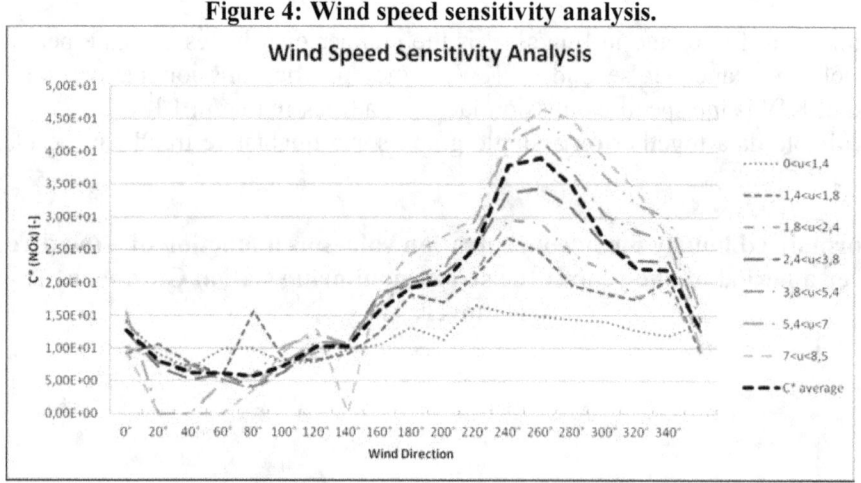

Furthermore, the wind speed sensitivity analysis reported in Figure 4 shows that the mean concentration trend is representative of all the ranges of velocity considered: hence, even though it is possible to identify 10 different wind speed classes throughout the entire year of measurements, one can assume that the mean wind speed at 42 m floats around 4.6 m/s. Following this reasoning the value of the friction velocity u^* was estimated as:

$$u^* = \frac{U_{42} \cdot k}{\ln\left(\dfrac{42}{z_0}\right)} = 0.254 \, \frac{m}{s}$$

Figure 4: Wind speed sensitivity analysis.

3.3. Computational domain

The computational grid represents a replica of the buildings located in Göttinger Strasse in Hanover; the area can be considered an example of a complex urban geometry, where traffic-induced turbulence and whirls around the buildings can affect the dispersion of pollutants emitted from cars. Its dimensions are L x W x H = 740 x 370 x 240 m³ and it is represented in Figure 5.

Figure 5: Computation domain.

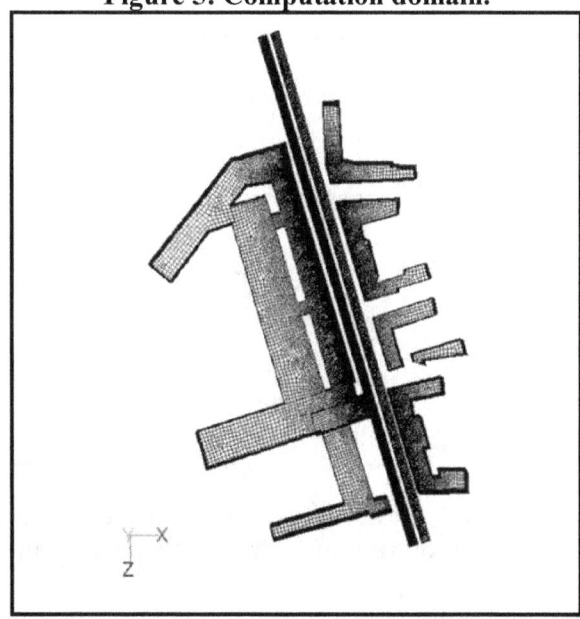

In this study, an unstructured hexahedral grid was generated by sweeping faces in the vertical direction, in order to allow full control over the grid quality, size and resolution [5]. The procedure was executed with the aid of the pre-processor Ansys GAMBIT 2.4.6, resulting in a hexahedral grid with about $6 \cdot 10^6$ cells. The range of the cell dimensions is broad: because of the extension of the domain, some parts of the geometry, which are less important for the flow field through the street canyon have a slightly coarser grid, with the width of the cells increasing towards the borders of the domain. Therefore, the whole grid was meshed using the concept of the meshed size function, which allows controlling the size of mesh elements, keeping the elements in the region surrounding the source small in comparison to those farther away. One specific guideline was also taken into account, in order to generate a grid according to the best practice for the CFD simulation of flows in urban environments [6]: the lateral and the top boundaries were set 5H away from the buildings, where H is the height of the target building, to allow a full flow development.

3.4. Analysis of computational data

Figure 6 shows the first complete set of calculations after convergence was reached; it can be noticed that, in the case of wind blowing from west (with respect to the street axis), the pollutant concentration at breath height is higher than in the case of wind blowing from east. This behavior is correctly predicted by the model. However, the model overestimates the concentration of NO for wind directions between 100° and 200° by up to a factor 3. Given the uncertainties related to both the experimental measurements and the input model parameters, this can be considered quite a fair agreement. The same figure also shows the influence of the u* value.

Figure 6: Comparison between field and computational data.

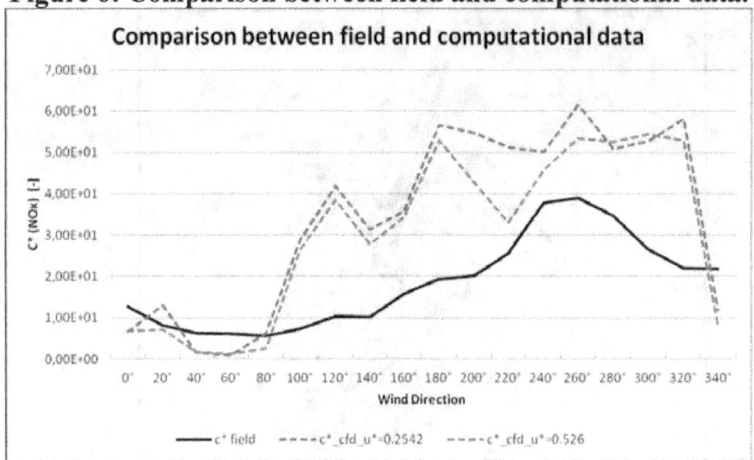

It can be seen that there are no large differences between results obtained using the u^* value suggested by Theodoridis and co-workers [7] and the one estimated from the average wind speed.

3.5. Grid sensitivity analysis

Turbulence models performance of flow around buildings is highly dependent on the mesh resolution. In order to understand if a refinement of the computational domain is needed, a grid sensitivity analysis was carried out by replicating the same computational domain with a coarser grid size obtained using the same size functions as in the first case, but different growth rate and maximum size of the cells.

Figure 7: Comparison between two different grid resolutions.

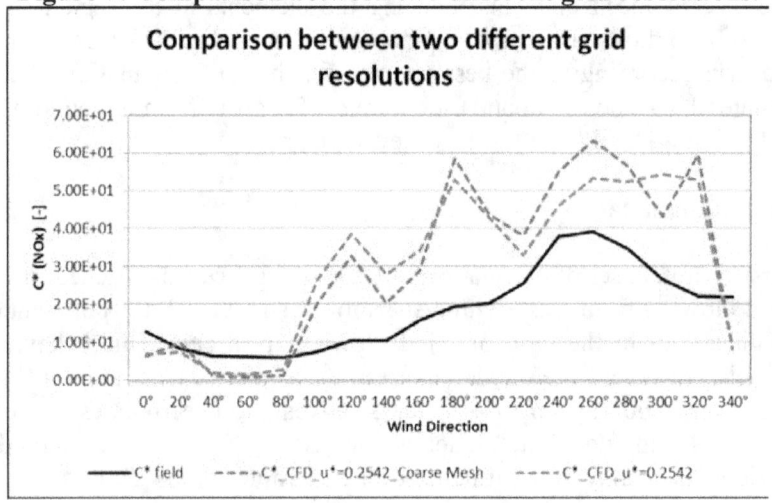

Figure 7 shows that no significant differences can be noticed comparing the first mesh with a coarser one.

3.6. Schmidt number sensitivity analysis

Another important input parameter is the Schmidt number, which is defined as the ratio of the turbulent momentum diffusivity v_t and the turbulent mass diffusivity D_t:

$$Sc_t = \frac{v_t}{D_t}$$

No universal value of Sc_t can be established and empirical values have been used in different studies: Flesch [8] reported estimated values of Sc_t ranging from 0.18 to 1.34, based on field observations under different atmospheric stability classes and wind conditions [9]. Regarding dispersion around buildings, the results with $Sc_t = 0.3$ show values of concentration, which are closer to the experiments. On the other hand, dispersion in street canyon is different from that around a single building, as argued in Hanna et al. [10] and Milliez and Carissimo [11] that suggest a higher value of Sc_t for the flow within a street canyons. In order to see if a modification of the Schmidt number produces changes on the predicted ground level concentrations, two other sets of calculations were performed, without modifying any other parameter but Sc_t, from its default value of 0.7 to 0.3 and then to 0.1, which are in the range of those used in previous studies. What can be noticed overall is that, by reducing the Sc_t, the dispersion of the pollutant increases, leading to smaller values of concentration, which are closer to the experimental findings. This can be explained because a lower Sc_t means that the turbulent mass diffusivity increases, hence leading to an easier mixing between pollutant and air within the street canyon. Figure 8 and Figure 9 show a comparison between field and computational data for a Schmidt number equal to 0.3 and 0.1, respectively:

Figure 8: Comparison between field and computational data for Sc_t=0.3.

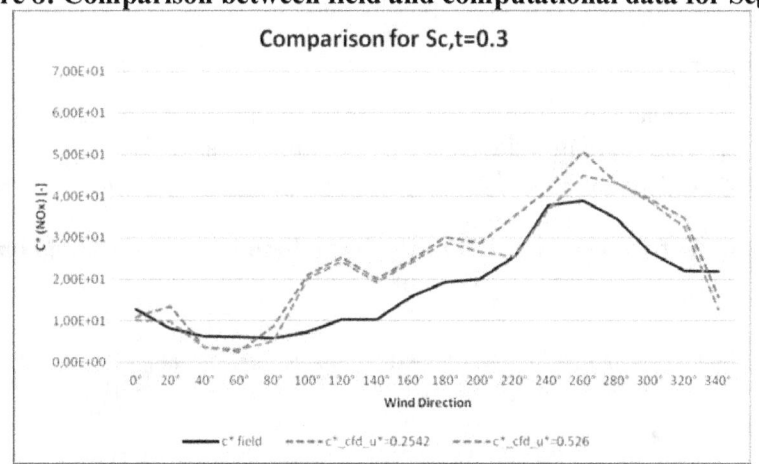

Figure 9: Comparison between field and computational data for Sc_t=0.1.

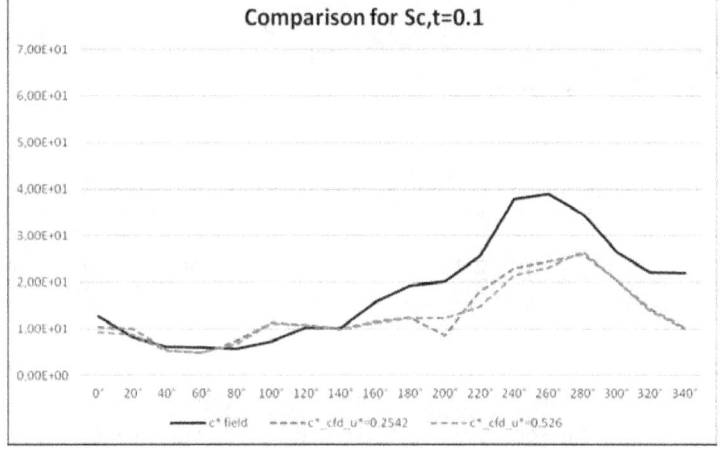

The two figures also confirm the low sensitivity of the model results to the u* value, in agreement with the results reported in Figure 6.

3.7. Extension of the domain

Another attempt was made in order to get further with the analysis and to see if the comparison between field and computational data could be even more improved: the domain of interest was increased, hence leading to the addition of more buildings, whose goal is to help the wind flow develop before it reaches the critical region, as shown in Figure 10.

Figure 10: Comparison between starting computational domain and extended computational domain.

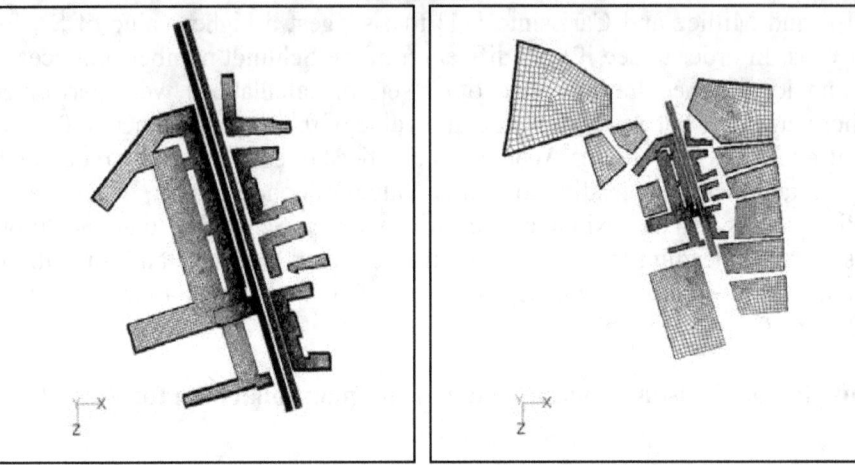

Despite the increase of the domain, the number of cells does not increase drastically due to the usage of size functions: the blocks of buildings added around Göttinger Strasse were meshed roughly, since no precise values of concentration are needed at such distant regions from the monitoring point.

Figure 11: Comparison between starting computational domain and extended computational domain.

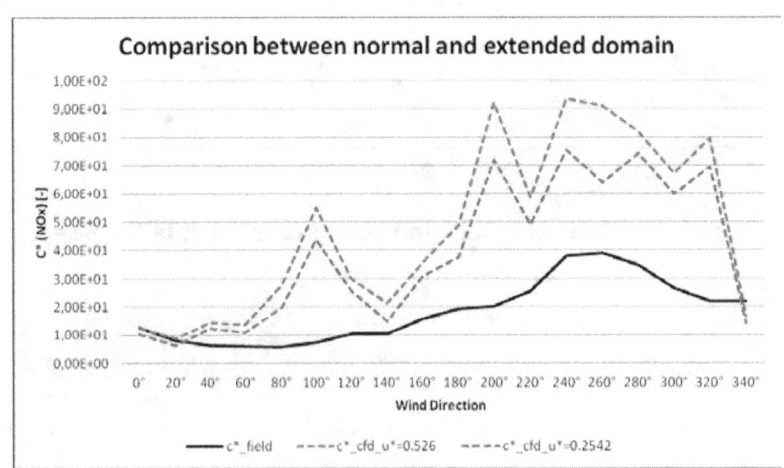

Figure 11 shows the influence of extending the computational domain. Overall, an increase in the dimensionless concentration can be observed, which might be due to the effect of the additional obstacles on the wind speed with a Sc_t value of 0.7.

Large changes are noticed if the Sc_t is modified, as previously shown, decreasing its value from the default option of 0.7 to 0.3 and to 0.1, as shown in Figure 12 and Figure 13.

The same figures also confirm the low sensitivity of the model predictions to the u* value.

Figure 12: Comparison between starting and extended computational domain (Sc$_t$=0.3).

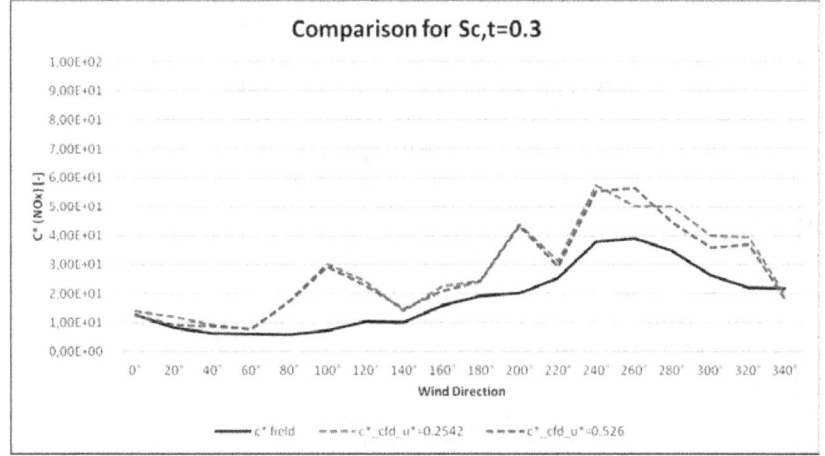

Figure 13: Comparison between starting and extended computational domain (Sc$_t$ = 0.1).

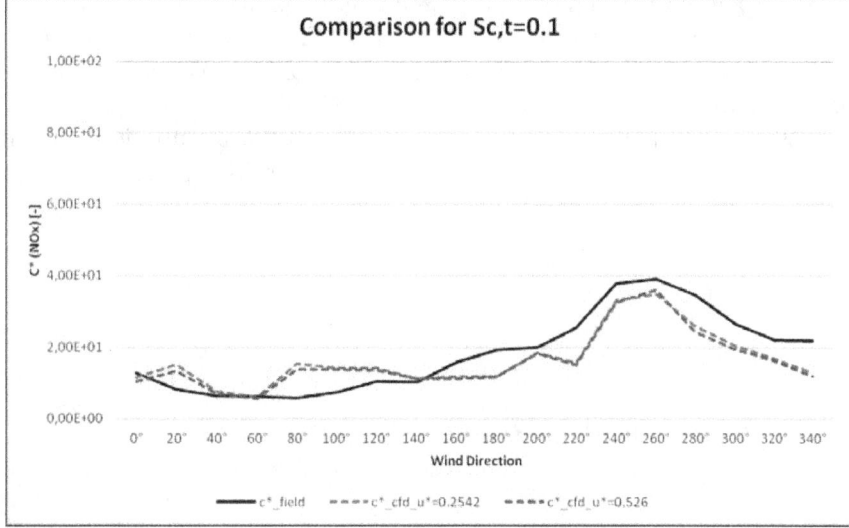

4. CONCLUSION

Purpose of this work was to validate a numerical model to predict pollutant dispersion in a urban street canyon (Göttinger Strasse, Hanover, Germany), by comparing its results to field data consisting of one-year measurements, carried out within an European project of pollution modeling in busy inner-city streets (TRAPOS: optimization of modeling methods for traffic pollution in streets). The simulations were performed with the commercial software FLUENT 12.1 in association with a pre-processor, GAMBIT 2.4.6, in order to generate the computational domain. Steady-state RANS simulations were conducted for the case under examination. All the CFD simulations were performed for neutral atmospheric conditions. The wind flow was simulated for 18 different directions, with a change of 20° each time: the qualitative agreement was found to be quite fair. Generally, the numerical results overestimate the concentration measured on the ground, but the concentrations trends were found to be similar. A Schmidt number sensitivity analysis was performed, leading to the conclusion that low values of Sc$_t$, such as 0.3 or 0.1, tend to improve the agreement of the predicted ground level concentrations inside the street canyon.

References

[1]   ANSYS Inc. "*ANSYS Fluent 12 User's guide*", Lebanon, NH, USA, (2009).
[2]   Blocken B., Stathopoulos T., Carmeliet J. "*CFD simulation of the atmospheric boundary layer: wall functions problems*", Atmospheric Environment, 41, pp. 238-252, (2007).

[3] Palmgren F., Berkowicz R., Ziv A,. Hertel O. *"Actual car fleet emissions estimated from urban air quality measurements and street pollution models"*, The Science of the Total Environment 235, pp. 101-109, (1999).

[4] Schatzmann M., Leitl B. *"Issues with validation of urban flow and dispersion CFD models"*, Journal of Wind Engineering and Industrial Aerodynamics, 99, pp. 169-186, (2011).

[5] Van Hooff T., Blocken B. *"Coupled urban wind flow and indoor natural ventilation modeling on a high-resolution grid: A case study for the Amsterdam Arena stadium"*, Environmental Modeling & Software, 25, pp. 51-65, (2010).

[6] Franke J., Hellsten A., Schluenzen H., Carissimo B., *"Best practice guideline for the CFD simulation of flows in the urban environment"*, (2007).

[7] Theodoridis G., Karagiannis V., Valougeorgis D. *"Numerical prediction of dispersion characteristics in an urban area based on grid refinement and various turbulence models"*, Water, Air and Soil Pollution 2, pp- 525-539, (2002).

[8] Flesch T.K. *"Turbulent Schmidt number from a tracer experiment"*, Agricultural and Forest Meteorology, 111, pp. 299-307, (2002).

[9] Tominaga Y., Stathopoulos T. *"Turbulent Schmidt numbers for CFD analysis with various types of flow field"*, Atmospheric Environment, 41, pp.8091-8099, (2007).

[10] Hanna S.R., Hansen O.R., Dharmavaram D. *"FLACS CFD air quality model performance evaluation with Kit Fox, MUST, Prairie Grass, and EMU observations"*, Atmospheric Environment, 38, pp. 4675-4687, (2004).

[11] Milliez M., Carissimo B. *"Numerical simulations of pollutant dispersion in an idealized urban area, for different meteorological conditions"*, Boundary Layer Meteorology, 122, pp. 321-342, (2007).

Consideration on the assessment of the environmental consequences and impacts during transport of radioactive materials (RAM)-A Safety Case

Gheorghe Vieru, AREN, Bucharest, ROMANIA

Abstract: The transport of Dangerous Goods-Class #7 Radioactive Material (RAM), is an important part of the Romanian Radioactive Material Management. The overall aim of this activity is for enhancing operational safety and security measures during the transport of the radioactive materials, in order to ensure the protection of the people and the environment. The paper will present an overall of the safety and security measures recommended and implemented during transportation of RAM in Romania. Some aspects on the potential threat environment will be also approached with special referring to the low level radioactive material (waste) and NORM transportation either by road or by rail. A special attention is given to the assessment and evaluation of the possible radiological consequences due to RAM transportation. The paper is a part of the IAEA's Vienna Scientific Research Contract on the State Management of Nuclear Security Regime (Framework) concluded with the Institute for Nuclear Research, Romania, where the author is the CSI (Chief Scientific Investigator).

Keywords: Transport, RAM, Environmental, Impact, Security.

1. INTRODUCTION

The IAEA defines nuclear security [1] as *"the means and ways of preventing, detecting and responding to sabotage, theft and unauthorized access to or illegal transfer of nuclear material and other radioactive substances, as well as their associated facilities"*. The IAEA works closely with Member States (MS) to establish and enhance the measures needed to control and protect nuclear and radioactive materials, as well as to prevent illicit nuclear materials trafficking [1].

On April 2010 the 47 MS attended the Nuclear Security Summit issued their Communiqué of the Washington Nuclear Security Summit, stating the they "reaffirm the essential role of the IAEA in the international nuclear security framework and will work to ensure that it continues to have the appropriate structure, resources and expertise needed to carry out its mandated nuclear security activities, in accordance with its Statute and its Nuclear Security Plans".

In the modern world the terrorism has renewed attention to security issued, prompting a profound re-thinking in the international approach to nuclear security. As a consequence Romania, as a MS, joined to the new realities in according with the IAEA Nuclear Security Plan.

Taking into consideration the above mentioned, the paper presents a methodology for risk assessment of the State Management of Nuclear Security in transport of Radioactive Materials (RAM) in Romania [2].

In the Figure 1 is presented the main routes for the transport of RAM.

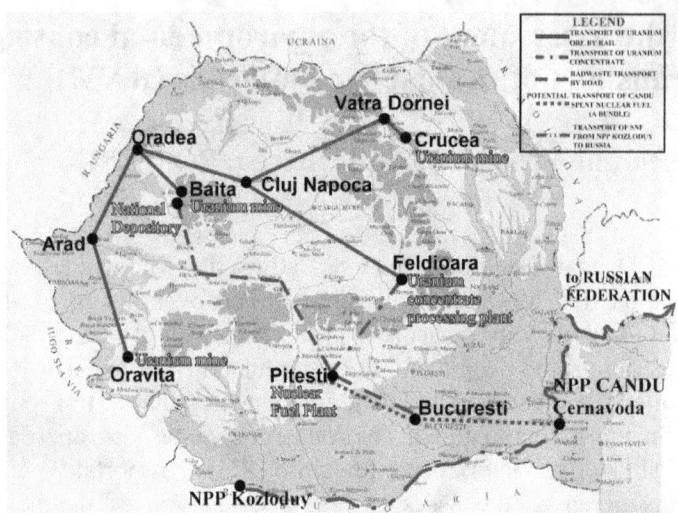

Fig 1. The Main Routes of RAM Transportation in Romania

2. DESCRIPTION OF THE METHODOLOGY TO BE USED

In order to determine a methodology to assess the radiological consequences due to a potential malicious act possible to be happen during the transport of RAM in Romania by using the INTERTRAN II and RADTRAN II computer codes [3].

3. IDENTIFICATION AND THE EVALUATION OF THE POTENTIAL RISKS DUE TO THE TRANSPORT OF RAM

3.1 TRANSPORT BY ROAD

In Figure 1 are shown the main routes for transport of the RAM in Romania. In order to evaluate the dose resulting from possible road accidents involving these radioactive shipments, based on the frequency of occurrence of accidents of specified severities the IAEA computer code INTERTRAN II has been used.
On the other hand for rail transport a probabilistic risk assessment method (PRA) has been adopted [4] for this work aimed to quantify the potential radiological consequences and the expected probability of occurrence of such sequences. Data to be used as input data to the computer code INTERTRAN II has been provided by postulate possible accidents scenarios [2] such as: transport hazards (fixed impact hazard, mobile impact hazard), malicious acts, potential threatens , accident frequencies by road.

Based on these there were calculated road accident probabilities such as:

probability of impact only: 0.421x10^{-5} per journey;
probability of impact and fire: 1.50x10^{-8} per journey;

It is also assumed that, following an impact and a malicious actor a terrorist attacks, the content may become available for dispersion.

The collective doses assessed are as follows:
dose to public along route: 0.25x10-5 person.Sv.y-1;
dose to public during stops: 0.37x10-8 person.Sv.y-1 ;
dose to truck crew: 0.47x10-5 person.Sv.y-1;
The total annual collective dose is: 0.72037x10-5 person.Sv.y-1;
The associated latent cancer fatality risk is estimated at 0.77x10-8 y-1.

3.2 TRANSPORTS BY RAIL

There are different kinds of operation contributing to the overall risk, such as: rail transport, rail road transfer activities handling and misoperation activities, etc. Transport and handling of possible accidents [3] or potential malicious acts may occur and pose a potential risk for the public and the environment.

Because the occurrence of such accidents is statistical in nature, the probability risk assessment (PRA) has been adopted in order to quantify the potential radiological consequences and the expected probability of occurrence of such accidental or potential malicious acts or terrorist attacks sequences. The potential radiological consequences have been calculated by using INTERTRAN II computer code.

The calculated radiological risks include [5]:

- RAM exposure to the public and transport personnel from routine (incident free) transport of the very low level radioactive material (uranium ore);
- transport accident and consequences of the potential malicious acts as well as terrorist attacks resulting in radiation exposure of the population and contamination of the environment.

The accidental sequences include steps such as: characterization and the type and quantity of shipment; determination, selection and description of the type, severity and probability of occurrence of transport and handling accidents; assessment of potential radiological consequences for the spectrum of wealth condition encountered along the rail route, consequences of potential malicious acts, landslide, etc.

The IAEA computer code INTERTRAN II has been used to determine the collective dose to population and transport personnel and the preliminary risk assessment results are:

- crew: 1.34×10^{-5} person Sv/y;
- members of the public: 1.78×10^{-5} person Sv/y;
- TOTAL: 3.12×10^{-5} person Sv/y

Radioactivity releases are not expected to occur in close proximity to a possible accident site at a probability level as low as 10^{-7}, i.e. a chance of 1 in 10 million for the total volume of the RAM to be transported. In case of the malicious acts, sabotage or terrorists attacks the radioactivity releases can increases significantly.

4, TRANSPORT OF SPENT NUCLEAR FUEL BY ROAD- A Safety case

A CANDU spent fuel bundle from NPP CANDU Cernavoda is intended to be transported, by road, to INR Pitesti (see Figure 2).

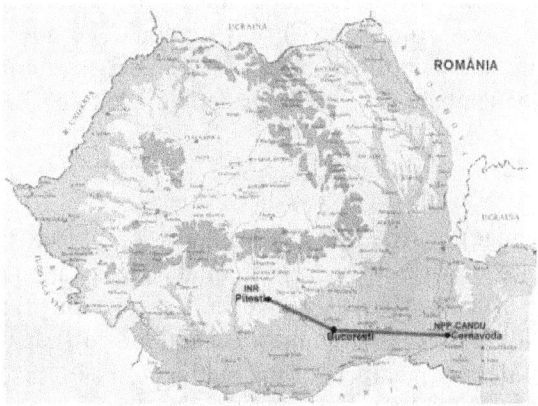

Fig 2. SNF route transportation from NPP Cernavoda to INR Pitesti

The scope of this transport is to be determined and analyzed the behaviour of the Romanian CANDU manufactured fuel (bundle) during burning. In order to evaluate the risk and the radiological consequences of routine as well as accident frequencies and radiological consequences due to the transport, the INTERTRAN 2 code has been used.

The assessment considers a possible future transport operation and specific details would be confirmed prior to a decision to undertake the movements. It is necessary to make assumption for the purpose of this assessment. An alternative route, for example, could include hazards excluded from this analyze. It is assumed also that the package will be transported on suitable trailer and the speed of the vehicle will be limited at 30 km/h (8.33 m s^{-1}). On the other hand, it is likely that factors such as operational controls and arrangements (high standards of driver training for example) will reduce the probability of many accident scenarios.

This analyze is referring to the off-site transport and consider also potential malicious acts, terrorist attacks or protest action.

The package contents: total products activity is approx. 696TBq, (18,800 Ci/bundle), respectively (activity is *945 Ci/kg* metallic uranium). It is also assumed that the true value may be up to *25%* larger than those quoted. Inside the package will be only 1 CANDU spent fuel bundle.
Population Density is required to estimate the consequences of an accidental release of radioactivity, the probability that an accidental release will occur in a particular population density area and the collective doses to the population from the normal transport of spent fuel.
Typical population densities for normal transport dose calculations have been determined to be: a) *urban*: 5,000 peoples/sq, km, b) *intermediate*: 230 peoples/sq.km, c) *rural*: 55 peoples/sq.km.

During incident free routine transport, the package external dose field will result in small radiation doses to exposed workers and members of the public. These doses are estimated using the IAEA computer code *INTERTRAN 2* [3].

RESULTS

The collective doses assessed (assuming 1 shipment/yr, 1 package/journey) are:

Dose to public alongside route	*1.43 x 10^{-3} (Sv/y)*
Dose to public sharing route	*1.78 x 10-3 (Sv/y)*
Dose to public during stops	*2.37 x 10-5 (Sv/y)*
Dose to package vehicle crew	*5.5 x 10-3 (Sv/y)*

The total annual collective dose to member of the public of *3.3 x10ˉ3 person Sv* can be compared with that they receive due to naturally occurring sources of radiation. The total number of the peoples exposed calculated from these areas and densities are about *140,256*. Assuming a risk factor of *0.067 Sv^{-1}*, the annual collective dose to member of the public is determined to be 2.3x10E-03.

RISK EXPECTATION VALUES

- The expectation value (or average) of risk, measured in terms of expected number of fatalities per year, is a convenient measure of risk. It suffers from the disadvantage that the averaging is performed over a wide range of consequences. However, the frequency associated with accidents involving more than one fatality is very small in this case and falls off rapidly for higher fatalities.
- The expected number of fatalities for members of the public per year associated with the proposed road transport operation are based on the expected number of fatalities for each scenario in urban, intermediate and rural population distribution and the probabilities of accidents occurring in each region.

The preliminary results are shown below:

- *radiological effects in accidents: 5x10-9;*
- *radiological effects in routine transport: 1.7x10-4*

5. CONCLUSIONS

The transport of RAM in Romania is a very sensible and complex problem taking into consideration the importance and the need of the security and safety for such activities. The Romanian Nuclear Regulatory Body set up strictly regulation and procedures according to the Recommendation of the IAEA Vienna and other international organizations. There were implemented the adequate regulation and procedures in order to keep the environmental impacts and the radiological consequences at the lower possible level and to assure the effectiveness of state nuclear security regime due to possible malicious acts in carrying out these activities including transport and the disposal site at the acceptable international levels. The levels of the estimated doses and risk expectation values [5] for transport and disposal are within the acceptable limits provided by national and international regulations and recommendations but can increase, significantly during potential malicious acts.

ACKNOLEDGEMENTS

The author wants to express deeply thanks to the IAEA Secretariat for the strong support, both technically and financially, under contract, in order to carry out the scientific research

REFERENCES:

[1] IAEA NUCLEAR SECURITY SERIES GUIDE-Security of Radioactive Material during Transport, (2007).

[2] G.Vieru, "Risk and Safety Evaluation in Radioactive Waste Transport in Romania", RAMTRANS, Vol. 10, No. 2, pp. 105-112, London, UK, (1999).

[3] G.Vieru, "The Identification and the approach of the risk and safety problems associated with the transport of radioactive materials in Romania", Romanian Nuclear Program, Internal Report, INR Pitesti, Romania, (2002).

[4] G.Vieru, "Safety criteria for the transport and storage of the Radioactive Materials, in Romania", Romanian Nuclear Program, Internal Report, INR Pitesti, Romania, (2001).

[5] G.Vieru, "The Safety of Transport of Radioactive Materials in Romania", RAMTRANS, Vol.14, No. 1, pp. 15-27, London, UK, (2003).

Health Effects of Technologies for Power Generation: Contributions from Normal Operation, Severe Accidents and Terrorist Threat

**S. Hirschberg[a*], C. Bauer[a], P. Burgherr[a], E. Cazzoli[b],
T. Heck[a], M. Spada[a] and K. Treyer[a]**
[a]Paul Scherrer Institute, Laboratory for Energy Systems Analysis, Villigen, Switzerland
[b]Cazzoli Consulting, Villigen, Switzerland

Abstract: As a part of comprehensive analysis of current and future energy systems we carried out numerous analyses of health effects of a wide spectrum of electricity supply technologies including advanced ones, operating in various countries under different conditions. The scope of the analysis covers full energy chains, i.e. fossil, nuclear and renewable power plants and the various stages of fuel cycles. State-of-the-art methods are used for the estimation of health effects. This paper addresses health effects in terms of reduced life expectancy in the context of normal operation as well as fatalities resulting from severe accidents and potential terrorist attacks. Based on the numerical results and identified patterns a comparative perspective on health effects associated with various electricity generation technologies and fuel cycles is provided. In particular the estimates of health risks from normal operation can be compared with those resulting from severe accidents and hypothetical terrorist attacks. A novel approach to the analysis of terrorist threat against energy infrastructure was developed, implemented and applied to selected energy facilities in various locations. Finally, major limitations of the current approach are identified and recommendations for further work are given.

Keywords: Power Generation, Health Effects, Normal Operation, Severe Accidents, Terrorist Threat, Life Cycle Assessment, Environmental Impact Assessment, Comparative Risk Assessment

1. INTRODUCTION

The goals of sustainability include minimization of negative health impacts of energy systems. Such effects may arise due to emissions of pollutants from the normal operation of power plants and the associated fuel cycles as well as from accidents, thus contributing to increased mortality and morbidity. In fact, health damages of power generation are major contributors to the corresponding external costs (e.g. [1]).

The health risks associated with energy supply are of high public interest and are frequently in the focus of attention in debates addressing the pros and cons of specific options. However, this is subject to major deficiencies and misunderstandings due to the lack of solid basis in terms of systematic comparisons of health effects caused by the normal operation on the one hand and by random or intentional accidents on the other hand. The scope of such comparisons should cover not only the power plants but also the full energy chains. Furthermore, proper attention has to be paid to the appropriate, balanced choice of reference technologies since the results are technology-specific.

In our previous work (e.g. [2]) we provided examples of comparisons of health risks associated with the portfolio of current and future electricity supply options of a major Swiss electric utility. This covered both normal operation and accidents. The present paper broadens the scope of the comparisons by including the terrorist threat, illustrates the impact of technological features and operational conditions, and reflects the new methodological developments and major extensions of the relevant databases.

---

* E-Mail: *stefan.hirschberg@psi.ch*

2. HEALTH IMPACTS OF NORMAL OPERATION

2.1. Estimates Based on Impact Pathway Approach

Health effects of normal operation are estimated using methods of Environmental Impact Assessment (EIA). The Impact Pathway Approach (IPA), allowing accounting for site-specific effects, is combined with detailed Life Cycle Assessment (LCA). The Impact Pathway Approach is based on methods developed in the European ExternE project series ([1], [3] and [4]). Methods and data for China refer to ([5] and [6]). The life cycle data are derived from the ecoinvent database ([7]), the most comprehensive LCA database worldwide.

Figure 1 shows health impacts of normal operation for different electricity generation technologies, different fuels, and different locations in Europe and China. The selected current (2010) European technologies represent very good environmental standards. For the Chinese case the typical, environmentally unfriendly technologies are contrasted to those having similar standards as the European ones. The health effects in terms of mortality are expressed in Years of Life Lost (YOLL) per kWh electricity produced (kWh$_e$). Large and small systems are considered. The biogas and natural gas combined heat and power (CHP) plants are of the order of 200 kW$_e$. For CHP, the environmental burdens are allocated to the generated electricity and heat according to the exergy allocation scheme. All results include contributions from the life cycle of the systems. Health effects due to climate change effects are not included.

Figure 1 Mortality in terms of Years of Life Lost (YOLL) per kWh electricity produced for different systems and different locations. Sources: Data for China plants from China Energy Technology Program ([5] and [6]); Swiss/European plants based on system choice in the Axpo project [8], adjusted to year 2010. (CH=Switzerland, FR=France, IT=Italy, DE=Germany, DK=Denmark, CC=Combined Cycle, CHP=combined heat and power, SOFC=solid oxide fuel cell, PV=photovoltaic, FGD=flue gas desulfurization).

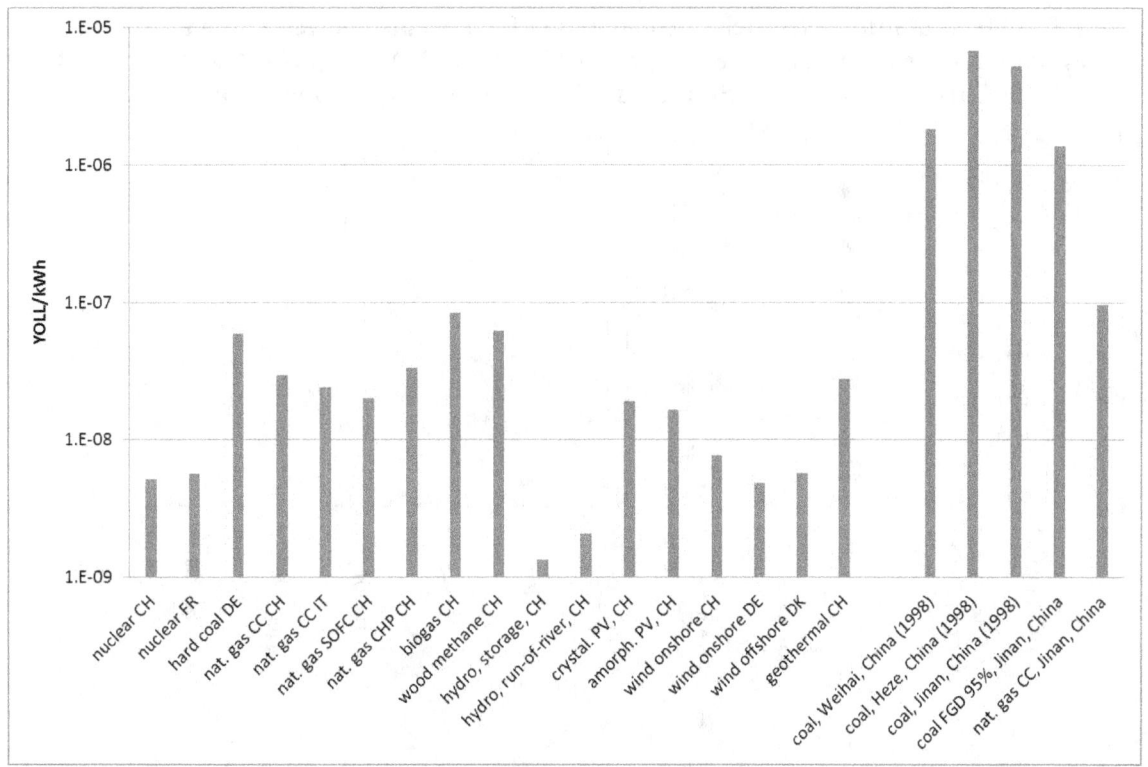

The health impacts due to normal operation of the selected systems vary over almost four orders of magnitude, depending on the technology but also strongly on the location of power plants. Each plantsite is characterized by specific population density, meteorological conditions and concentration of species relevant for chemical transformations leading to production of secondary particulates.

Hydro power plants show the lowest health damages per kWh electricity. Most other renewables and nuclear are in the middle range. Coal power plants without emission reduction measures, located in high population areas in China, yield the highest health damages per unit of electricity produced among the selected systems; use of clean coal technologies and modern combined cycle gas plants leads to large reduction of health effects.

2.2. Estimates Based on Life Cycle Assessment

The impacts of electricity production during normal operation can alternatively be assessed by means of Life Cycle Assessment. Human Health Damage is an endpoint in different Life Cycle Impact Assessment (LCIA) methods, expressed in Disability Adjusted Life Years (DALYs; DALY = Years of Life Lost + Years Lived with Disability). Depending on the LCIA method chosen and the social perspective taken, the influence of impact categories on the total human health impacts (HHI) varies significantly. The technologies chosen are based on ([7]- [11]). As opposed to the IPA the LCIA approach does not allow for representation of site-specific dependencies.

Figure 2 shows the total HHI of a number of European electricity producing technologies and the associated energy chains under three social perspectives used in the LCIA context, i.e. Hierarchist (H), Egalitarian (E) and Individualist (I). By normalizing to the technology with the highest impact within each of the three perspectives, the influence of the value choice on the ranking is reflected in the results. All three perspectives result in a clear difference in HHI of nuclear and renewables (N&R) on the one side and hard coal and lignite on the other side. In the Egalitarian perspective, impacts from natural gas are in the same range as those from N&R. With this perspective, the effect of the implementation of carbon capture and storage (CCS) is not as clearly advantageous as with the (H) and (I) perspectives.

Figure 2 Total HHI of future electricity production in Europe, normalized to the technology with highest impacts in each of the three perspectives [9]. NGCC: Natural Gas Combined Cycle; CCS: Carbon Capture and Storage, EPR: European Pressurized Reactor.

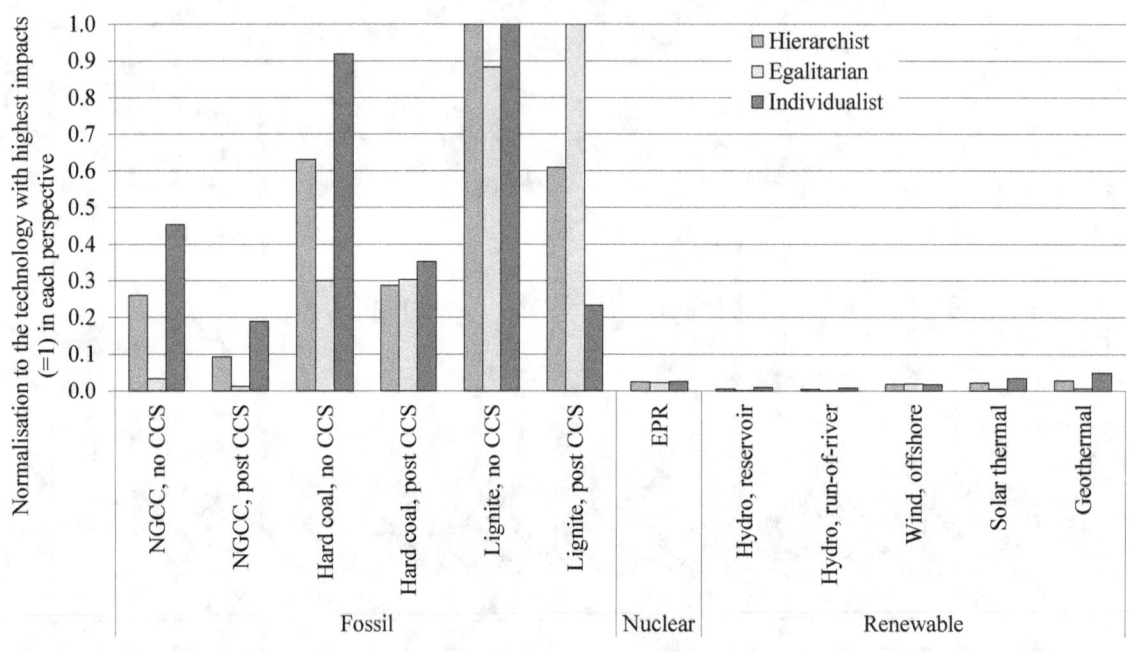

Whereas the (H) and (I) perspectives are mostly dominated by impacts due to climate change and, to a lower extent, by particulate matter formation, most of the impacts according to the (E) perspective originate from long-term effects of (ground)water pollution.

Figure 3 shows the comparison of DALYs for the same technologies and the associated energy chains as those covered in Figure 1 where IPA was used as the basis for the estimation. The most commonly used (H) perspective was employed. Table 1 summarizes the results for all three perspectives.

Figure 3 Impacts on human health from normal operation for electricity production, calculated with ReCiPe Endpoints (Hierarchist, Europe H/A) in DALY/kWh. "Other categories" include Human toxicity, ionizing radiation, photochemical oxidant formation, and particulate matter formation. All inventory data are for current European technologies and are based on [8].

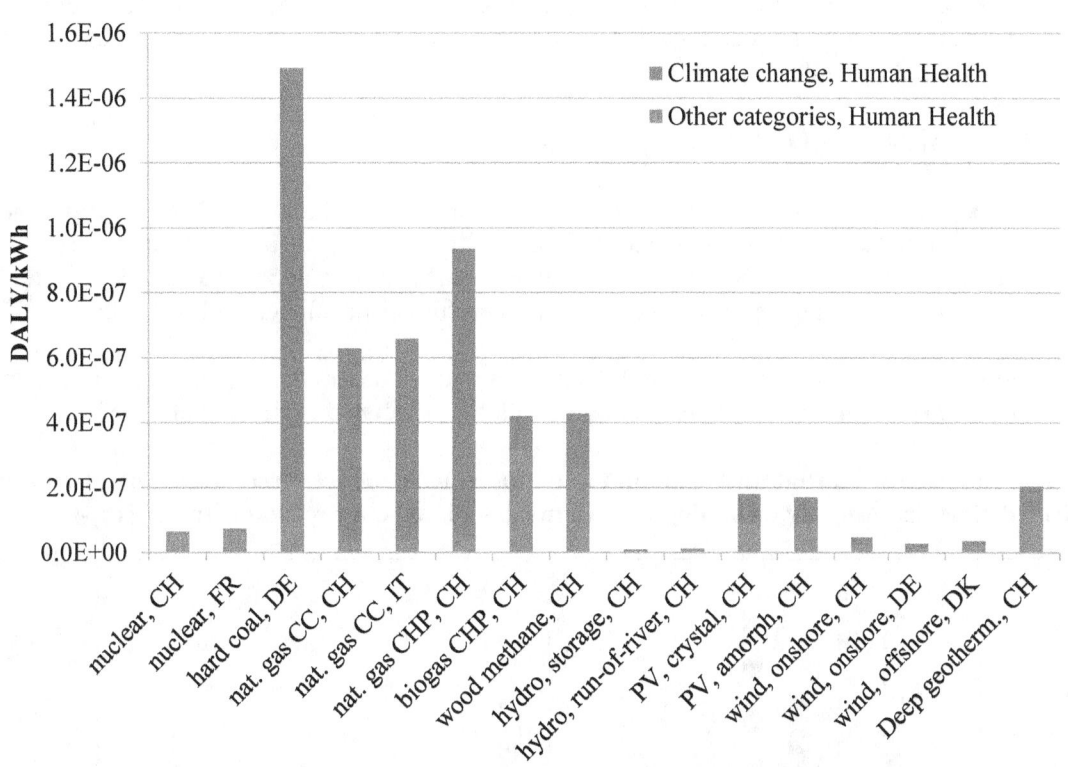

Table 1 Total human health impacts of different electricity producing technologies in nano-DALY/kWh for results calculated with ReCiPe and in nano-YOLL/kWh for results calculated with IPA. Abbreviations: ReCiPe, R (H) = Hierarchist, R (E) = Egalitarian, R (I) = Individualist; CC = Climate Change; IPA = Impact Pathway Approach; CCS = Carbon Capture and Storage; EPR = European Pressurized Reactor.

		Fossil						Nuclear	Renewables				
		Hard coal, no CCS	Hard coal, post CCS	Lig-nite, no CCS	Lig-nite, post CCS	NGCC, no CCS	NGCC, post CCS	Nuc-lear, EPR	Hydro, reser-voir	Hydro, run-of-river	Wind, off-shore	Solar thermal	Geo-thermal
Total human health, with CC	R(H)	1449	660	2300	1399	598	213	56	11	10	42	49	64
	R(E)	13819	13956	40738	46122	1527	570	1039	38	35	851	245	288
	R(I)	1121	429	1220	284	553	230	31	11	9	21	42	60
Total human health, w/o CC	R(H)	319	390	1000	1192	30	32	49	5	6	30	18	30
	R(E)	11117	13434	37497	45636	146	177	1024	23	24	824	179	208
	R(I)	60	77	99	91	27	28	25	4	5	10	14	27
IPA		59	-	-	-	27	-	5	1	2	6	-	28

It should be noted that the results of IPA and LCIA are not directly comparable. The estimation methods are much different with LCIA involving subjective elements related to the various social perspectives while IPA is based on natural sciences and allows simulation of site-specific effects. The health impact estimators have different scopes, i.e. YOLLs derived using IPA are a subset of DALYs generated using LCIA. The estimates based on LCIA cover not only health impacts of major pollutants but also the highly uncertain ones caused by the climate change; the latter are not included in IPA-estimates.

Overall, the IPA-estimates are more robust than those generated using LCIA. In spite of the differences in approaches a closer examination of the numerical results shows some consistency between the relative results for the estimates of health effects caused by pollution as obtained using the two methods. For the European technologies analyzed in this work these impacts are lowest for hydro and highest for coal, biogas and synthetic natural gas from wood, with nuclear, natural gas and other renewables in the middle range.

3. HEALTH IMPACTS OF SEVERE ACCIDENTS

The results presented here build on evaluations based on our database ENSAD (ENergy-related Severe Accident Database) and applications of simplified Level III Probabilistic Safety Assessment (PSA) for nuclear power plants. ENSAD and the framework for the comparative assessment of severe accidents was originally established in 1998 [12] and then further developed and refined ([13]-[17]).

Figure 4 shows the expected severe accident fatality rates and maximum consequences (square points) assessed for selected electricity supply technologies with the associated energy chains ([18]-[21].

Figure 4 Severe accident fatality rates and maximum consequences (red points) assessed for selected electricity supply technologies with the associated energy chains (after [18]-[21]).

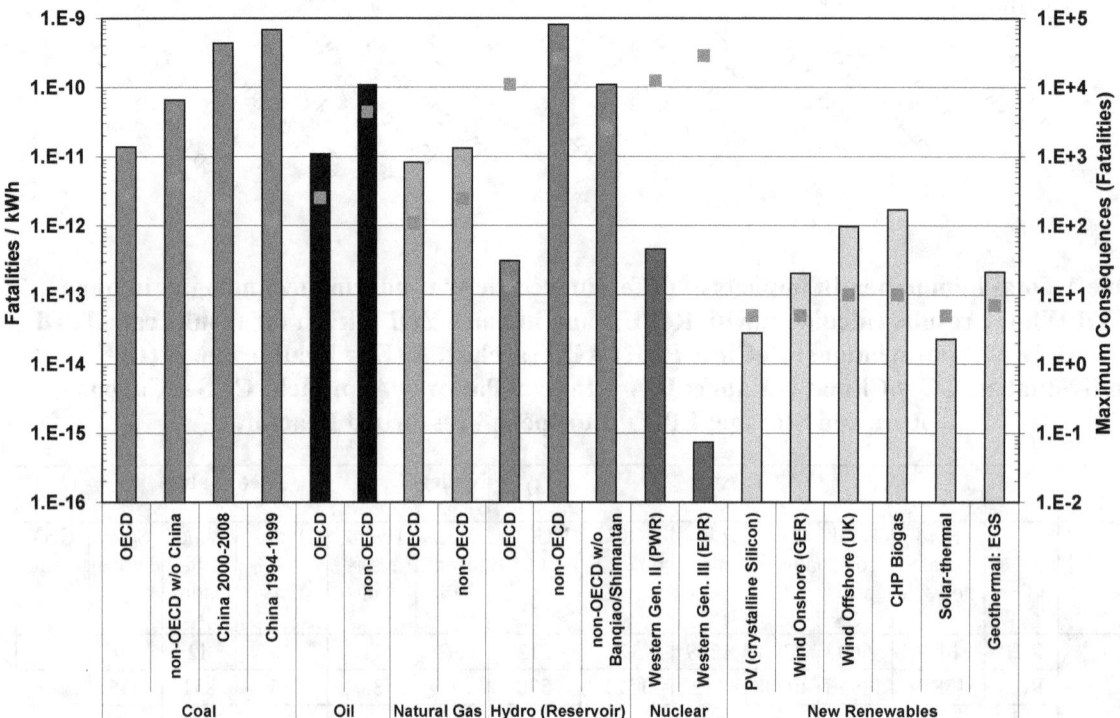

The results for fossil options are exclusively based on historical evidence according to ENSAD. The same applies to hydro with the exception of the high value of maximum consequences for OECD (red square point) corresponding to simulated consequences for a specific Swiss dam at a site characterized by relative ely high population density downstream from the dam. For nuclear energy a simplified Level III PSA was applied to a specific GEN II plant in Europe and to hypothetical GEN III plant in the same location; the maximum consequences include the dominant latent fatalities. For new renewables a combination of limited historical experience, literature data and expert judgment were used; the maximum credible consequences of accidents could in some cases (e.g. solar PV) be much higher than indicated. For further details on the methodological approach we refer to ([18]-[21]).

Non-OECD fatality rates are clearly higher than those for OECD. Fossil energy chains and non-OECD hydro have much higher fatality rates than the other options. Nuclear and hydro accidents may, however, have very large consequences. This is further illustrated by frequency-consequence curves below. The experience-based maximum consequences of accidents with new renewables are small. Further work exploring hypothetical accident scenarios for example in the manufacturing of solar cells are needed.

In order to facilitate comparison of the estimates for normal operation and severe accidents, respectively, it is worth noting that one premature fatality caused by air pollution roughly corresponds to 10 (chronic) YOLLs.

Figure 5 compares frequency-consequence curves for full energy chains in OECD countries for the period 1970 - 2008. The curves for coal, oil, natural gas and hydro are based on historical accidents and show immediate fatalities. For the nuclear chain, we extended Level II Probabilistic Safety Assessments (PSA) for a representative Generation II plant and a generation III plant (EPR) by conducting simplified Level III PSA; both immediate and latent fatalities are covered by these results. It should be noted that the Fukushima accident is not yet included since a reliable assessment of its consequences is still an open issue.

Figure 5 Comparison of frequency-consequence curves for full energy chains in OECD countries for the period 1970 - 2008 (after [20] and [21]).

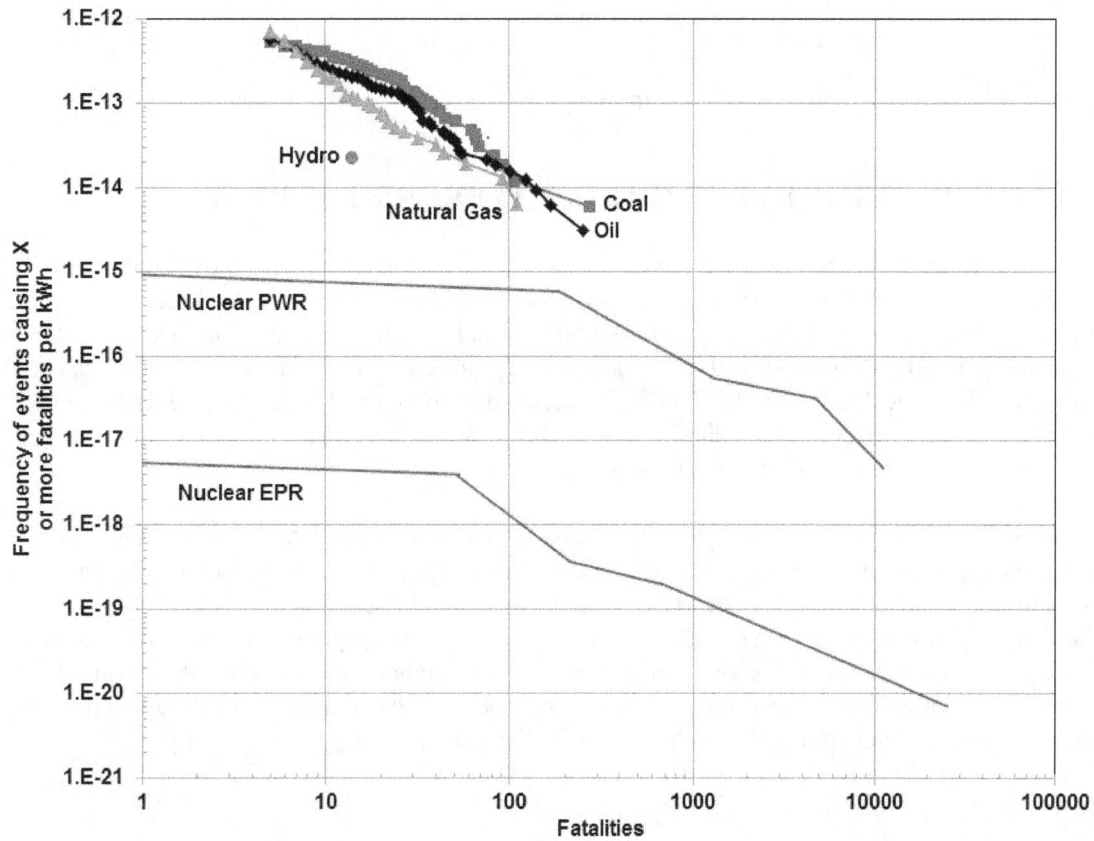

Corresponding frequency-consequence curves for non-OECD countries are shown in Figure 6. The curves for coal, oil, natural gas and hydro are based on historical accidents and show immediate fatalities. For the nuclear chain, the results represent immediate and estimated range of latent fatalities caused by the Chernobyl accident.

Figure 6 Comparison of frequency-consequence curves for full energy chains in non-OECD countries for the period 1970 - 2008 (after [20] and [21]).

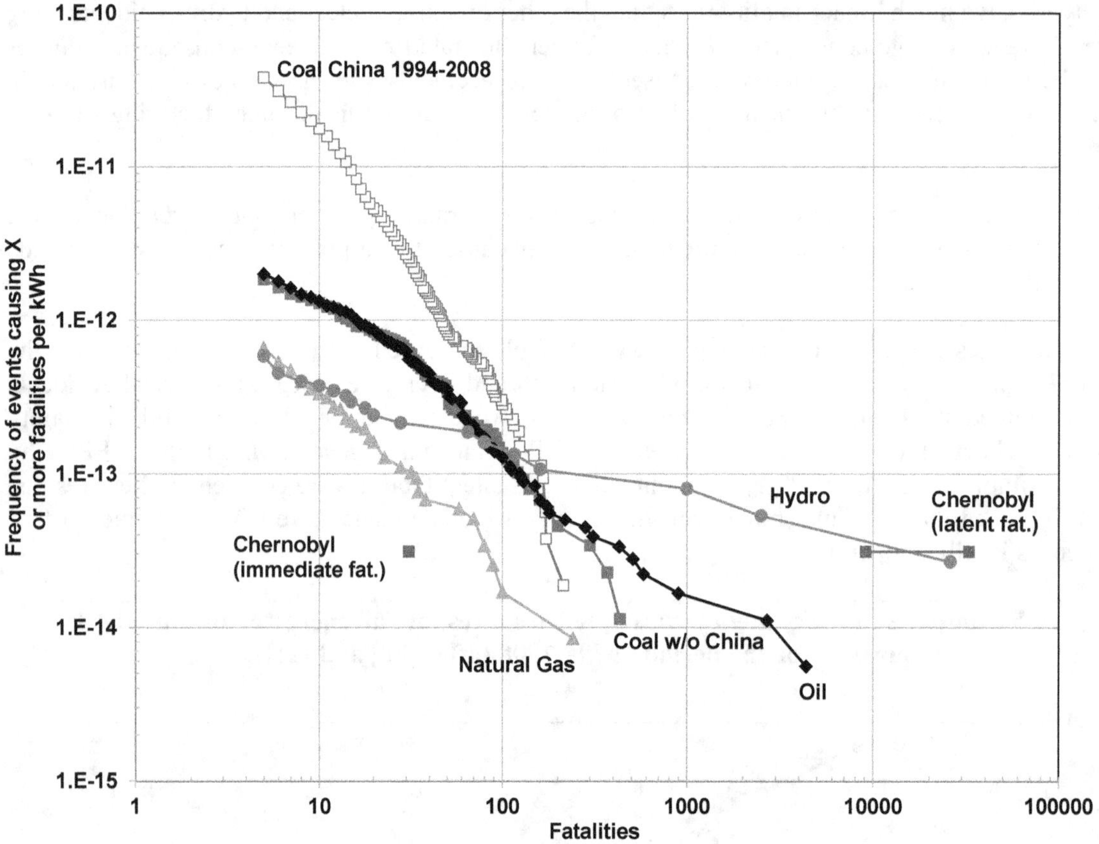

4. HEALTH IMPACTS OF HYPOTHETICAL TERRORIST THREAT

Within the EU-project SECURE [22] we developed, implemented and applied a novel methodology for the analysis of terrorism threat to energy infrastructure facilities with the potential for catastrophic consequences following a terrorist attack [23]. The targets include oil refineries, liquefied natural gas (LNG) terminals, hydropower dams and different types of nuclear power plants that rely on current (EPR) as well as future technologies (HTR – High Temperature Reactor; LMR – Liquid Metal-cooled Reactor). For each type of energy installation a specific location in China, Europe and the US was defined, where possible representing a real facility.

The developed framework allows integration of diverse expertise ranging from political sciences and intelligence on the motivation of terrorists to military knowledge on scenario planning to physical assessment of consequences. The framework also addresses the challenge of the large differences in the reliability of information in the different areas. While consequences can be modeled with relatively high confidence, the motivations of terrorists can be judged only within large error limits. The resulting large variation of uncertainty in the quantification of those aspects is addressed through a consistent treatment of uncertainty through all steps in the model.

The risk is calculated based on three factors:

- The probability that an attack is planned based on historical evidence of attractiveness of a target and evidence of terrorist activity in the considered country
- The probability that a certain scenario can be implemented based on the necessary resources, time, know-how and countermeasures in place
- The consequences in terms of fatalities, injured and land contamination.

The reasoning behind this approach is that a terrorist will, more or less formally, follow the same evaluation: consequences of an attack should be maximized, but this aim has to be weighted against the success probability, the planning effort and the financial and personnel means available.

Several different concepts were integrated into the framework: The scenario quantification is based on fault/event tree logic. The "initiator frequency" of terrorist attacks, i.e. the probability that a given target is chosen per year is treated with Bayesian frequency updating. Uncertainties in the quantification process are addressed by using fuzzy logic, i.e. uncertainty functions that are evaluated by Monte Carlo analysis. This allows the systematic and formalized integration of expert judgment with a physical analysis of the consequences and attack scenarios to generate a complete picture of the probability that an attack can be successfully executed and of the likely resulting consequences.

Figure 7 and 8 show respectively the estimated fatality risks and frequency-consequence curves for the analyzed energy infrastructure.

Figure 7 Risk of immediate and delayed fatalities due to hypothetical terrorist attacks [23].

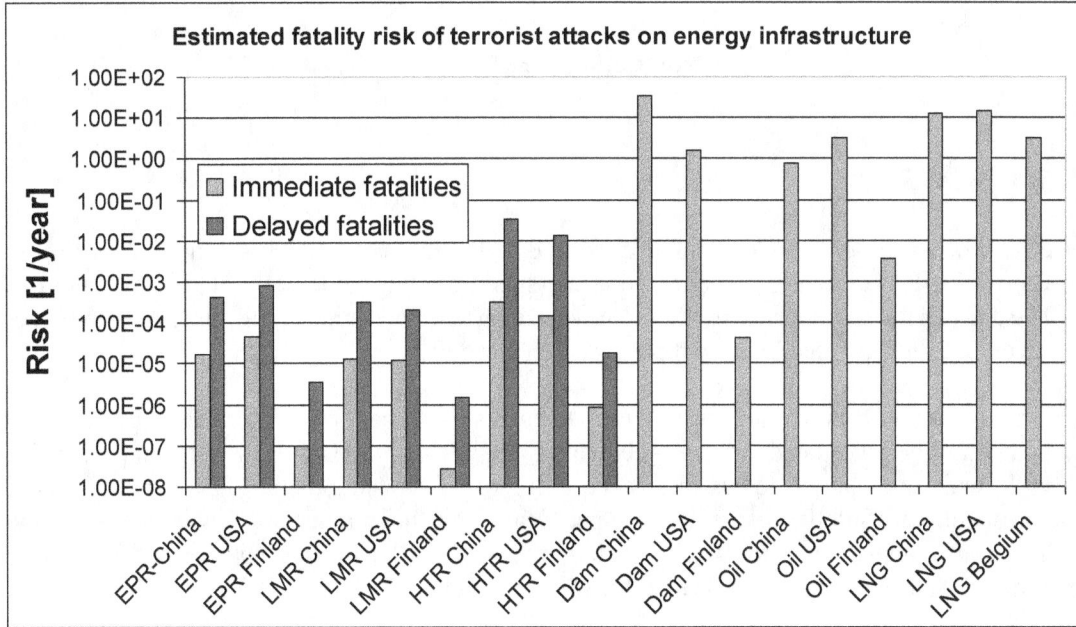

The results are strongly technology-, country- and site-specific. The risk to oil refineries and LNG terminals may be substantial though maximum consequences are much more limited than for hydro and nuclear. Countermeasures on site may reduce the impact of a terrorist attack but will not ensure the total elimination of threats. Risks from attacks to dams are potentially very large in cases with high density of population down-stream from the plant (China). It should be noted that the analyzed Chinese dam is the largest world-wide electricity generation plant. However, the chance that a catastrophic accident can be induced by a terrorist attack is much smaller than for oil and LNG installations. Finally, the chance that a terrorist attack would cause very large consequences at the examined nuclear installations is extremely small, and comparable to the corresponding hypothetical risks associated with random severe accidents at these plants.

Figure 8 Frequency-consequence curves for hypothetical terrorist attacks on energy infrastructure [23].

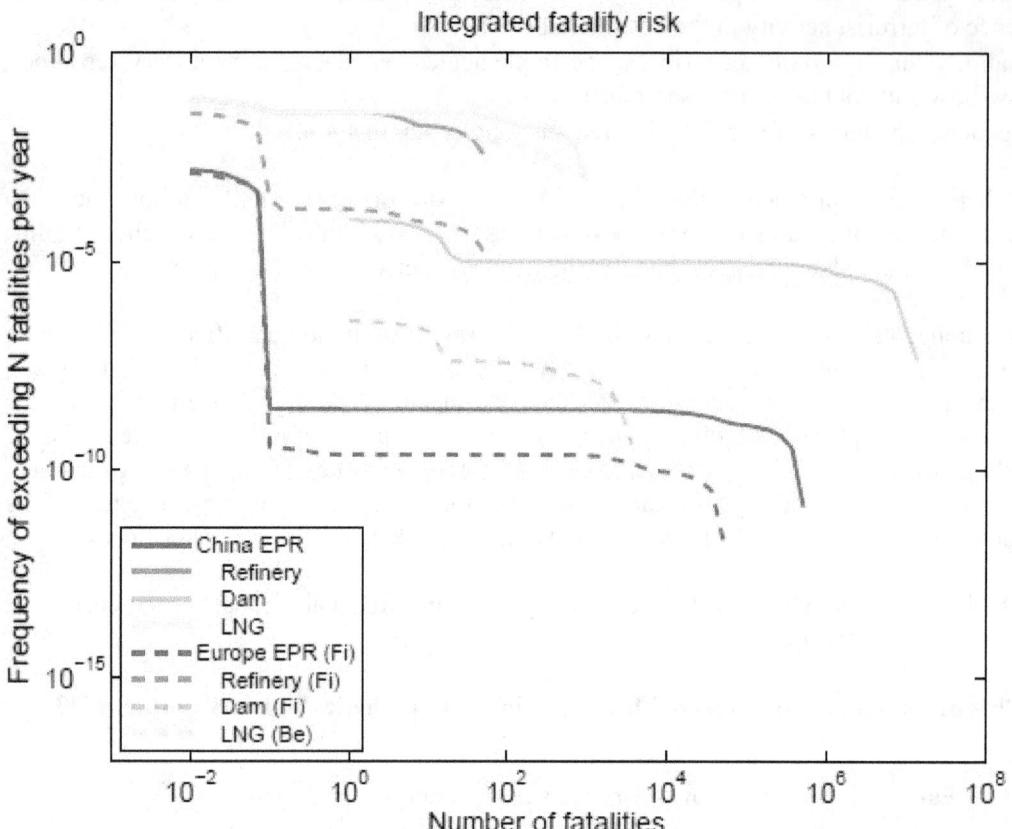

5. CONCLUSIONS

Health effects of normal operation are estimated using detailed Life Cycle Assessment (LCA), Impact Pathway Approach (IPA) allowing accounting for site-specific effects, and combination of these two methods. The LCA part is supported by the ecoinvent database, the largest and most detailed LCA database, developed and operated by us and our research partners.

Estimation of health effects caused by severe accidents is based on historical experience as represented in our ENergy-related Severe Accidents Database (ENSAD), the most comprehensive database world-wide covering accidents in the energy sector. This is supplemented by the results of full scope Probabilistic Safety Assessments. Specifically for new renewables we use a hybrid approach including statistics (e. g. for wind), modeling (e. g. for solar PV), proxies (e. g. partially for geothermal, biogas) and expert judgment (e. g. for solar thermal or wave and tide).

A novel approach to the analysis of terrorist threat against energy infrastructure was developed, implemented and applied to selected energy facilities in various locations. It considers a number of factors including: attractiveness of specific objects as targets for an attack, implementation scenarios depending on resources, time, know-how and countermeasures, and estimation of consequences.

On the basis of our work employing the above approaches and carried out during the last 20 years a number of conclusions can be drawn with regard to the various types of risks associated with the energy supply. This paper focused on health risks represented by mortality.

Our work demonstrates that both in the context of normal operation and severe accidents it is highly essential to cover the full energy chains.

The contributions to health risks other than those caused by direct emissions from the operation of power plants, may in some cases be dominant but the relative shares vary a lot between the different options. For the normal operation such contributions (e.g. burdens related to material inputs and component manufacturing), are particularly important for solar PV, solar thermal, wind and hydro options. The rest of the energy chains along with the construction of power plants can also have quite high significance for specific burdens associated with fossil and nuclear fuel chains as well as the production of biomass.

The two approaches used for the estimation of health impacts from normal operation are much different both in terms of scope and the underlying methodology. While there are large numerical differences between the technology-specific estimates obtained using the two approaches, there are also some parallels in terms of technology ranking. Thus, renewables (with the exception of biogas) and nuclear mostly exhibit very good performance with hydro being the best option; coal ranks mostly worst while performance of natural gas is mixed.

In the context of severe accidents the fuel extraction, processing and transports within the fossil energy chains and hydro in non-OECD countries are most accident prone. The lowest fatality rates apply to hydro and nuclear in OECD countries though in both cases events with very low frequency can lead to quite extreme consequences. Generally, the fatality rates due to accidents in non-OECD countries are substantially higher than in OECD-countries and exhibit a number of accidents with very large consequences not experienced within OECD.

Overall, the fatality rates due to normal operation are much higher than the corresponding rates due to severe accidents.

In spite of large uncertainties the first-of-its-kind analysis of the terrorist threat indicates that the frequency of a successful terrorist attack with very large consequences is of the same order of magnitude as can be expected for a disastrous accident in the respective energy chain. This is primarily due to the fact that centralized large energy installations are hard targets and relatively easy to protect, requiring sophisticated attack scenarios to cause significant damage and lasting impacts. Historic preference of terrorists for fatalities implies lower risk compared to soft targets, which are much more vulnerable and do not necessitate mobilization of very large resources by the terrorists.

Further work on health effects associated with energy technologies should strive for dealing with limitations of the current work thus reducing the uncertainties. There is a need to improve the consistency of the analysis by fully consequent choice of reference technologies and the associated fuel cycles when carrying out the various types of analysis presented here. The analysis scope should be extended both geographically and in terms of covering future technologies. Novel approaches to the treatment of spatial dependencies in LCIA should be considered along with accounting for health effects associated with climate change when using IPA.

In the treatment of severe accidents it is desirable to extend the use of PSA to better reflect the quite heterogeneous safety performance of nuclear and hydro by extended use of PSA. More extensive analysis is needed for renewables handling large amounts of toxic materials, in particular solar PV.

The main merit of the current exploratory study is that it provides a structured methodology for quantitative assessment of terrorist threats against energy infrastructure. Such a framework has not been available until now. The framework was implemented and applied to selected facilities in specific locations. The numerical results should be seen as indications and depend on the judgments made by risk analysts engaged in the project. Full scale implementation would call for engagement of a variety of intelligence and technology specialists to provide more robust judgments and address the credibility of the postulated scenarios.

References

[1] European Commission, *"ExternE - Externalities of energy - Methodology 2005 update"*, P. Bickel, R. Friedrich (eds.), European Commission, Brussels, ISBN 92-79-00423-9 (2005).

[2] S. Hirschberg, C. Bauer, P. Burgherr, E. Cazzoli, R. Dones, T. Heck and W. Schenler, *"Treatment of risks in sustainability assessment of energy systems"*, Invited paper, PSAM 9, 9th International Probabilistic Safety Assessment & Management Conference, 18-23 May 2008, Hong Kong, China.

[3] European Commission, *"ExternE - Externalities of energy, Volume 7: Methodology 1998 update"*, M. Holland, J. Berry, D. Forster (eds.), European Commission, Brussels. ISBN 92-828-7782-5 (1999).

[4] NEEDS: New Energy Externalities Developments for Sustainability (2009), http://www.needs-project.org/.

[5] S. Hirschberg, T. Heck, U. Gantner, Y. Lu, J.V. Spadaro, W. Krewitt, A. Trukenmüller, and Y. Zhao, *"Environmental impact and external cost assessment"*. In: Integrated Assessment of Sustainable Energy Systems in China - The China Energy Technology Program", Book Series: Alliance for Global Sustainability Series, Vol. 4, pp. 445-586, Kluwer Academic Publishers, Dordrecht/Boston/London ISBN 1-4020-1198-9 (2003).

[6] T. Heck and S. Hirschberg, *"China: economic impacts of air pollution in the country"*, in J.O. Nriagu (ed.), "Encyclopedia of Environmental Health", Burlington, Elsevier, Vol. 1, pp. 625–640 (2011).

[7] Ecoinvent database v2.2 (2010), www.ecoinvent.ch.

[8] S. Roth, S. Hirschberg, C. Bauer, P. Burgherr, R. Dones, T. Heck and W. Schenler, *"Sustainability of electricity supply technology portfolio"*, Ann. Nucl. Energy, Vol. 36, pp. 409-416 (2009).

[9] K. Treyer, C. Bauer and A. Simons, *"Human health impacts in the life cycle of future European electricity generation"*, Paper in review. Submitted to Energy Policy, special issue "Nuclear Power & Sustainable Development".

[10] K. Volkart, C. Bauer and Boulet, C., *"Life Cycle Assessment of Carbon Capture and Storage in Power Generation and Industry in Europe"*, International Journal of Greenhouse Gas Control, Vol. 16, pp. 91-106 (2013).

[11] A. Simons, and C. Bauer, *"Life cycle assessment of the EPR and the influence of different fuel cycle strategies"*, Proceedings of the Institution of Mechanical Engineers, Part A: Journal of Power and Energy, Vol. 226(3), pp. 427-444 (2012).

[12] S. Hirschberg, G. Spiekerman and R. Dones, *"Severe accidents in the energy sector"*, PSI Report No. 98-16, Villigen PSI, Switzerland: Paul Scherrer Institut (1998).

[13] S. Hirschberg, P. Burgherr, G. Spiekerman, E. Cazzoli, J. Vitazek and L. Cheng, *"Comparative assessment of severe accidents in the Chinese energy sector"*, PSI Report Nr. 03-04, March 2003, Paul Scherer Institut, Villigen, Switzerland (2003).

[14] S. Hirschberg, P. Burgherr, G. Spiekerman, E. Cazzoli, J. Vitazek and L. Cheng, *"Assessment of severe accident risks"*, In: Integrated assessment of sustainable energy systems in China. The China Energy Technology Program - A framework for decision support in the electric sector of Shandong province (eds. Eliasson, B. & Lee, Y.Y). Alliance for Global Sustanaibility Series Vol. 4, Kluwer Academic Publishers, Amsterdam, The Netherlands (2003).

[15] P. Burgherr, S. Hirschberg, A. Hunt and R.A. Ortiz, *"External costs from major accidents in non-nuclear fuel chains"*, Work Package 5 Report prepared for European Commission within Project NewExt on New Elements for the Assessment of External Costs from Energy Technologies, Paul Scherrer Institut, Villigen, Switzerland (2004).

[16] P. Burgherr, S. Hirschberg and E. Cazzoli, *"Final report on quantification of risk indicators for sustainability assessment of future electricity supply options"*, NEEDS Deliverable n° D7.1 – Research Stream 2b. NEEDS project "New Energy Externalities Developments for Sustainability", Brussels, Belgium (2008).

[17] P. Burgherr, P. Eckle, S. Hirschberg and E. Cazzoli, *"Final report on severe accident risks including key indicators"*, SECURE Deliverable n° D5.7.2a, SECURE project "Security of

Energy Considering its Uncertainty, Risk and Economic Implications", Brussels, Belgium (2011).

[18]   P. Burgherr, *"Accidents and risks"* (chapter 9.3.4.7). In: J. Sathaye, O. Lucon, A. Rahman, J. Christensen, F. Denton, J. Fujino, G. Heath, S. Kadner, M. Mirza, H. Rudnick, A. Schlaepfer, A. Shmakin, 2011: Renewable Energy in the Context of Sustainable Energy. In IPCC Special Report on Renewable Energy Sources and Climate Change Mitigation [O. Edenhofer, R. Pichs-Madruga, Y. Sokona, K. Seyboth, P. Matschoss, S. Kadner, T. Zwickel, P. Eickemeier, G. Hansen, S. Schlömer, C. von Stechow (eds)], Cambridge University Press, Cambridge, United Kingdom and New York, NY, USA (2011).

[19]   P. Burgherr, P. Eckle and S. Hirschberg, *"Comparative risk assessment of severe accidents in the energy sector based on the ENSAD database: 20 years of experience"*, In: Safety, Reliability and Risk Analysis: Beyond the Horizon (Steenbergen, R.D.J.M., van Gelder, P.H.A.J.M., Miraglia, S., Vrouwenvelder, A.C.W.M. (Eds.)). pp. 2111-2117. Taylor & Francis Group, London, UK (2014).

[20]   P. Burgherr, S. Hirschberg and M. Spada, *"Comparative assessment of accident risks in the energy sector"*, In: Handbook of risk management in energy production and trading (Eds.:Raimund M. Kovacevic, Georg Ch. Pflug, Maria Teresa Vespucci). Springer Science+Business Media New York (2013)..

[21]   P. Burgherr and S. Hirschberg, *"Comparative risk assessment of severe accidents in the energy sector"*, Energy Policy (in press), http://dx.doi.org/10.1016/j.enpol.2014.01.035 .

[22]   SECURE: Security of Energy Considering its Uncertainty, Risk and Economic Implications (2010); http://www.secure-ec.eu/.

[23]   P. Eckle, E. Cazzoli, P. Burgherr and S. Hirschberg, *"Analysis of terrorism risk for energy installations"* , Confidential Report, SECURE project "Security of Energy Considering its Uncertainty, Risk and Economic Implications", Brussels, Belgium (2011).

Metal remediation of acid mine drainage using a hybrid system of microalgae reactor

Young-Tae Park [a], Hongkyun Lee [a], Hyun-Shik Yun [a], Jaeyoung Choi [a*]

[a]Korea Institute of Science and Technology- Gangneung Institute, 679 Saimdang-ro, Gangneung 210-340, South Korea

Abstract: Acid mine drainage(AMD) contains high concentrations of heavy metals and has become a serious environmental problem. A pipes inserted microalgae reactor(PIMR) was constructed to cultivate microalgae and purify AMD. The effects of metal concentration, pH and sulfate after pretreatment on the removal of iron and microalgae growth were investigated. Batch studies showed that PIMR and microalgae can adsorb iron with an uptake of 63.21 ± 9.8 mg/L iron. Microalgae growth was measured by optical density (OD) and dry cell weight (DCW); OD and DCW were 3.96 and 1.54g/L respectively. Continuous studies also proved that PIMR can be used for metal remediation and microalgae cultivation.

Keywords: Pipes inserted microalgae reactor, Acid mine drainage, Microalgae, Metal removal

1. INTRODUCTION

Acid mine drainage (AMD) is one of the major sources of heavy metals. AMD occurs when sulfide minerals, such as pyrrhotite (FeS) and chalcocite (Cu_2S), are exposed to air and water. Pyrite (FeS_2) is one of the most common sulfide mineral that reacts in the presence of water and oxygen to yield sulfuric acid and iron [1].

Microalgae are photosynthesis microbial cell factories that convert carbon dioxide to potential biofuels, foods supplements, animal feed, and high-value bioactive materials. Microorganisms have been used for bioremediation of pollutants under in-situ and ex-situ conditions. Furthermore, microalgae are able to remove a xenobiotic by a biosorption process using living or dead nitrogen fixing biofertilizers [2-5].

This study investigated the feasibility of using AMD as a source of minerals for microalgae cultivation. We also investigated the microalgae's effectiveness in terms of heavy metal removal. For this purpose, a two- step process was developed. In the first step, AMD is neutralized by chemical materials, and iron in AMD is precipitated. However, iron colloids exist in the effluent. This may result in the effluent being a reddish color, which is sometimes considered to be an aesthetic problem [6]. In the second step, a pipes inserted microalgae reactor (PIMR) is introduced to remove iron colloids and cultivate microalgae by supplying minerals simultaneously.

---

* Corresponding author.

Tel.: +82-33-650-3701; Fax: +82-33-650-3729; E-mail address: jchoi@kist.re.kr

2. Materials and Methods

2.1. Microalga strain and growth media

The microalga selected for this study, *Nephroselmis sp.* KGE 8 was isolated from a heavy metal-rich environment in an abandoned coal mine located in South Korea. This microalga was selected because of its tolerance to heavy metals; however, its adaptability to heavy metal was not considered in this study. Bold's basal medium (BBM) was selected to incubate the microalgae [7]. The components of BBM are shown in Table 1.

Table 1. The components of BBM media for incubate microalgae

BBM media components	
KH_2PO_4	175 mg/L
$CaCl_2*H_2O$	25 mg/L
$MgSO_4*7H_2O$	75 mg/L
$NaNO_3$	250 mg/L
K_2HPO_4	75 mg/L
NaCl	25 mg/L
H_3BO_3	11.42 mg/L
Microelement stock solution	1 ml
Solution 1	1 ml
Solution 2	1 ml

Microelement stock solution components	
$ZnSO_4*7H_2O$	8.82 g/L
$MnCl_2*4H_2O$	1.44 g/L
MoO_3	0.71 g/L
$CuSO_4*5H_2O$	1.57 g/L
$Co(NO_3)_2*6H_2O$	0.49 g/L

Solution 1	
Na_2EDTA	50 g/L
KOH	3.1 g/L

Solution 2	
$FeSO_4$	4.98 g/L
H_2SO_4	1 ml/L

2.2. PIMR and culture conditions

PIMR consisted of an open tank, 2000mm × 600mm × 1100mm, constructed from transparent tempered glass, with 30 acrylic pipes inserted at regular intervals (Fig. 1). The pipes could deliver light evenly although the incubated microalgae stuck to the reactor and interfered with light distribution. The light sources were sunlight and LED sticks. The LED sticks (147 $\mu mol/m^2/s$) were inserted into the pipes. Sunlight provided light from 10:00 to 16:00. The LED sticks were used from 16:00 to 10:00. The microalgae reactor without pipes could contain 1 ton of water with one input pipe for air flow to agitate the microalgae and provide carbon dioxide.

Figure. 1 : Schematic diagram of the pipes inserted microalgae reactor (a) and the size of PIMR (b)

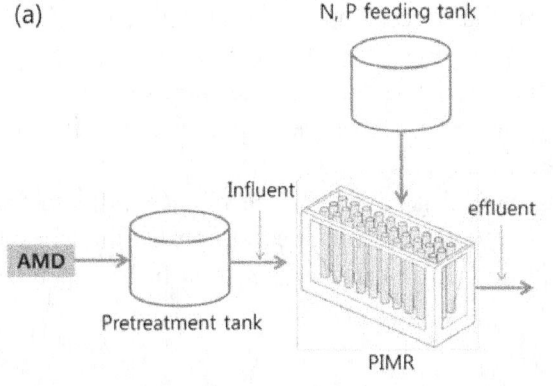
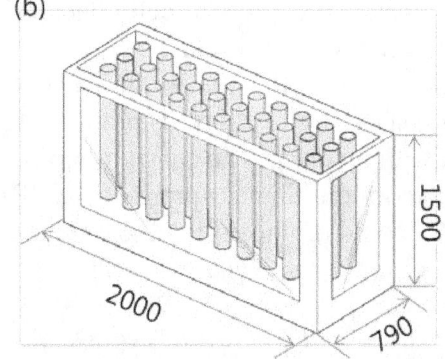

2.3. Chemical treatment for AMD pretreatment

The selected abandoned coal mine, the Yeong-Dong mine, is located 200km from Seoul, along the northeastern border of the Tae-back Mountain. AMD was contaminated with 217.8 mg/L Fe and 5.7 mg/l Mn.

Calcium hydroxide ($Ca(OH)_2$) and magnesium hydroxide ($Mg(OH)_2$) were used to pretreat AMD. Depending on amount of $Ca(OH)_2$ and $Mg(OH)_2$ required and the reaction time, effectiveness of pretreatment were determined. For pretreatment of 1000 L AMD, air was injected into the reactor at a rate of 100 L/min. When the chemical reaction was complete, the treated AMD was transferred to a settling tank and the supernatant was stored.

2.4. Microalgae growth measurement.

Optical density (OD) and dry cell weight (DCW) were used to measure microalgae growth. OD was measured at 680 nm using a spectrophotometer (HS-3300; Humas, Daejeon, Korea). Microalgae growth by DCW was determined by: biomass productivity (P), as expressed in Eq. (1).

$$P = M_b - M_{b0}/T - T_0 \qquad (5)$$

where M_b and M_{b0} are microalgae biomass at time T and starting time T_0 respectively

2.5. Analysis of various parameters.

To measure metal (Fe, and Mn) concentration, samples were collected and digested using organic matter and added sulfuric acid. The samples were analyzed by inductively coupled plasma optical emission spectroscopy (ICP-OES 730; Varian Inc. Palo Alto, CA, USA).

To investigate the effect of anions in AMD, the sample was diluted 1/10, 1/100 and analyzed for sulfate (SO_4) by ion chromatography (850 Professional IC; Metrohm, Herisau, Switzerland). The phenanthroline method was employed to quantify ferrous iron [8]. All experiments were performed in duplicate and results were expressed as the mean value.

3. Results and Discussion

3.1. Batch test of *Nephroselmis* sp. KGE 8 in PIMR

Microalgae cultivation was performed in PIMR to examine the effects of initial concentrations of cations and anions in the pretreated AMD. Four types of pretreated AMD (Table 2) were used initially as culture media at lab scale. (For PIMR equal light energy and cell concentrations were initially provided to the cultures, which were cultivated for 25 days.) Time courses of cell growth and biomass productivity relative to the type of pretreatment influent are shown in [9]. For cell concentration in the second type of pretreatment influent, OD and DCW were 3.96 and 1.54g/L, respectively. In the control tanks, where microalgae were incubated in BBM media, the respective measurements were lower, i.e., OD was 3.54 and DCW was 1.38g/L. Biomass productivity (P) for the second type of pretreatment influent increased with increasing Fe concentration until 8.52 ±2.4.

The maximum biomass productivity (P) was 0.0602 g/L/day. These results indicate that Fe in AMD could promote biomass productivity. However, for the second and fourth types of pretreatment influents, higher Fe concentrations could inhibit microalgae growth.

Table 2. The change of AMD characteristics after pretreatment by Calcium hydroxide (Ca(OH)₂) and Magnesium hydroxide (Mg(OH)₂) [9]

	Initial AMD	1st pretreatment	2nd pretreatment	3rd pretreatment	4th pretreatment
Fe	237.8 ±12.5 mg L^{-1}	4.64 ± 0.5 mg L^{-1}	8.52 ± 2.4 mg L^{-1}	24.21 ± 2.7 mg L^{-1}	20.51 ± 9.8 mg L^{-1}
Fe^{2+}	187.3 ± 8.6 mg L^{-1}	0.5 ± 0.1 mg L^{-1}	1.2± 0.2 mg L^{-1}	4.1 ± 0.9 mg L^{-1}	3.2 ± 1.4 mg L^{-1}
Mn	5.7 ± 1.5 mg L^{-1}	3.9 ± 0.7 mg L^{-1}	4.8 ± 0.5 mg L^{-1}	5.3 ± 0.6 mg L^{-1}	5.4 ± 0.5 mg L^{-1}
SO₄	320.4 ± 24.3 mg L^{-1}	214.6 ± 19.6 mg L^{-1}	252.9 ± 16.2 mg L^{-1}	294.2 ± 17.1 mg L^{-1}	311.7 ± 14.3 mg L^{-1}
NO₃	< 0.1 mg L^{-1}	< 0.1 mg L^{-1}	< 0.1 mg L^{-1}	< 0.1 mg L^{-1}	< 0.1 mg L^{-1}
PO₄	< 0.1 mg L^{-1}	< 0.1 mg L^{-1}	< 0.1 mg L^{-1}	< 0.1 mg L^{-1}	< 0.1 mg L^{-1}
T-N	< 0.1 mg L^{-1}	< 0.1 mg L^{-1}	< 0.1 mg L^{-1}	< 0.1 mg L^{-1}	< 0.1 mg L^{-1}
T-P	< 0.1 mg L^{-1}	< 0.1 mg L^{-1}	< 0.1 mg L^{-1}	< 0.1 mg L^{-1}	< 0.1 mg L^{-1}
pH	3.7	7.1	6.2	6.0	5.7

3.2. The removal of Fe in PIMR

Figure 2a, 2b, 2c shows that variation in total Fe, Fe (II), and SO₄ concentration in effluent. Fe could be removed through various processes that included (i) absorption of microalgae for cell growth [10], (ii) biosorption of metal ions on microalgae [11], and (iii) precipitation of metals inside the biological reactor [12]. A direct relationship was observed between the initial Fe concentration and the amount of Fe taken up; the higher the initial Fe concentration, the larger the amount of Fe taken up. The maximum Fe loading capacity of Nephroselmis sp. KGE 8 was found to be 59.92 mg/g for the fourth type of pretreatment wastewater (Fig. 2a). However, the pH in PIMR ranged from 5.4–7.1 (Fig. 2d) and precipitation could occur [13]. Therefore, it is significant that the PIMR system can remove both the insoluble Fe that remains in suspension and Fe precipitates in a sludge system.

The ferrous iron was removed immediately until 4.1±0.9 mg/L in the third pretreatment influent because ferrous iron oxidation occurred because of air input to PIMR. For the fourth pretreatment influent, ferrous iron was removed at 5 h reaction time in PIMR. Although the sulfate concentration varied during the reaction time, the Fig. 2c indicates that except for the first type of influent, sulfate concentration did not change appreciably, and 20% of the sulfate was removed when the first type of influent was used in PIMR. Some previous studies have reported the importance of sulfate reduction processes to AMD remediation [14]. In particular, it has been reported that dissimilatory sulfate reduction is an important mechanism in AMD purification [15].

Figure 2. Variation of influent (t=0) and sample in PIMR. (a) Concentration of total iron, (b) Concentration of ferrous (Fe^{2+}). (c) Concentration of Sulfate, (d) pH

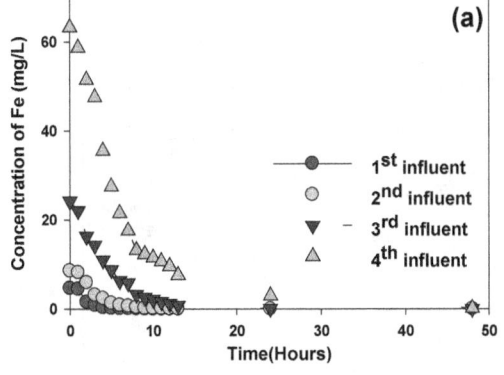
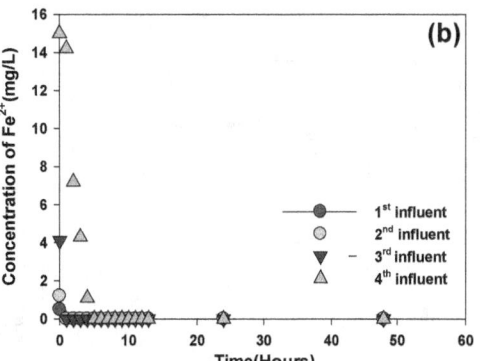

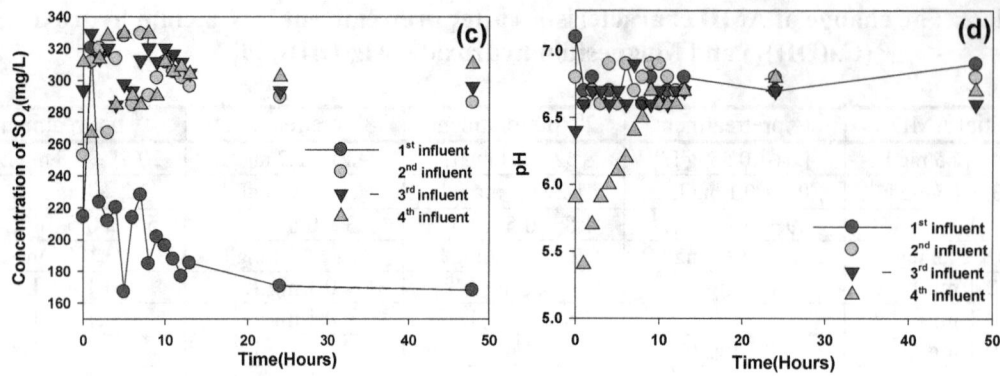

3.3. Continuous operation of PIMR

On the basis of preliminary batch results for microalgae growth and efficient removal of metals (Fig.2), we employed the second type of pretreatment effluent as the influent to supply PIMR. Continuous Fe removal and microalgae cultivation were investigated by supplying the pretreatment influent and medium with 50 mg/L nitrate and phosphate based on the composition of BBM media. The flow rate for the second type of pretreatment influent used in PIMR was 47L/h. This influent was supplied with 50 mg/L nitrate and 50 mg/L phosphate from the supply tank. Retention time was 21.27hours.

3.4. Variation in pH and removal of mineral.

The results of the pH profile are graphically illustrated in Fig. 3a. The results indicate that the pH in PIMR was maintained from 6.4 to 6.8 over the lasting 120 hours. This pH condition could be optimal for microalgae growth and Fe removal.

Figure 3b and 3c shows the influent and effluent concentrations for both total Fe and ferrous iron. Fe concentrations in the influent were 8.21 - 8.64 mg/L and in the effluent were < 0.1 mg/L. Iron oxide particles were observed in the influent but not in the effluent.

There was 1.4 - 1.0 mg/L of ferrous iron in the influent. Ferrous iron was removed from the effluent. Ca, Mg, and Mn concentrations were 20.02 - 20.48 mg/L, 39.86 - 40.47 mg/L, and 3.02 - 3.38 mg/L, respectively (Fig. 3d). Although Ca, Mg, and Mn were taken up by the microalgae, their concentrations did not change because Ca and Mg were used for pretreatment and Mn occurs in AMD and accumulated in PIMR. Sulfate concentration was 267.13 - 329.42 mg/L and unlike the results for the batch test, did not change significantly.

Figure 3. pH profile (a) on effluent and removal of iron (b), ferrous (c) and calcium (Ca), magnesium (Mg), manganese (Mn) and sulfate (d) and variation of optical density (OD) and dry cell weight (DCW) (e) during continuous PIMR operating in 127 hours

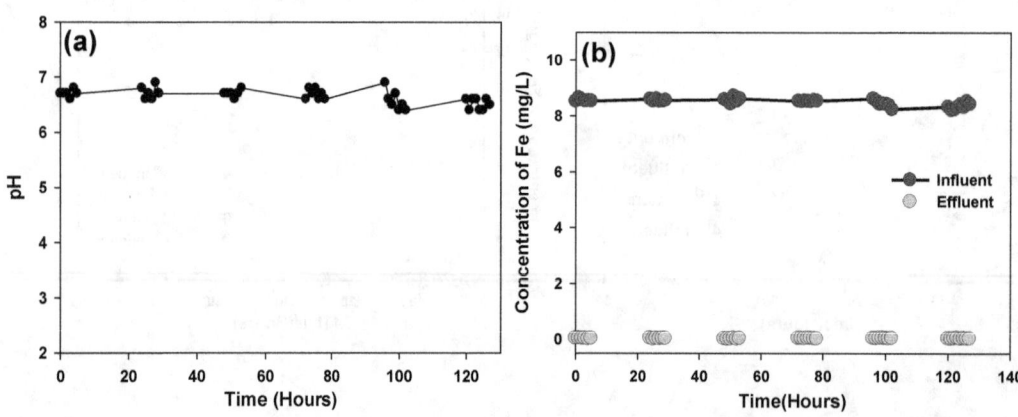

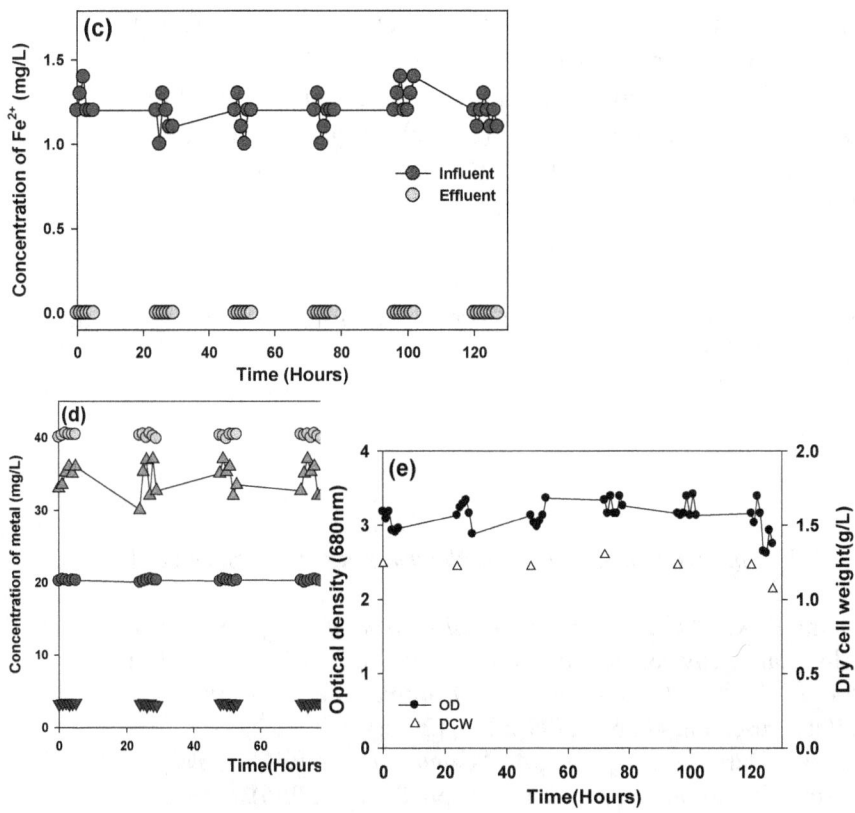

3.5. Cell growth in continuous cultivation.

The time courses of biomass concentration for continuous cultivation are shown in Fig. 3e. In the batch test, OD and DCW for *Nephroselmis* sp. KGE 8 cultivation were 3.96 and 1.54g/L respectively. In the continuous cultivation mode, microalgae growth was maintained at OD 2.64 - 3.41 and DCW 1.07 - 1.30g/L for 25 days. OD decreased because few microalgae can persist in the effluent. However, the cells can be maintained during continuous growth. This study demonstrated that compared with the batch test, the continuous growth mode can achieve proximity harvest cell.

4. Conclusions

A PIMR containing pretreatment system including Ca and Mg, was developed and employed for microalgae-mediated heavy metal remediation. It reduced the initial high Fe and Mn concentrations released from AMD and supplied PIMR. The hybrid system was combined with a pretreatment system, and PIMR enhanced heavy metal reduction in AMD. It was economical in improving bioremediation and enhancing microalgae production.

A PIMR system was developed and operated for microalgae cultivation. The pipes in PIMR allowed effective light penetration and distribution. Moreover, PIMR could be used efficiently for Fe removal from AMD and microalgae cultivation. Batch studies showed that PIMR and microalgae can adsorb Fe with an uptake of 63.21 ±9.8 mg/L. Continuous studies also proved that PIMR can be used for metal remediation and microalgae cultivation.

Acknowledgements

The project was financially supported by the Korea Institute of Science and Technology(Gangneung) (Grant. 2Z03860).

References

[1]   Singer, P.C., Stumm, W.. " *Acidic mine drainage: the rate-determining step*", Science, 167, pp. 1121- 1123, (1970).

[2]   Kalin, M., Wheeler, W., Meinrath, G.. "*The removal of uranium from mining waste water using algal/microbial biomass*", Journal of environmental radioactivity, 78, pp 151-177, (2004).

[3]   Munoz, R., Guieysse, B.. "*Algal–bacterial processes for the treatment of hazardous contaminants: a review*", Water research, 40, pp. 2799-2815, (2006).

[4]   Spolaore, P., Joannis-Cassan, C., Duran, E., Isambert, A.. "*Commercial applications of microalgae*", Journal of bioscience and bioengineering, 101, pp. 87-96, (2006).

[5]   Walker, T.L., Purton, S., Becker, D.K., Collet, C.. "*Microalgae as bioreactors*", Plant cell reports, 24, pp. 629-641, (2005).

[6]   Kimball, B.A., Callender, E., Axtmann, E.V.. "*Effects of colloids on metal transport in a river receiving acid mine drainage, upper Arkansas River, Colorado, USA*", Applied Geochemistry, 10, pp. 285-306, (1995).

[7]   Fabregas, J., Dominguez, A., Regueiro, M., Maseda, A., Otero, A.. " *Optimization of culture medium for the continuous cultivation of the microalga Haematococcus pluvialis*", Applied Microbiology and Biotechnology, 5, pp.530-535, (2000).

[8]   Muir, M.K., Andersen, T.N.. "*Determination of ferrous iron in copper-process metallurgical solutions by the o-phenanthroline colorimetric method*", Metallurgical and Materials Transactions B, 8 , pp. 517-518, (1977).

[9]   Young-Tae Park, Hongkyun Lee, Hyun-Shil Yun, Kyung-Guen Song, Sung-Ho Yeom, Jaeyoung Choi. " *Removal of metal from acid mine drainage using a hybrid system including a pipes inserted microalgae reactor*", Bioresource Technology, 150, pp.242-248 (2013).

[10]   Liu, Z.Y., Wang, G.C., Zhou, B.C.. "*Effect of iron on growth and lipid accumulation in Chlorella vulgaris*", Bioresource technology, 99, pp.4717-4722, (2008).

[11]   Çetinkaya Dönmez, G., Aksu, Z., Öztürk, A., Kutsal, T.. "*A comparative study on heavy metal biosorption characteristics of some algae*", Process Biochemistry, 34, pp.885-892, (1999).

[12]   Katsou, E., Malamis, S., Loizidou, M.. "*Performance of a membrane bioreactor used for the treatment of wastewater contaminated with heavy metals*", Bioresource technology,102, pp. 4325-4332, (2011).

[13]   Spiteri, C., Regnier, P., Slomp, C.P., Charette, M.A.. "*pH-Dependent iron oxide precipitation in a subterranean estuary*" Journal of geochemical exploration, 88, pp. 399-403, (2006).

[14]   Hedin, R., Hammack, R., Hyman, D.. "*Constructed Wetlands for Wastewater Treatment: Municipal, Industrial and Agricultural*", Lewis Publishers, 1989, Chelsea Michigan.

[15]   Vile, M.A., Wieder, R.K.. "*Alkalinity generation by Fe (III) reduction versus sulfate reduction in wetlands constructed for acid mine drainage treatment*", Water, Air, and Soil Pollution, 69, pp. 425-441, (1993).

Experiences from Developing and Implementing Shutdown Fire PRA at Forsmark NPP

Erik Cederhorn[a*], Maria Frisk[a]
[a*] Risk Pilot AB, Stockholm, Sweden
[b] Risk Pilot AB, Stockholm, Sweden

Abstract: The cold shutdown mode has earlier been considered as a safe mode without a significant risk for a major accident. However during the last few decades knowledge has improved regarding risks during shutdown mode. Many activities are on-going during this period and the risk of fire occurrence may be affected. Due to an increased number of plant activities the integrity of the fire compartments may not be intact and this could lead to more extensive fire spreading. At the same time important barriers may be unavailable due to maintenance and a fire event could become critical. Time available for recoveries before fuel is exposed in the reactor pressure vessel after an initiating event i.e. fire event, which results in loss of residual heat removal, is in many cases significantly longer than 24 hours.

Area event analyses for shutdown mode generally tend to produce quite conservative results, which is why efforts have been made to increase realism in the analyses by using of improved methods. In order to increase realism dependencies between plant risk and maintenance activities, i.e. different combinations of safety system alignments, during the shutdown period have been studied in detail. This has had an impact on the estimation of both fire ignition frequencies and probabilities for fire spreading between different compartments.

This paper will discuss the methodology applied to the fire PRA at Forsmark NPP during the cold shutdown period, with focus on fire frequency analysis and fire scenario analysis. The implementation of fire analysis in the PRA and lessons learned from this will also be addressed.

Keywords: PRA, PSA, shutdown mode, Fire PRA, Fire PSA

1. INTRODUCTION

This paper will discuss the methodology applied to the fire Probabilistic Risk Assessment (PRA) at Forsmark NPP during the cold shutdown period, with focus on fire frequency analysis and fire scenario analysis. The paper will also address the implementation of the fire analysis in the PRA and lessons learned from this.

The cold shutdown period has earlier been considered as a safe mode without a significant risk for a major accident. However during the last few decades knowledge has improved regarding risks during shutdown mode. Many plant activities are ongoing during this period and the risk of fire occurrence may be affected. Due to the increased number of plant activities the integrity of fire compartments may not be intact and this could lead to a more extensive fire spreading. At the same time important barriers may be unavailable due to maintenance and a fire event could become critical.

Fire event analyses for the shutdown period generally tend to produce quite conservative results, which is why efforts have been made to increase realism in the analyses by using improved methods. In order to increase realism in the analyses dependencies between plant risk and maintenance activities, i.e. different combinations of safety system alignments, during the shutdown period has been studied in detail. This has had an impact on both fire ignition frequencies and probabilities for fire spreading between different compartments.

The conditions during the shutdown period at Forsmark NPP must be known in order to enable the analysis. In order to perform the analysis it is relevant to know for example conditions during the different phases of the cold shutdown and also what initiating event that should be analysed.

1.1 Initiating Event
The definition of an initiating event, at Forsmark NPP, is a disturbance in the nuclear power plant that requires one or more automatic or human initiated actions to bring the nuclear plant to a "safe" and "stable" mode. Loss of manual or automatic action can cause risk of a continuing process that may lead to release of radioactive materials to the environment.

During the cold shut down mode the initiating event is defined as an event that causes loss of residual heat removal. The different ways in which residual heat removal is maintained during the different phases is described in the section below.

1.2 The phases of during cold shut down mode
At Forsmark NPP the analysis has been divided into six phases. The systems operating for residual heat removal differs depending on phase and this affects how a fire event should be analysed. In phase 1-3 the residual heat removal in the reactor pressure vessel, RPV, is performed by residual heat removal system (321).

These are the different phases during the cold shutdown mode:
- In phase 1 the reactor lid is mounted but the filling of the RPV has not started.
- In phase 2 the reactor lid is still mounted and the process for filling the RPV has started. At the end of this phase the reactor lid has been dismounted.
- In phase 3 the reactor lid is dismounted and the filling process of the pool above the RPV has begun.
- In phase 4 the pools in the reactor service room are filled with water. In this phase it is assumed that the majority of all maintenance is ongoing. During this phase the residual heat removal is performed by the residual heat removal system (321) and the fuel pit cooling and cleaning system (324), combined.
- Phase 5 see phase 2, the only difference is that the draining process of the RPV has begun.
- Phase 6 see phase 1.

The safety systems are divided into four independent trains A-D. During cold shut down mode it is assumed that two trains is unavailable for maintenance. In Forsmark 1 and 2 it is only possible for the combination of A and C or B and D to be unavailable at the same time. In Forsmark 3 all combinations of two trains can be unavailable at the same time.

Table 1   Phase classification during cold shutdown

	Reactor lid	Water Level in RPV / Reactor pool	Operational system for residual heat removal system
Phase 1	Mounted	Normal	The residual heat removal system (321) is cooling RPV
Phase 2	Mounted	Top filling/water level above steamlines.	The residual heat removal system (321) is cooling RPV
Phase 3	Dismounted	Reactor pool is empty	The residual heat removal system (321) is cooling RPV
Phase 4	Dismounted	Reactor hall pools are met	The residual heat removal system (321) and the fuel pit cooling and cleaning system (324) is cooling RPV and reactor halls pool together.
Phase 5	Dismounted	Reactor pool is empty	The residual heat removal system (321) is cooling RPV
Phase 6	Mounted	Normal	The residual heat removal system (321) is cooling RPV

The environment in the containment is assumed not to be inert during the whole shut down period and fires inside the containment can therefore occur.

3. METHOD
This chapter aims to give a short introduction and background to the method chosen for the analysis at Forsmark NPP. The standard method for the cold shutdown period at Forsmark is retrieved from reference [1]. The method used in this analysis is partly based on reference [2]. But due to limited resources the method could not be used completely.

3.1 Identify critical equipment
Critical components are those components that directly or indirectly are included in the safety related systems and also are included in the PRA model. When the critical components have been identified the failure mode in case of fire needs to be determined. Only active equipment which need power supply is assumed to be affected by the fire. Passive objects such as a heat exchanger can be assumed to be unaffected by a fire.

3.2 Mapping of the electrical system
Mapping of the electrical system is a very important step in the fire analysis. The electrical systems are built with circuits that form networks and branches with cables, loads, cabinets and breaking points. The electrical networks are widespread throughout the building and therefore sensitive to area events such as fire.

The power supply to various consumers is hierarchical. From an overhead power input the electrical system branches out to supply voltage to different loads over switches, fuses and cables.The cable routes can be identified when the electrical system for all objects included in the PRA model is completely mapped. Cable routs can be described by listing all the rooms which cables go through. Cable paths should be mapped for each fuse and breaker and room dependencies are set. It is important that all of the cables, as fire can causes blown fuses, can be identified. In general cables are placed in open cable trays, but cables can also be placed in fireproof boxes. However this is not taken into account in the fire analysis. The objects need different kind of power supply like for example control voltage and power supply. All these voltages should be mapped. Hot shorts are not required to be mapped in this analysis and this could be non-conservative.

3.4 Identify fire events
Fire can be assumed to occur in all rooms in the power plant. Rooms containing objects included in the PRA model are analysed. Fire spreading from other rooms into these rooms must be taken into account.

3.3 Screening
A screening of fires in fire compartments or fire cells which implies an initiating event, the screening criterion, i.e. fires that causes loss of residual heat removal, should be done. A fire should be assumed to destroy all the electrical system in the fire compartment or fire cell. This approach can be sensitive to errors in electrical mapping which is fundamental to the screening process. This approach does not consider if a fire event causes a degraded barrier, only events that lead to an initiating event will be taken into account.

3.4 Data analysis

Frequency for fire occurrence should be calculated for each room included in the fire compartments or fire cells that have been screened out. The frequency depends on the type of room and is calculated from fire statistics from the Swedish and Finnish NPPs countries. The statistics are based on fires during shut down periods. In the statistics there is information about whether the fire occurred due to on-going work in the room. The risk for fire during on-going work could be much higher, in order of twice the risk.

For example assume that power supply train A is unavailable because of maintenance. This means that the all rooms that contain at least some equipment for train A the room gets a higher fire frequency, this is relevant in the PSA-model because the room could also contain equipment from train B. In rooms that only contain equipment from train B, which is not shut down for maintenance, it is assumed that no work is in progress and the lower frequency should be applied.

If the fire compartment or fire cell contains equipment from the train which is unavailable for maintenance the frequency for on-going work should be applied. So depending on what train combination that are unavailable for maintenance the frequency varies. The frequency should be allocated to the different phases in proportion to its length.

3.4.1 Fire extinguishing
Manual and automatic extinguishing is not modeled explicitly in the PRA model. However successful firefighting efforts could be considered when calculating fire occurrence frequencies.

3.4.2 Fire spreading
It is possible for the fire to spread inside the fire compartment or fire cell. A probability for the fire to spread to the next room can be applied.

3.4.3 Impact of manual operation
Fire could have an impact on manual operations, Post-incident actions (Category C), it could for example affect the place where local maneuvers are performed or the information at the main control room could be affected. Therefore the probability for failure of manual operations should be re-estimated during an on-going fire.

3.5 Detailed analysis
If needed a more detailed analysis can be done, focusing on analyzing the compartments that give high core damage frequency.

4. ANALYSIS
In this chapter the fire analysis performed at Forsmark NNP is described. An extensive work of mapping of the electrical system, according to the described method, has been performed and applied in the full power PRA model. Figure 1, below, illustrates the entire analysis process. This chapter will explain and describe the different parts of the process.

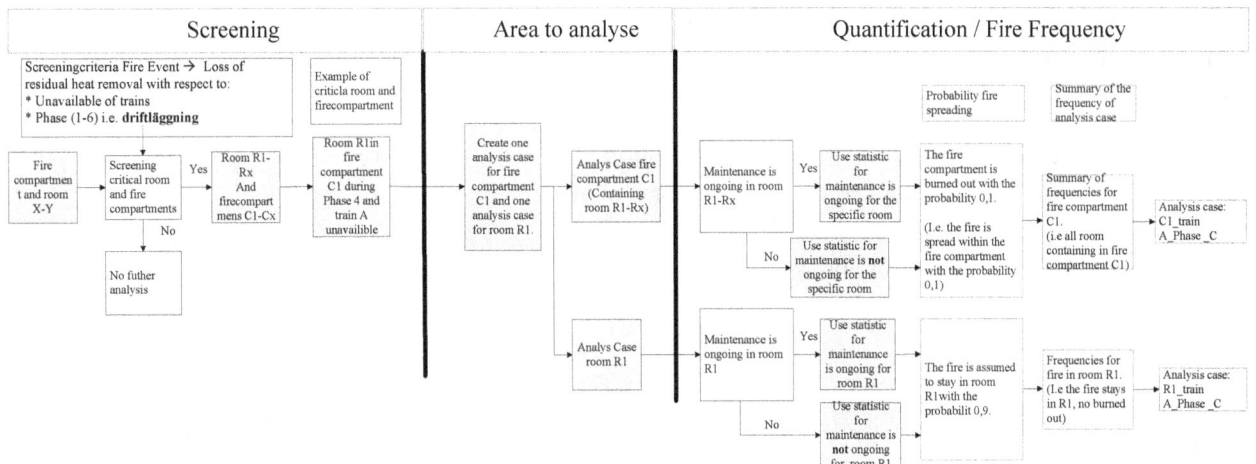

Figure 1: Fire analysis process

4.1 Screening process

A screening was done for fire in fire compartment and fire in room that leads to initiating event, i.e. loss off residual heat removal, see figure 2. To be considered a critical compartment a fire need to cause loss of residual heat removal if the whole compartment is burnt out and to be a critical room fire needs to cause loss of residual heat removal if the whole room is burnt out. One critical compartment could consist of several critical rooms.

Since there are different conditions, depending on phase during the cold shutdown and what train combination is unavailable, the screening was done for each train combination unavailable during every phase. When a list of fire compartments and rooms needed to be analysed further was done the analysis cases was created and described in the next section.

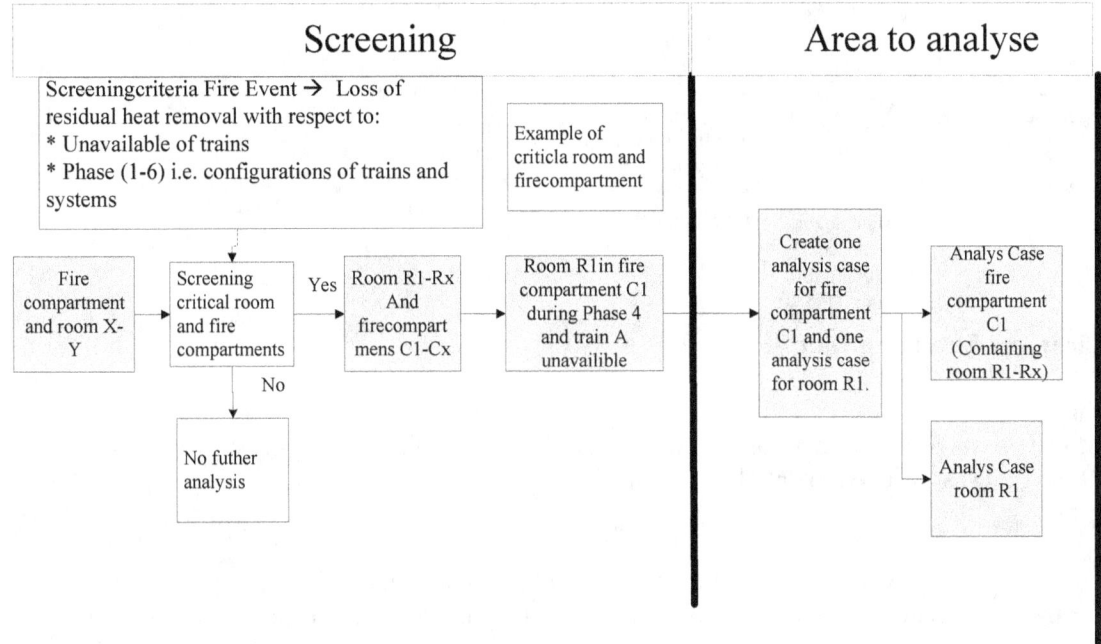

Figure 2: Screening process

4.1 Analysis cases and frequency for fire occurrence

For each critical room or compartment analysis cases was created for the critical conditions i.e. phase and unavailable train combinations.

The frequency for fire occurrence in each room or fire compartment is calculated depending on several parameters described in the method. The process is described in figure 3.

- Type of room
- On-going work/no work on-going
- Length of the phase

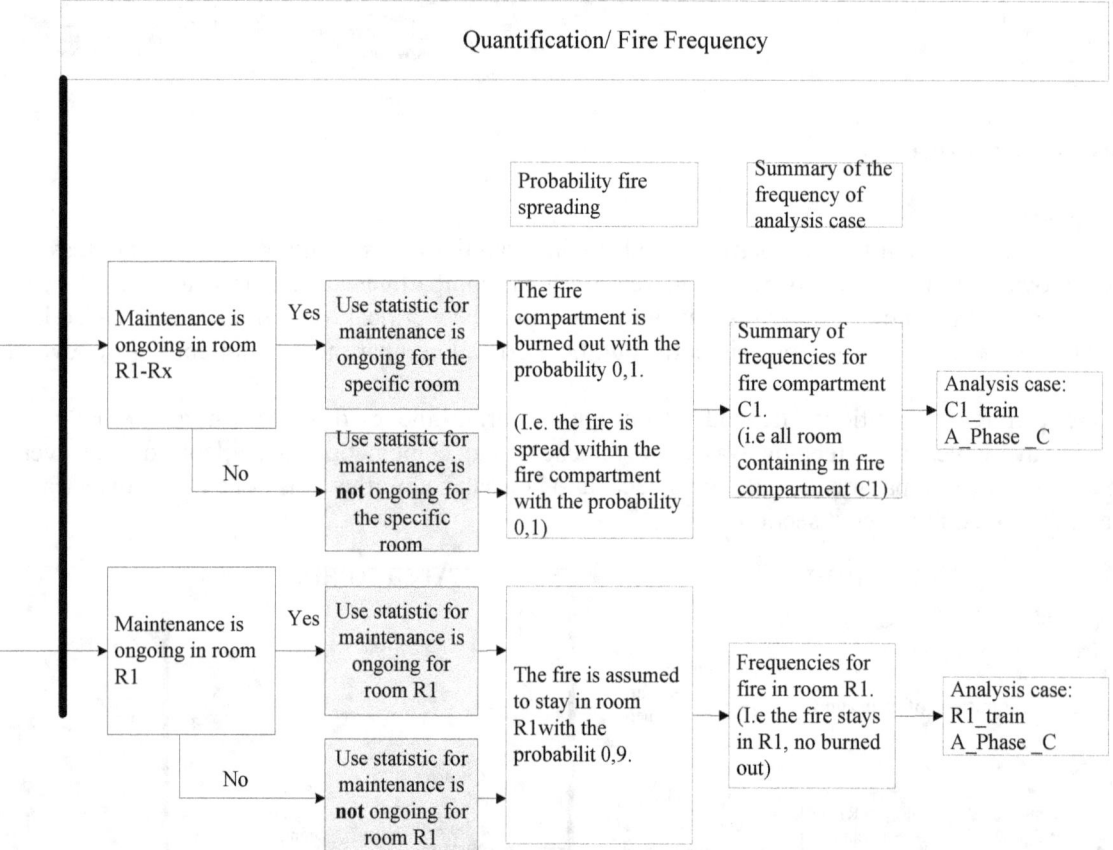

Figure 3: Quantification / Fire Frequency

4.1.1 Fire spreading

Fire in fire compartment was assumed to burn out the whole compartment with a probability of 0.1 and with a probability of 0.9 the fire was assumed to only burn out the room.

4.2 Walk downs

Walk downs have been performed with purpose to find deficiencies in the mapping of fire compartments. The main reason for the deficiencies in the mapping of compartment is that the layout in some cases does not correspond to reality.

5. RESULTS

In order to increase realism, dependencies between plant risk and maintenance activities, i.e. different combinations of safety system alignments, during the shutdown mode have been studied in detail. This has had an impact on both fire occurrence frequencies.

Fire during the cold shutdown period leading to loss off residual heat removal gives a core damage frequency in the magnitude of 1E-6 and is 4,5% of the total core damage frequency for Forsmark 3. Fire during power operation gives a core damage frequency in the magnitude of 1E-7 and is 3% of the total core damage frequency for Forsmark 3. Fire is not a dominating initiating event. But the risk of getting a critical fire is greater during cold shutdown mode than during power operation.

The risk for fire during cold shutdown period is much higher that during power operation. The reason for that is the risk increase because of on-going maintenance.

6. CONCLUSION

The results of our analysis indicate that the most critical phases with respect to a fire event are the phases when upper head (reactor vessel lid) still is mounted. If a fire event occurs during any of these phases it leads to loss of residual heat removal and the time of recovery is quite short.

Due to an increased number of plant activities during the cold shutdown period the integrity of fire compartments may not be intact and this could lead to an even more extensive fire spreading. At the same time important barriers may be unavailable due to maintenance and a fire event could become critical. The risk for a fire to occur and to be critical is more probable during cold shutdown than during power operation. Therefore it is very important to analyse this and implement this in the PRA studies. Available time for recoveries before fuel is uncovered in the reactor pressure vessel after an initiating event, i.e. fire event that result in loss of residual heat removal, is in many cases significantly longer than 24 hours. The reason for this is that during phase 4 when most of the maintenance is ongoing all pools are filled with water and a large water volume must be boiled off before fuel is uncovered. This means that other aspects related to fire events during the cold shutdown period might be more relevant. For example consequences originating from the spent fuel pool after a fire event or combinations of fire event that could cause a leakage from RPV. It seems that the only phases a fire event could lead to exposed fuel within 24 hours are during phase 1, 2 and 3.

The frequency of fire occurrence differs significantly between the rooms where maintenance is on-going and rooms where maintenance is not on-going. On the other hand the probability of successfully fire extinguishing is significantly higher in rooms where work is in progress. Fire frequencies are generally higher during the cold shutdown mode compared to operating mode.

During our work with fire analysis the points below with possible further development were found.
- More detailed analysis of how human error is affected by a fire.
- Secondary events caused by fire leading to combined events, for example LOCA caused by a fire.
- Secondary fire event caused by the initiating event.
- Fire when all fuel is unloaded from reactor pressure vessel and put into the spent fuel pool.
- A more detailed analysis of how human error is affected by a fire.

Reference

[1]   Probabilistic Safety Assessments of Nuclear Power Plants for Low Power and Shutdown Modes, IAEA, Vienna, 2000 IAEA-TECDOC-1144, ISSN 1011-4289
[2]   Fire PRA Methodology for Nuclear Power Facilities, EPRI/NRC-RES, vol 2 detailed methodology, EPRI 1011989, NUREG/CR-6850

Fire PSA and insights

F. Nicoleau[a], F. Corenwinder[a], G. Georgescu[a]

[a] Institute for Radiological Protection and Nuclear Safety (IRSN), Fontenay-Aux-Roses, France

Abstract: IRSN (TSO of French Nuclear Safety Authority) develops simplified Fire Level 1 probabilistic safety assessments (PSAs) for nuclear power plants (NPPs) in order to establish his own independent opinion on the assumptions and results of the licensee Fire PSAs (EDF). IRSN Fire PSAs are extensions of the IRSN in-house developed NPP Level 1 PSAs for internal events.

The licensee and IRSN studies are similar in scope; however the objectives and some main assumptions may be different. The licensee objectives are to answer to the Safety Authority requests to perform complete PSA studies as a complementary approach of the deterministic studies of the fire risks. The IRSN study objectives are to provide an independent verification of the licensee study and also to allow further PSA applications in the framework of technical instruction of safety issues. In particular, IRSN main goal is to focus on the most critical equipment and compartments in terms of fire-related risks.

The paper gives two examples of specifics insights obtained regarding the licensee PSAs in the field of Fire.

The first example is related to the ongoing third periodic safety review of 1300MWe NPPs. The second example deals with IRSN review of the licensee Fire PSA for the commissioning of the French EPR reactor (at Flamanville 3).

Keywords: Fire PSA, Periodic Safety Review.

1. INTRODUCTION

The periodic safety review procedure is a periodic process implemented for every reactor. In France, the periodic safety reviews occur every ten years and concern all reactors of a given serie (e.g. 900 MWe or 1300 MWe or 1450 MWe reactors).

In that context, IRSN, the French Institute for Radiological Protection and Nuclear Safety, which is the technical support of the French Nuclear Safety Authority (ASN), develops his own PSA to assess the PSA developed by the licensee.

IRSN began to develop and use level 1 probabilistic safety assessment (PSA) for French Nuclear Power Plants in the 90's. In the frame of its missions, the in-house development of PSA allowed gaining valuable knowledge on nuclear safety. In the same time, a deep independent analysis of the PSAs developed by the licensee (EDF) was performed. Since 2002 [2], PSA review became an important part of the periodic safety reviews of the operating plants.

PSA development program is still in progress at IRSN and at the licensee. These developments aim to introduce new knowledge and to extend their scope, in order to increase the possible fields of applications. Both organizations are working in parallel on PSA developments. The licensee objective is to establish reference PSAs for each plant series. IRSN objective is to obtain independent insights, precious to evaluate and point out needs for further developments. Comparisons between these two independent PSAs highly contribute to the quality of the studies.

In the context of the third decennial visit for the French 1300MWe nuclear power plants, IRSN developed a fire level 1 PSA for 1300MWe reactors. IRSN main goals were to gain knowledge in order to be able to evaluate assumptions and results of the licensee Fire PSA. The study is an extension of IRSN in-house 1300MWe NPPs Level 1 PSA, for internal events. The development of a Fire PSA is necessary due to the importance of fire on the risk of core damage.

A lot of information was exchanged between the licensee and IRSN during the development of the project. The licensee and IRSN studies are similar in scope and use the same principles; however the objectives and the main assumptions may be different (for example: damage temperature considered for the equipment, fire source characteristics…). In particular, IRSN main goal is to identify and to quantify preponderant accident sequences leading to core melt. The study will therefore focus on the most critical equipment and compartments in terms of fire-related risks. IRSN objectives are also to provide an independent verification of the licensee study and to use Fire PSA applications in the framework of technical instruction on specific safety issues.

IRSN also reviewed the licensee Fire PSA for EPR reactor commissioning (at Flamanville 3). The assessment of the EPR Fire PSA was very particular. Especially because EPR is currently under construction and a lot of data are still missing: this is the case of railway cables and several components which are not localized in fire areas. Moreover, fire action procedures for operator are not developed yet. However the method used for the Fire PSA is globally the same as the one used for 1300MWe NPPs.

2. FRENCH SPECIFIC CONTEXT

Regarding nuclear industry, France represents a unique situation with a rather large fleet of Nuclear Power Plants (58 in operating, 1 in construction) which are all built by the same manufacturer (AREVA) and operated by the same licensee (EDF). This nuclear fleet is standardized in 3 PWR series - soon 4 with EPR – (900MWe: 34 plants, 1300MWe with two types of plants named P4 and P'4: 20 plants, 1450MWe: 4 plants; EPR: 1 plant). The plants of each PWR serie are almost identical in design and operation, excepted EPR. The standardized series has real advantages in terms of experience feedback. In the specific field of PSA, the situation is particularly favorable for data collection, and moreover a single PSA (at least for level 1 PSA and internal initiating events) is sufficient for a whole PWR serie of plants. In fact, three PSAs - 4 with EPR - are sufficient to cover all operating plants, for internal events analysis; for Fire PSA only three Fire PSAs - soon 4 with EPR - are sufficient to cover all operating plants. Since few years, IRSN has begun to develop also PSA for internal hazards in order to increase its capability to assess similar studies developed by the licensee and because those studies are important for safety. Regarding PSA for external events, developments are still ongoing especially for seismic hazards and other external hazards inducing long term loss of offsite power and heat sink.

Concerning IRSN Fire PSA, two models have been developed. The first development of Fire PSA started in the 90's and concerned French 900MWe. This study was achieved in 2007. It was a very complete study, developed as recommended in the international practice [1]. It will be updated in 2014 to take into account new data and new experience feedbacks. Moreover, the model will be completely reviewed and implemented with Risk Spectrum tool® in order to facilitate sensitivity studies.
The second development of Fire PSA started in 2005 and concerned French 1300MWe reactors. The general method adopted by IRSN for his 1300MWe Fire PSA is similar to the one used for 900MWe reactors Fire PSA. Nevertheless, the lessons learned from the development of the 900MWe reactors Fire PSA as well as the progress in computer tools have led to some improvements, such as the selection of critical compartment (the most critical compartments for the 900MWe NPPs, were essentially localized in the electrical building) and the development of the study only with RiskSpectrum® tool by linking events trees (event trees of fire scenarios and internal event level 1 event trees).

3. PERIODIC SAFETY REVIEW

3.1. Generality

The periodic safety review procedure is a periodic process implemented for a given reactor type, which in order to take into account operating experience and updated knowledge. For PSA, the review is mainly divided in 2 steps.
In the first step, the periodic safety review procedure aims to demonstrate the conformity of the "*reference plant situation*" with the "*safety reference system*". The "safety reference system" consists

of all the safety rules, criteria and specifications applicable to a reactor type resulting from the safety analysis report. The "reference plant situation" consists of the state of the installation and its operating conditions. Any observed deviations are corrected or justified.

In application of the general procedure, PSAs are used during the periodic safety review to assess the core damage frequency and its change compared with the assessment made at the end of the previous review, including the analysis of the potential changes in system characteristics and in operating practices.

In addition, the identification and the analysis of the main contributors to the core damage frequency (for example analysis of the predominant functional sequences) are achieved in order to highlight potential weak points for which design and operation changes should be studied. They can be ranked using PSA results to define priorities. In particular, the analysis must take into account the frequency of the sequences, possible consequences on containment integrity and uncertainties.

During the first step of the periodic safety review, the reference PSA is updated, with the most recent operating experience (identification of frequency of initiating events, equipment reliability data, plant operating states…), updated plant design and operation. It also includes new knowledge about the plant behavior, obtained from the most recent studies.

After the review of all conservative assumptions of the PSA, this analysis results either in a status quo or in an indication of the usefulness or the needs of implementing design or operational modifications. Following the periodic safety review, a new version of the reference PSA is produced taking into account the plant changes or modeling improvements decided during the review process.

The use of PSA for periodic safety review is done accordingly with the French PSA Basic Safety Rule [2].

Regarding Fire PSA, the periodic safety review or the anticipated safety review for commissioning of EPR is divided into two phases based on the two steps described above with specificities for Fire PSA. During the first phase, the licensee develops a Fire PSA and, then, IRSN compare this PSA with his own study. The objective of the Fire PSA development is to allow gaining valuable knowledge on risks due to fire on nuclear plant and to identify main contributors, risk of fire in different compartments… IRSN Fire PSA leads to several requests of changes from French Safety Authority (ASN) to the licensee. For the requests, endorsed by ASN, the licensee should propose solutions (design or operational improvements) at the end of the first phase of the periodic safety review.

In the second phase (after the licensee solutions proposal), the licensee study is updated, as IRSN Fire PSA, and is finally used to verify the improvement associated to the changes decided.

3.2. IRSN PSA Development for Periodic Safety Review

For the third periodic safety review of the 1300MWe French plants, the licensee updated his Internal Events PSA and also developed a Fire PSA, an Internal Flooding PSA and a Fuel Pool PSA. In order to prepare the review of the licensee studies, IRSN updated his own 1300MWe internal events PSA and developed a 1300MWe Fire PSA.

The paragraph 5.1. describes the Fire PSA use in the frame of the third periodic safety review of 1300MWe plants.

4. THE ANTICIPATED SAFETY REVIEW OF EPR

The anticipated safety review for the EPR for the commissioning is particular because the review is divided into several steps depending on the deadline of the commissioning application.

Some requests proposed by IRSN will be taken into account by EDF for commissioning application and other will be achieved for another deadline corresponding to end of the commissioning tests report; the deadline is decided regarding the potential effect of the request on core damage frequency.

For the anticipated safety review of EPR for its commissioning, the licensee developed an Internal Events PSA a Fire PSA, an Internal Flooding PSA and an Explosion PSA. In order to prepare the review of the licensee studies, IRSN only developed an Internal Events PSA.

The paragraph 5.2. presents the IRSN assessment of the Fire PSA for the anticipated safety review of EPR for its commissioning.

5. IRSN ASSESSMENT

5.1. The third Periodic Safety Review for the 1300MWe NPPs

For The third Periodic Safety Review for the 1300MWe NPPs, IRSN firstly, as a preparatory work, developed his own Fire PSAs, which consist of two different models for the two types of 1300MWe plants, in order to be able to better distinguish the specificities of each design. For P'4 type plant, an adapted method of the international practice was implemented by IRSN: the *"qualitative screening"* [1] was reduced at the selection of compartments containing important safety equipment which are the most important contributors to the core damage frequency (estimated by importance calculation with Risk Spectrum® tool [3]) or at the selection of compartments which are adjacent to a compartment containing equipment important to safety. For one compartment, only one type of component was taken into account for the source fire characteristics. For P4 type plant, a very simplified model was developed by IRSN for few fire areas, based on the conclusion of the P'4 plant Fire PSA. Fire areas, taken into account, were selected considering the results of the type P'4 in terms of core damage frequency (CDF) due to a fire.

The licensee presented a "reference" Fire PSA which is common for the two types of 1300MWe plants (P4 and P'4). IRSN considers that the two types of 1,300MWe NPP are different due to compartment geometries, different types of component contained into the compartments, localization of compartments in the buildings and different types of adjacent compartments. The fire areas are not the same too. For Fire PSA, those elements have consequences on the result for fire simulation and on the list of components lost after a fire. If the compartments are different due to dimension, geometry and due to the combustible they contain, the failure time could be different. All these reasons led IRSN to ask the licensee to develop two Fire PSAs: one by type of 1,300 MWe plant. This conclusion of the periodic safety review was approved by ASN who asked to the licensee to develop a fire PSA for the 1,300MWe reactor type P4 for the second step of the periodic safety review.

The reference study performed by the licensee pointed out the need to change the type of the manual command on the Main Control Room (MCR) of the pressurizer safety valves to avoid the spurious signal leading to their opening in case of the failure of I&C cabinet due to fire.

A first conclusion of IRSN assessment was that during the first step of Fire PSA development, hypothesis more or less conservative, as well as parameters values with various uncertainties are used: it's very important to analyze the effects of those choices on the PSA results and to identify the possible cliff-edge effects and the needs for R&D.

IRSN estimated that the use of Fire PSA approach proposed by the licensee was acceptable and consistent with requirements of the French basic safety rules for PSA. Regarding the licensee conclusions, a particularly deep verification of the Fire PSA developed by the licensee was performed by IRSN, based mainly on the use of the IRSN Fire PSA models. The licensee and IRSN studies were developed by using the same computer code Risk Spectrum®. Some differences exist between the two studies. They include, among others, the following aspects:

- reliability data about fire damper and fire door,
- characterization of fire,
- damage temperature,
- fire spreading between fire areas which contains components of the two electrical trains,
- human reliability analysis (HRA); IRSN and the licensee didn't use the same method,
- departure of fire on the current part of cable,

- development of two Fire PSA for IRSN: one by type of 1,300MWe NPPs.

Those differences led to several recommendations due to the potential impact on the core damage frequency, obtained at the end of the first step of the periodic safety review.

The licensee proposed to take into account most of the recommendations, in the updated model of his Fire PSA developed for the second step of the periodic safety review, but he maintained his position for the damage temperatures.

The damage criteria are important parameters of Fire PSAs. These criteria correspond to the failure of equipment. In case of fire, these criteria should be linked to temperature, smoke concentration, humidity, etc. In the licensee Fire PSA, the damage criteria taken into account are associated to a temperature threshold defined for each component.

The licensee considered the value of damage temperature equal to 95°C for electronic equipment (only I&C cabinet) and 137°C for electrical component including all electrical cabinets.

IRSN considered that the value considered for electrical cabinet was not acceptable because (i) electrical cabinet contain electronic cards and (ii) the value considered is not the recommended value of damage temperature in international practices [1].

Regarding international R&D, different values of damage temperature are proposed but there is a lack of knowledge on electronic and electric cards contained in electrical cabinets. For that reason, IRSN decided to set up specific R&D programs in this area: the objective was to quantify the damage temperature and to investigate the impact of the smoke on the components. Three components of electrical cabinet were tested: two electronic devices (named "electronic card") and one circuit breaker (IRSN considered that they are potentially the most affected components of electrical cabinet). Two experimental programs were defined [3].

The first experimental program, called "CATHODE", took place in an experimental small-scale compartment named SIROCCO. The objective of those tests was to define the empirical temperatures of the three components. The results were not sufficient to conclude because the experimental program was not in real condition of fire. For example, there was not soot taken into account. To confirm the results, it was necessary to pursue the experimental program to test components in real conditions of fire.

The second experimental program, called "CATHODE Suies" ("Suies" stands for soot), was performed in a real-scale experimental compartment. The objective of the tests was to obtain elements of answer concerning the damage criteria of component of electrical cabinets in real conditions of fire: first, in terms of temperature and second, in terms of soot concentration. The effect of soot was particularly studied. Four tests were performed between June 18 and October 8, 2009 in the DIVA facility. One type of component of the first SIROCCO experiment was tested in DIVA facility: the electronic cards. They were placed at two different heights (two tests at 1.80 meters and two tests at 0.55 meter) in the compartment containing the electrical cabinet, in a fire. The electronic cards were lost when they were positioned at a height of 1.80 meters but were still operating at 0.55 meter. The first analysis of the tests showed that the electronic cards were lost at a value of temperature lower than the value of temperature found in the first SIROCCO experimental program (upper than 100°C): the value of the damage temperature, obtained in DIVA facility, was superior or equal to 65°C. Another conclusion was that the electronic cards did not work temporarily when some conditions on temperature and soot are reached: a combination of values of two "damage criteria" (temperature and soot) could cause the relay's malfunction. At the beginning of 2014, new series of experiments will be performed to check this assumption and to quantify the values of temperature and soot for which the components are lost.

Considering the results of the experimental programs, IRSN considered a damage criteria of 65°C in his Fire PSA for electrical and I&C cabinet. This value corresponds to the temperature of an area with hot smoke which leads at the failure of components.

The damage criteria were used in fire simulation to estimate the time at which the component fails called the "failure time". It occurs when the area around the component reaches the damage criteria.

For fire simulation, IRSN relies on SYLVIA code (a two-zones fire model), a software system for simulating fire, ventilation and aerosol contamination phenomena developed at IRSN. SYLVIA code estimates pressure, temperature and concentration in carbon which allow to estimate the failure time for various safety related components, in the critical compartment. It also gives similar information in the adjacent compartment, in case of fire spreading.

To estimate if a component is lost during the fire scenario a comparison is done between the failure time and the duration of fire. If the failure time is lower than the fire duration and if damage criteria are reached: the component is lost. In the other case, the component is available. This method gives the list of components or cables which are lost.

Then, a sensitivity study is performed, taking into account two different values for the damage temperature, to evaluate their effects on core damage frequency:

- for the reference study, the value of damage temperature of electrical cabinets is equal to 65°C, it is assumed to be conservative,
- for the sensitivity study, the value of damage temperature of electrical cabinets is equal to 95°C (value taken by the licensee).

Experimental results and sensitivity studies were presented to the licensee. A conclusion was: If the damage temperature increases of 30°C, less component are lost in the fire compartment and in the adjacent compartment. The increase of damage temperature has an important impact on core damage frequency as less initiating events are induced and less equipment important for safety are lost.

In conclusion of the periodic safety review, ASN asked the licensee to change the damage temperature or to perform a sensitivity study on the damage temperature in his Fire PSA for electrical cabinet.

5.2. The Anticipated Safety Review for EPR for commissioning

For EPR reactors, the safety demonstration was significantly improved. EPR design is based on the "technical guidelines for the conception and the construction of the next generation of nuclear reactors with pressurized water", established in 2000, after the French-German experts assessment of EPR safety options. These guidelines mentioned in particular that "the demonstration of safety for the nuclear power plant of the next generation must be made in a determinist way, completed by probability methods and works of research and development suited".

They also mention that "a significant reduction of the global frequency of core damage must be obtained for the nuclear power plant of the next generation. The implementation of improvements of the in-depth defense of these NPP should lead to the obtaining of a global frequency of core damage lower than 10^{-5} per reactor year (/r.y.) by taking into account uncertainties and all the types of failures and hazards. "

Within the framework of the reactor EPR-FA3 commissioning, the licensee achieved a level 1 PSA relative to internal fire. IRSN did the assessment of this Fire PSA.

The Fire PSA results are consistent with the general safety objectives (especially to obtain a global frequency of core damage lower than 10^{-5}/r.y.). For IRSN, this result supports the EPR design regarding fire risks, in particular the separation of electrical train in four different buildings (one by electrical train and fire areas within every building).

IRSN considers that the method, the main assumptions and the data used by the licensee are suitable. Nevertheless, IRSN highlights that the study does not take into account electrical cables and piping of hydrogen, neither as ignition equipment nor as potential targets.

As the reactor is under construction, the localization of cables was unknown when Fire PSA was developed by the licensee. Additional work is then needed to consider the risk due to cables and the study will be completed by the end of the commissioning application. This study concerns fire risks in the containment annulus and, more generally in all compartments with cables of different electrical train.

Furthermore, IRSN identified needs of further development concerning spurious orders in case of fire and evaluation of frequencies for fire departures.

The licensee committed himself to updating his Fire PSA for the end of the commissioning tests report, by taking into account these aspects.

More specifically, during its assessment, IRSN pointed out that the licensee first based his EPR Fire PSA on deterministic principle (fire barriers are always fire-resistant: the spreading of the fire is not possible). For example, a fire door is considered to resist to any fire. Therefore, the licensee didn't take into account, in his fire PSA, the probability of the failure of the fire door during a fire. These assumptions will be checked by IRSN during the deterministic studies assessment.

5. CONCLUSION

It is important to note that a PSA development program is still in progress at IRSN and at the licensee. The developments aim to improve PSA quality and to extend their scope, in order to increase the field of applications. Both organizations are working in parallel on PSA developments. These two independent works, which could be considered as a particularly deep external review, highly contribute to the quality of the studies.

The third periodic safety review for the French 1300MWe nuclear power reactors is performing with the particularity to extend PSA to internal hazards like fire and flooding. The development of an IRSN Fire PSA made it possible to conduct in-depth analysis of the licensee study. IRSN estimated that the use of Fire PSA approach proposed by the licensee was acceptable and consistent with requirements of the French basic safety rules for PSA. Some plant improvements were identified. They concerned the change of the type of the manual command on the Main Control Room (MCR) of the pressurizer safety valves to avoid the spurious signal leading to their opening in case of the failure of I&C cabinet due to fire. The conclusion of the periodic safety review led ASN to ask the licensee to develop a fire PSA for the 1,300MWe reactor type P4 because some elements, important for a Fire PSA, are different between the two types of 1,300MWe plant (geometry, fire barriers, localization of component…) and to change the damage temperature or to do a sensitivity study on the damage temperature in his Fire PSA for electrical cabinet. The study, performed by IRSN, highlighted the importance of the values of the damage criteria considered in Fire PSA and the need to perform experimental programs in this field.

Concerning EPR, the Fire PSA results are consistent with the general safety objectives. These results support EPR design against fire risks, in particular the separation of electrical train in four different buildings (one by electrical train) and the measures of separation into fire area within every building. Additional studies will be performed by the license by the end of the commissioning application, especially to take into account risks associated to fires in compartments including cables of different electrical train.

References

[1]   EPRI/NRC-RES, NUREG/CR-6850, "Fire PRA Methodology for nuclear Power Facilities".
[2]   French PSA Basic Safety Rule 2002-01.
[3]   Article PSAM10, "Fire PSA for French 1300MWe NPPs".

Complex investigation of Fire PSA dominant scenario related to direct flame contact with safety related pipes

Shahen Poghosyan[*a], **Tsolak Malakyan**[a], **Gurgen Kanetsyan**[a] **and Armen Amirjanyan**[a]

[a] Nuclear and Radiation Safety Center, Yerevan, Armenia

Abstract: Fire risk is one of the complex problems and potentially serious challenges to the safety of Nuclear Power Plants (NPPs). Fire PSA is a powerful and systematic tool which can reveal critical safety issues from the point of view of fire. A detailed fire PSA study performed for Unit 2 of the Armenian NPP (ANPP) shows that overall fire risk is driven by several fire scenarios. However before applying the results in a safety-related decision making process, it is important to verify the robustness of conclusions related to the identified risk contributing factors. Observation shows that the results received for the confinement oil fire scenario imply a need to implement substantial modernization activities. On the other hand, the approach used for oil fire modeling in confinement is considered conservative and the results obtained are considered to have considerable associated uncertainty. The aim of the current paper is to present a more accurate complex investigation of the oil fire scenario in the ANPP confinement building in order to create an adequate basis for further plant modernization decisions. The aim of the current paper is to present a more accurate complex investigation of the oil fire scenario in the ANPP confinement building in order to create an adequate basis for further plant modernization decisions.

Keywords: PSA, Fire, flame, pipe rupture, direct contact

1. INTRODUCTION

Fire risk is one of the complex problems and potentially serious challenges to the safety of Nuclear Power Plants (NPPs). Probabilistic safety analysis (PSA) performed for VVER reactors shows that fire could contribute up to 50% of overall core damage frequency (CDF) [1]. Armenian NPP Unit 2 is not an exceptional case, fire risk contributes about 20% of overall CDF for full power operational modes. Fire PSA performed for Armenian NPP Unit 2 allowed to reveal several specific fire safety issues for first generation VVER-440 reactors [2]. According to Fire PSA results, the most dominant contributor is large oil fire scenario in confinement area (more than 30% of overall fire-induced CDF).

Oil located in main coolant pumps (MCP) oil system's pipelines is considered as a source for large fire. It was assumed that fire could start in case of oil leakage from MCP oil system and its contact with hot surfaces. Fire scenario analysis evaluated possible impact on discharge pipes of emergency make-up system (EMS) which are passing through A-013/2 compartment in confinement area (see Figure 1). According to state of the art fire PSA approach fire induced pipe ruptures are typically could be neglected from the analysis [3]. However it was decided to considered mentioned scenario due to the following factors:
- existence of water stagnation zones at the closed valves on the pipes (see Figure 1)
- high pressure inside the pipe (P=125atm)
- direct flame contact with pipes and potentially large amount of combustible material (oil from MCP system)
- direct core damage in case of EMS pipes failure (due to LOCA through the pipes breaks and simultaneous unavailability of EMS system)

Analysis of mentioned scenarios was performed by fire simulation using CFAST two-zone code. It was concluded that pipeline metal temperature reaches critical temperature when integrity of the pipe could not be credited. Meanwhile CFAST has several limitations which does not allow to fully address

---

* *E-mail: s.poghosyan@nrsc.am*

mentioned fire scenario. Particularly CFAST does not allow to correctly model direct flame contact effect taking into account radiation heat transfer. In addition performed analysis considered pipelines as a thermally thick target neglecting existence of water inside the pipe.

Figure 1: EMS discharge pipeline geometry in compartment A-013/2

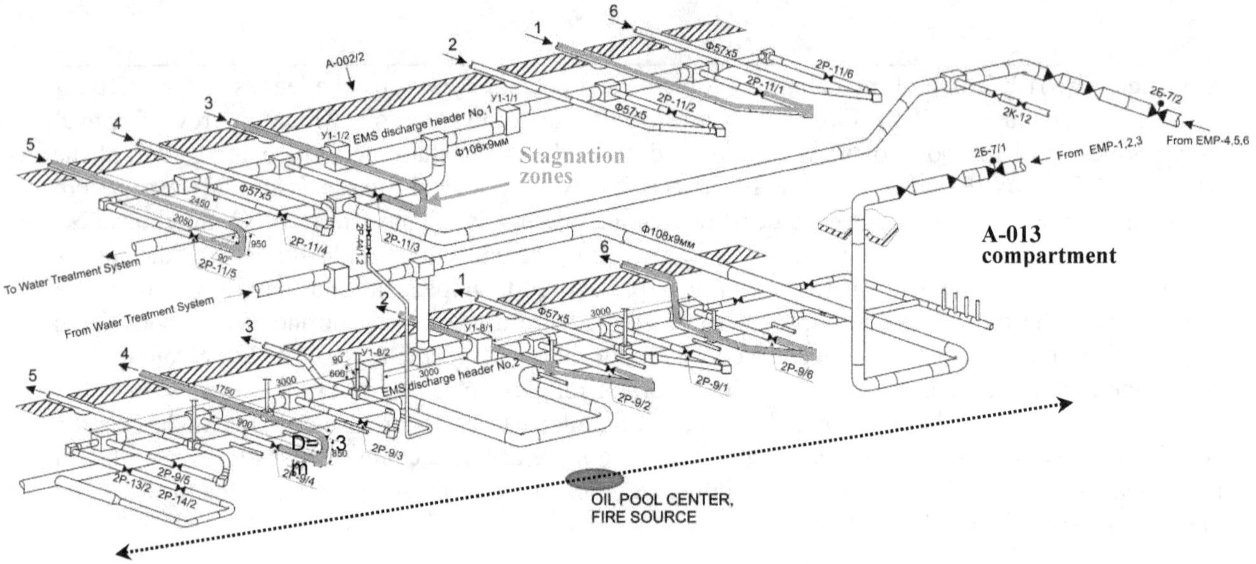

Results received for the confinement oil fire scenario require implementation of substantial modernization activities. On the other hand, the approach used for oil fire modeling is considered to be conservative and the results obtained are considered to have considerable associated uncertainty. Basically, following aspects considered to be questionable:

- Heat transfer from fire to the pipe
- Strength analysis for given pipe metal temperature

The aim of the current paper is to present more accurate complex investigation of the oil fire scenario in the ANPP confinement building taking into account overall heat flux from the fire flame, geometrical characteristics of the pipe and pressure in the pipeline in order to create an adequate basis for further plant modernization decisions.

2. ANALYSIS OF HEAT TRANSFER FROM FIRE TO THE PIPE

RELAP5 was used for analyzing of heat transfer phenomena from fire flame to EMS pipeline. EMS pipeline was modeled in RELAP5, and integrated into plant-specific RELAP5 model. Geometrical characteristics of modeled EMS pipeline segment are introduced in the table 1.

Table 1: Geometrical characteristics

Parameter	Size in mm
Total length of EMS pipeline	68000
HPI pipeline length covered by fire	5000
Inner diameter	47
Outer diameter	57

Taking into account that EMS pipeline is connected to cold leg of main circulation loop, the temperature of coolant was obtained based on steady-state calculation of integral ANPP model. The results of steady-state calculation showed that initial temperature of coolant equals 268 °C.

EMS pipeline were modeled as "pipe" element consisting of 15 volume elements with 1.7E-3m² cross-sectional area (see Figure 2). Each of 13 volume elements has length of 5m; remaining two volume elements have correspondingly 2 and 1 meter of length.

Figure 2: RELAP5/MOD3.3 nodalization of ANPP Unit 2 primary circuit

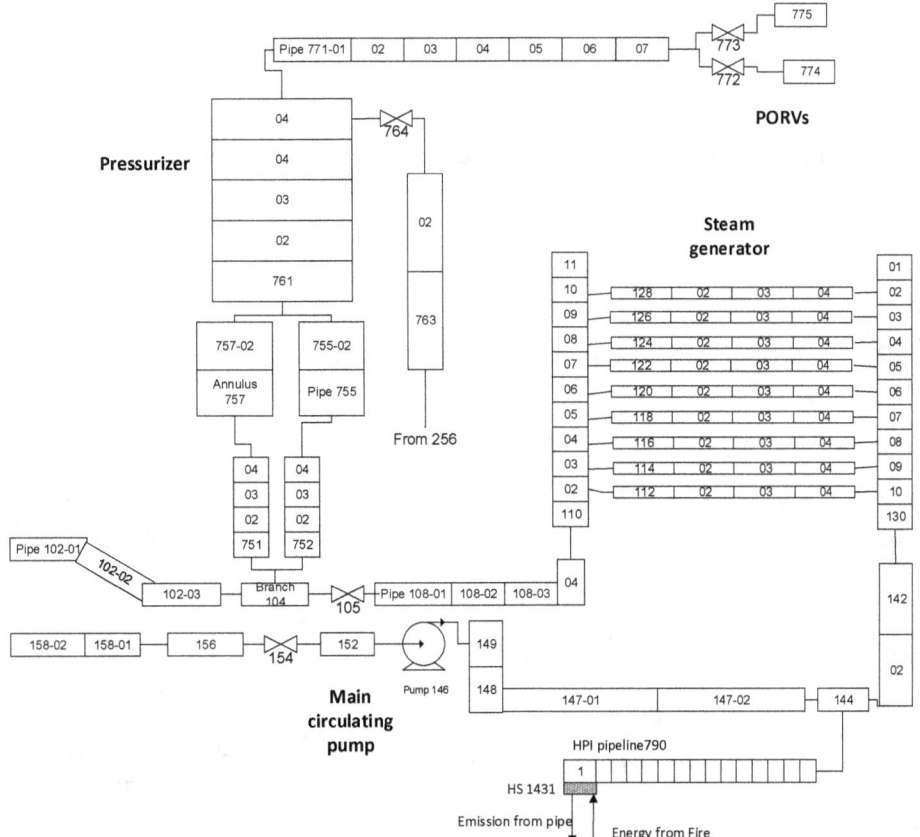

Maximum time-average heat fluxes from the flame to an object (size of the object was small relative to flame size), located in the flame, was taken 120kWt/m² [4] which includes radiation and convective heat fluxes.

Radiant flux emitted from pipe was modeled with Stephan-Boltzmann equation:

$$q_r = \varepsilon \sigma T^4 \qquad (1)$$

where ε is the emission coefficient of steal, σ is Stephan-Boltzmann constant, T is the temperature of pipe.

In order to calculate the emission from the pipe specific metal characteristics were used. The material of the pipeline is 0X18H10T for which ε=0.85 emission coefficient is taken from [5]. Heat transfer from pipe wall to water was modeled by "Heat Structure" component (see Figure 2), element HS1431 simulates pipe wall. Duration of simulation was taken 1000 seconds.

From the calculation results, it can be noticed that at the volume which interacts with fire water temperature reaches saturation point in 43 second after fire initiation (see Figure 3 and 4). Due to continuous heat flux from fire to the pipe the water in mentioned volume starts to evaporate and fully transfers to the steam on 52nd second after fire initiation.

Starting from this time point, temperature of steam rapidly increases and reaches value of 707°C by 400th second. After 400th second radiant flux emitted from pipe becomes comparable with fire heat transfer to the pipe (see Figure 3 and 4). This effect is conditioned by the fact that pipe metal temperature increase rate starts to reduce.

At the end of calculation, temperature of considered steam reached 807°C (see Figure 3). As it could be seen from Figure 5 temperature of pipe wall has the same behavior as steam temperature and at the end of calculation the considered pipe segment walls temperature reaches 949°C.

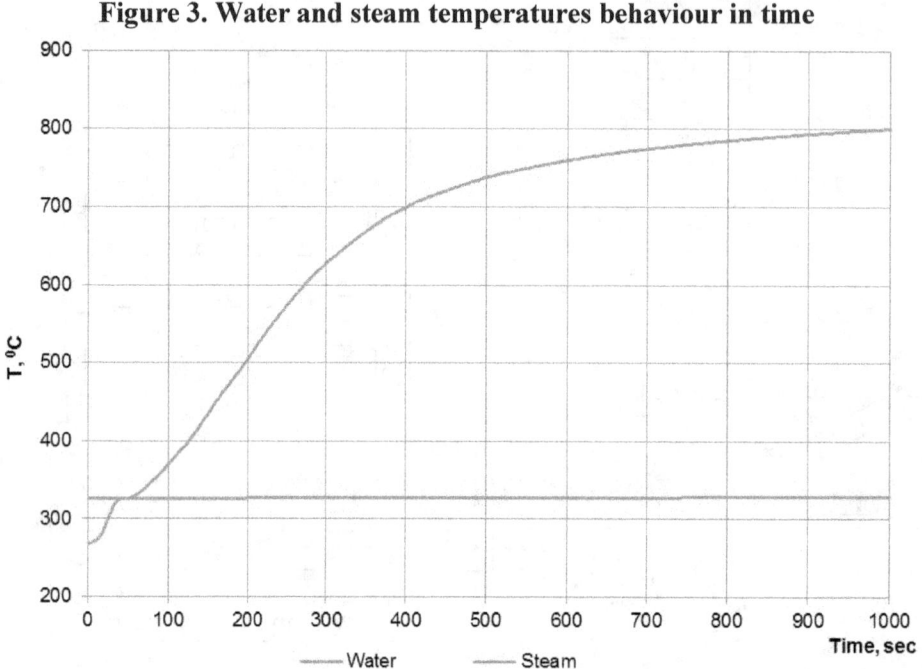

Figure 3. Water and steam temperatures behaviour in time

Figure 4. Radiant flux emitted from pipe and fire energy flux in time

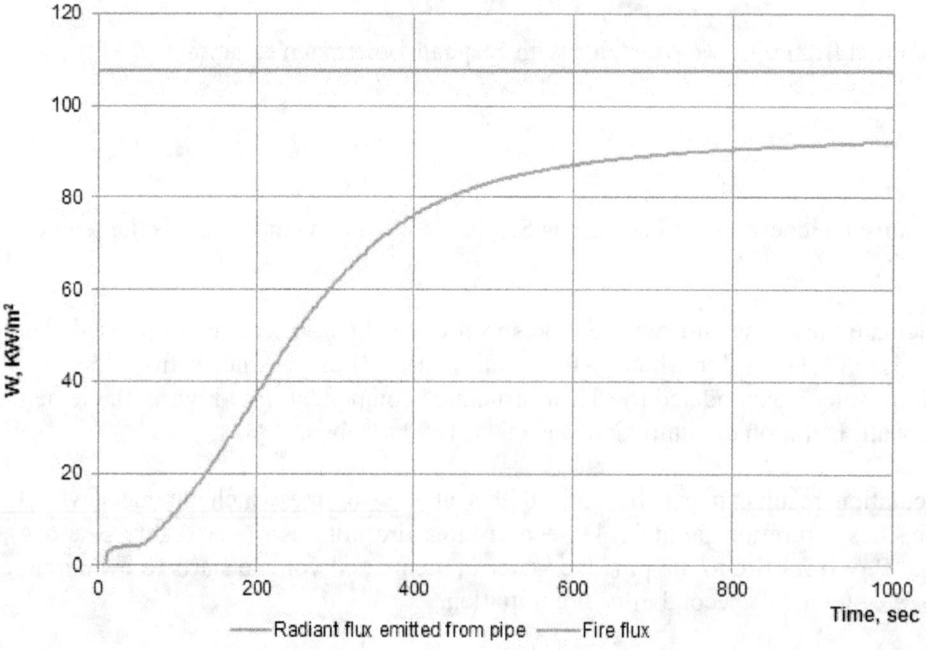

Figure 5. Pipe wall temperatures behaviour in time

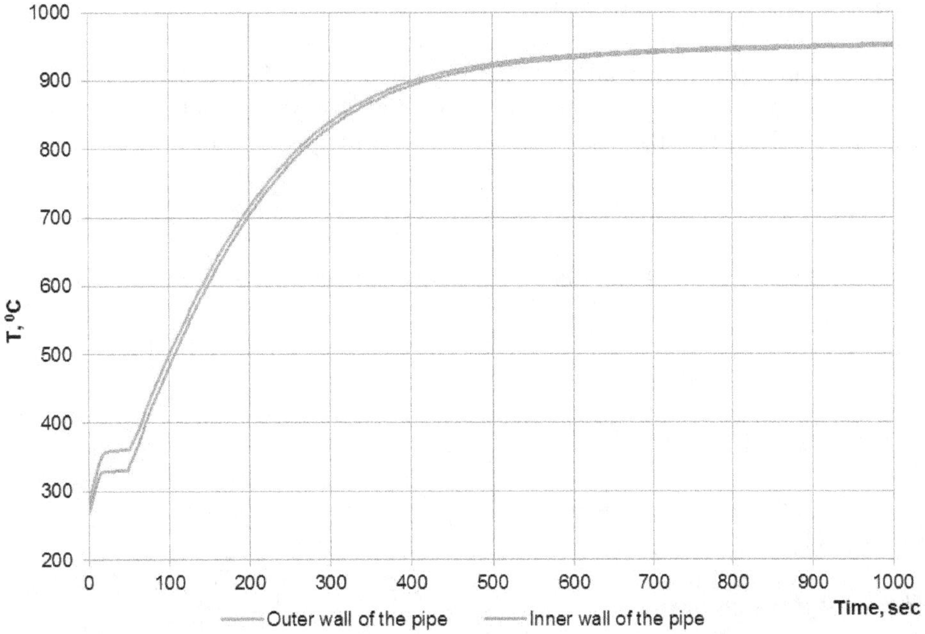

3. PIPE STRENGTH EVALUATION

Pipe strength evaluation was implemented using PNAE G-002-86 standards [6] applicable for pipe metal type (0X18H10T). According to [6] allowable stress (σ) is calculated using following equation:

$$\sigma = \frac{p(Dm_3 + Sm_2)}{Sm_1} \qquad (2)$$

where p – pressure inside the pipe (MPa), D – inner diameter of the pipe (mm), S – wall thickness (mm), m_1, m_2 and m_3 are parameters which depends on the pipe geometry. According to [6] for cylindrical pipe m_1=2, m_2=m_3=1. By putting in the equation all of the pipe characteristics presented in Table 1 and corresponding pipe pressure once could obtain allowable stress– σ=71.3MPa.

Following equation was used in order to calculate critical temperature of the pipe metal when σ=71.3MPa is reached from temporary resistance point of view

$$R_m = n_m \sigma \qquad (3)$$

where R_m is allowable for particular temperature, n_m –strength margin coefficient which equals 2.6 for temporary resistance limit [7]. Taking into account that for considered EMS pipes allowable stress equals σ=71.3MPa, obtained R_m is 185.38 MPa. According to [7] R_m=185.38 MPa corresponds to 750°C of pipe metal temperature (see Figure 6).

Following equation was used in order to calculate critical temperature of the pipe metal when σ=71.3MPa is reached from elasticity limit point of view

$$R_{p0.2}^T = n_{0.2} \sigma \qquad (3)$$

where $R_{p0.2}^T$ is elasticity limit for particular temperature, $n_{0.2}$ – strength margin coefficient which equals 1.5 for elasticity limit [7]. Taking into account that for considered EMS pipes allowable stress equals σ=71.3MPa, obtained $R_{p0.2}{}^{T}$ is 106.95 MPa. According to [7] R_m=106.95 MPa corresponds to pipe metal temperature close to 1000°C (see Figure 7).

As it is presented on Figure 6 pipe metal temperature reaches 750°C value approximately in 223 seconds after fire initiation. Therefore it could be concluded that for considered fire scenario pipe integrity could not be credited. For elasticity limit pipe metal temperature does not reach the critical $R_{p0.2}{}^{T}$ value.

Figure 6. R_m dependency on pipe metal temperature

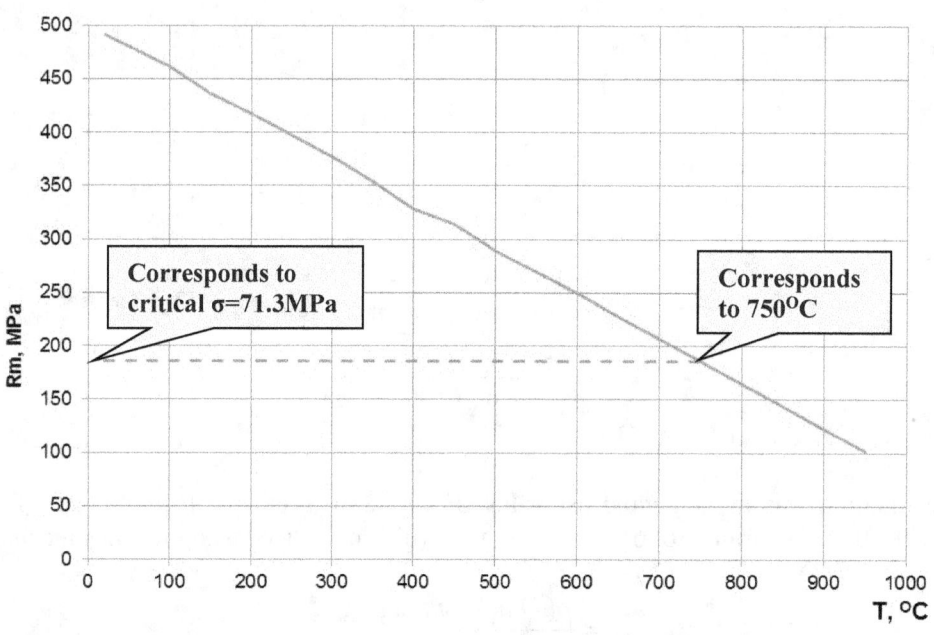

Figure 7. $R_{p0.2}{}^{T}$ dependency on pipe metal temperature

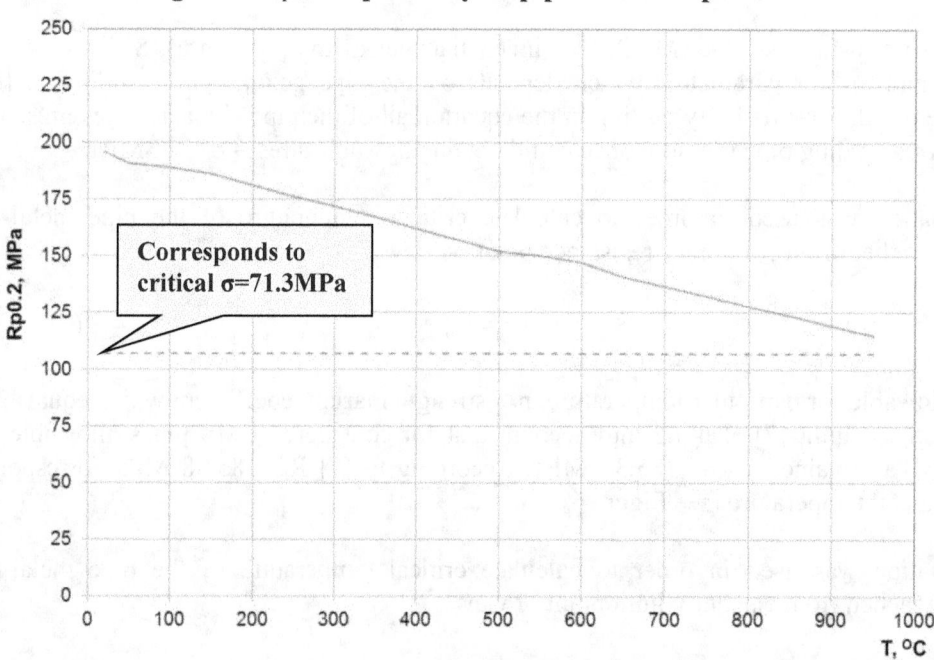

4. CONCLUSION

Fire PSA performed for Armenian NPP Unit 2 revealed that the most dominant contributor is large oil fire scenario in confinement area. The risk of dominant fire scenario is conditioned by impact on discharge pipes of emergency make-up system (EMS). Previous analysis of this fire scenario was performed by fire simulation using CFAST two-zone code that has several limitations which does not allow to fully trust obtained results. The main problems of CFAST: limitation in modelling of direct flame contact effect and neglecting of pipe water effect. Since elimination of considered fire scenario requires implementation of substantial modernizations, it was decided to spend efforts for more accurate investigation of the scenario in order to assure credibility of obtained results.

Investigation implies analysis of heat transfer from fire to the pipe and evaluation of pipe metal temperature. In addition strength analysis for given pipe metal temperature was provided. Heat transfer analysis from fire to the pipe was done taking into account both radiation and convective heat fluxes. Analysis shows that pipe temperature exceeds $949^{\circ}C$ in 1000 seconds after fire initiation.

Pipe strength evaluation was implemented using PNAE G-002-86 standards [6] applicable for pipe metal type (0X18H10T). Calculation shows that EMS pipe's maximal allowable tension– $\sigma=71.3MPa$ is reached in 223 seconds after fire initiation.

The overall conclusion is that pipe integrity could not be credited for considered oil fire scenario. Therefore it was recommended to develop and introduce measures to decrease risk of fires in confinement room A-013/2. Particularly following actions could be suggested:
- increase reliability of MCP oil pumps disconnecting electrical scheme in order to decrease amount of oil spill in the compartment
- create possibility for fire detection in A-013/2 compartment
- decrease likelihood of oil pipe rupture (i.e. pipe cover installation)

References

[1]   R. Virolainen et al., *"Comparison of PSA Practises and Results Summary Report of the Second Phase Activities Co-operation Forum for VVER Regulators (VVER Forum) Working Group on the Use of PSA. Final Report."*, STUK, Helsinki, Finland, 2009
[2]   Sh. Poghosyan, G. Kanetsyan and A.Amirjanyan, *"Specific fire safety issues for first generation VVER-440 reactors"*, Proceeding of PSAM11& ESREL2012 joint conference, Helsinki, Finland, June 25-29, 2012
[3]   *"Fire PRA Methodology for Nuclear Power Facilities (NUREG/CR-6850), Volume 2: Detailed Methodology"*, US NRC, Washington, USA, 2005
[4]   *"SFPE handbook of fire Protection Engineering (3rd edition)"*, National Fire Protection Association, Massachusetts, USA, 2002
[5]   I. Kikoin, *"Handbook on physical quantities"*, ATOMIZDAT, Moscow, USSR, 1976
[6]   *"Equipment and pipelines strength analysis norms for nuclear power plants (PNAE G-7-002-86)"*, ENERGOATOMIZDAT, Moscow, USSR, 1989

Fire Risks of Loviisa NPP During Shutdown States

Sami Sirén[a]*, Ilkka Paavola[a], Kalle Jänkälä[a]
[a] Fortum Power And Heat Oy, Espoo, Finland

Abstract:

Fire PRA for all 15 shutdown states of Loviisa NPP has been performed. The fire PRA for power operation and the internal event PRA for shutdown have been used as a basis for the analysis, reducing the time needed for investigating cable routing and potential of fire-induced initiating events. The hot states are mostly modeled using applicable power operation fire scenarios. For the cold states, 342 fire scenarios have been created and integrated with the PRA model. Fire frequencies have been estimated with an empirical Bayesian method using both plant data and international data. The importance of moving from conservative modeling towards best estimate is underlined in the shutdown fire PRA. The real availability of systems instead of the minimum requirements in Technical Specifications has been taken into account to decrease the conservatism related to maintenance activities. Fires inside the control building during cold states dominate the risk. (The shutdown fire risk is relatively small,) but it would be hundredfold without the backup RHR system.

Keywords: Fire PRA, Shutdown PRA.

1. INTRODUCTION

Loviisa nuclear power plant in southern Finland consists of two almost identical VVER units, commissioned in 1977 and 1980. Unit 1 (PRA was started in 1985 and the) internal event PRA for power operation was completed in 1989. Since then the PRA has been continuously updated and expanded to cover new initiating event groups, such as severe weather and flooding events. The fire PRA for power operation was first completed in 1997. It was and still is a major contributor to the overall core damage risk. As the PRA was developed, fires during shutdown became the last missing part of the level 1 PRA that already included internal events, floods, severe weather and seismic events for both power operation and shutdown and fire events for power operation.

Even though both shutdown risks and power operation fire risks contribute very significantly to the overall plant risk, fires during shutdown had been estimated to be much less important. However, to achieve completeness of the level 1 PRA and thus improve the accuracy of PRA applications such as the risk informed inspection of Technical Specifications, it was decided to develop a fire PRA for shutdown states. The focus was on getting a best estimate quantification result while making use of the already finished parts of Loviisa PRA as much as possible to reduce the amount of work needed and to keep the risk model simple and manageable. The starting point for the shutdown fire PRA is described in Chapter 2. Chapters 3 and 4 include the description of the shutdown fire PRA development and the results. Conclusions can be found in Chapter 5.

2. LOVIISA PRA MODEL

The basis for the shutdown fire PRA model is in the internal event PRA for shutdown and in the fire PRA for power operation. Both are very significant contributors to the annual core damage frequency of Loviisa NPP.

---

* sami.siren@fortum.com

2.1 Internal Events Shutdown PRA

The shutdown PRA model of Loviisa NPP consists of 15 plant operating states (POSs), as shown in Table 1. Each POS represents a distinct phase in an average refueling shutdown. Initiating events, success criteria, reliability data etc. are separate for each POS. Currently, internal initiating events during shutdown contribute 29 % of the annual core damage frequency (CDF). The shutdown fire PRA was not considered very significant because most of the core damage frequency in shutdown states is caused by events that are unlikely to be caused by fires, e.g. drops of heavy loads.

Table 1: Plant Operating States in Loviisa Shutdown PRA

POS	Description	Avg. duration (h)
B	Low power and sub-criticality	2.1
C	Hot standby	8.6
D	Hot shutdown	7.7
E	Hot shutdown, residual heat removal system is water solid	14.3
F	Cold shutdown	8.8
G	Cold shutdown, primary circuit open	38.4
H	Refueling shutdown, procedures before refueling	39.9
I	Refueling shutdown	106.5
J	Refueling shutdown, procedures after refueling	74.1
K	Cold shutdown, assembly of the reactor	65.9
L	Cold shutdown, pressurization of the primary circuit	76.0
M	Hot shutdown	61.7
N	Hot standby	40.2
O	Startup	13.4
Q	Power operation, power increase and turbine startup	6.9

2.2 Power Operation Fire PRA

Fire PRA for Loviisa NPP unit 1 power operation was developed in the late 1980s and 1990s. Fire PRA analysis methods used for other NPPs at the time were varied, but little detailed information on the methodology was available. Furthermore, many methods concentrated on systems thought to be the most critical to safety and did not consider cabling systematically. Therefore, a method was developed for Loviisa fire PRA.

The analysis contained identification of the fire-induced initiating events (IEs), estimation of different fire frequencies and estimation of conditional core damage (CD) probabilities for the fire events. IE fault trees were used to identify possible fire-induced IEs in rooms or groups of adjacent rooms. [1]

In the first phase of the analysis, all cabling and equipment inside the ignition room was assumed to fail due to the fire. Internal event PRA models were used to calculate the conditional core damage frequency. The possibility of the fire spreading to adjacent rooms through doors and openings was estimated based on the fire load and its location. Automatic extinguishing systems were taken into account in the fire spreading scenarios, but not when considering the fire damage in the ignition room.

In the second phase, the most significant fire scenarios were examined more closely to remove conservatism. The total fire frequency of a room was allocated to separate ignition sources and more detailed fire scenarios were created to consider their respective damage potential based on the location of fire loads and critical equipment. Extinguishing systems were taken into account also in the ignition room in case of equipment located outside the flame area. Various fire simulation codes were used when needed. Less important fire scenarios were left as is.

As a result of the fire PRA, weaknesses were identified and several plant modifications were made to address them. Fire protection covers were installed for critical cabling, sprinkler system was extended and high pressure hydraulic oil pipelines were covered to prevent jet fires. A new backup residual heat removal (RHR) system, shared by both units, was installed to mitigate the impact of large turbine hall oil fires.

After its completion the power operation fire PRA has been continuously updated due to plant and risk model modifications. It currently consists of 241 ignition rooms and 718 analysis cases, some of which include more than one fire scenario, and contributes 29 % of the annual core damage frequency.

3. SHUTDOWN FIRE PRA DEVELOPMENT

The objective in the shutdown fire PRA development was to combine the methodology and fire data from the power operation fire PRA and the plant response models and reliability data from the existing shutdown PRA. By fully integrating the shutdown fire model into the existing living PRA model makes it easier to keep it up to date.

3.1 Identification of Initiating Events

The fire PRA is based on the initiating events identified in the internal event PRA. The principles of the method include:

- Multiple simultaneous initiating events of the internal event PRA were considered possible due to fire
- Outside the control building the IEs are modeled with fault trees taking into account possible equipment and cable failures due to fire
- Inside the control building the analysis is more coarse and based on conservative assumptions about cable routes and potential fire damage

The principles used in the fire PRA for power operation were applied also in the shutdown fire PRA. In addition, the initiating events included in the shutdown fire PRA were screened using several criteria, such as:

- Only IEs applicable for the POS in question are considered
- If an IE cannot be caused by fire alone, it is excluded
- If fire as the cause of the initiating event does not have any effect on the consequences and the IE is more likely to happen due to other reasons, it is excluded

Most of the IEs are the same as during power operation, but some are only relevant in shutdown states (e.g. loss of residual heat removal system) or triggered differently than during power operation (e.g. loss of off-site power during the maintenance of main transformers). The identified initiating events for shutdown fire PRA in each POS are shown in Table 2.

In the hot POSs B-D, M-O and Q, the remaining initiating events after the screening are all included in the fire PRA for power operation. Also, the plant response is similar to power operation and the fire risk was expected to be small due to the short time spent in these POSs. Therefore, the fire scenarios for these POSs were adopted from power operation fire PRA, only making minor POS-specific adjustments.

Table 2: Applicable Initiating Events in Each Plant Operating State

Initating event		Relevancy by plant operating state														
Acronym	Description	B	C	D	E	F	G	H	I	J	K	L	M	N	O	Q
LDCP	Loss of DC Power	x	x	x	x	x	x	x	x	x	x					x
LIRV	Loss of Instrumentation Room Ventilation	x	x	x	x	x	x	x	x	x	x					x
LMFW	Loss of Main Feed Water	x	x													x
LOOP	Loss Of Offsite Power	x	x	x	x	x	x	x	x	x	x					x
MLOCA	Medium Loss Of Coolant Accident	x	x	x	x								x	x	x	x
PLOCA	Pressurizer Loss Of Coolant Accident	x	x											x	x	x
PLRR	Partial Loss of RR (residual heat removal)				x	x	x	x		x	x					
PLSW	Partial Loss of Service Water				x	x	x	x	x	x	x					
RT	Reactor Trip	x														x
TLFW	Total Loss of Feed Water	x	x	x												x
TLRR	Total Loss of RR (residual heat removal)				x	x	x	x		x	x					
TLSW	Total Loss of Service Water	x	x	x	x	x	x	x	x	x	x					x
XSLOCA	Very Small Loss Of Coolant Accident	x	x	x	x											x

For the cold POSs F-L, new fire scenarios were developed. Criteria for triggering the applicable IEs were, where needed, redefined to match the plant configuration, e.g. Loss Of Offsite Power was redefined because during the shutdown states, the offsite power is supplied from the 110 kV grid instead of the 400 kV grid. Equipment and cable failure combinations leading to the IEs were then identified.

The hot shutdown state E has some properties of both the hot and cold POSs. Loss of coolant accident is still considered possible due to loss of primary coolant pump sealing water, but loss of RHR system can also be a problem. Therefore, it was modeled using the hot POS procedure, but adding the fire scenarios related to IEs PLRR, PLSW and TLRR from the cold POS analysis.

3.2 Cable Routing and Plant Walk Downs

After identifying the systems and cabling related to triggering each IE, the IEs were linked to individual rooms and areas by carrying out plant walk downs. Individual cable routes were investigated on a room-level accuracy and all the rooms containing those systems or cables were assessed for relevant characteristics, such as fire loads, fire propagation and suppression possibilities. The same was done with safety systems needed to prevent core damage after the IE. Normally this would be very time consuming, but for the shutdown fire PRA, most of the needed data was already available as part of the extensive work carried out earlier for power operation fire PRA. For the cold POSs, only five new ignition rooms were identified that were not part of the power operation fire PRA.

3.3 Fire Frequency Estimation

Shutdown fire events from the plant and international fire databases [2,3] were used for the fire frequency estimation. The fire events were first distributed among 'hot POSs before refueling', 'cold POSs' and 'hot POSs after refueling' and then among 17 room types, e.g. 'process rooms' or 'cable spreading rooms'. A screening was then done using the criteria in the fire PRA for power operation with two exceptions:

- Only fires during a refueling shutdown were included
- Fires related to maintenance activities were excluded, except in areas where the fire could affect systems in the other redundancy, e.g. in the turbine hall

Fire frequencies were estimated using an empirical Bayesian method [4]. No fire events were allocated for 'hot POSs before refueling' so fire frequencies for power operation were used instead. For fires during 'hot POSs after refueling', there were six fire events and the overall fire frequency was about 3.5 times that of power operation. However, to get a more realistic distribution among different types of rooms, power operation frequencies for each room type multiplied by 3.5 were used instead. For the 'cold POSs', there was sufficient data and the fire frequencies for each room type were estimated normally. The overall fire frequencies used in the analysis are presented in Table 3.

Table 3: Overall Fire Frequencies in Each Plant Operating State

POSs	Description	Fire Events	Overall Fire Freq. (1/h)
P	Power operation	344	7.1E-06
B...E	Hot POSs before refueling	0	7.1E-06
F...L	Cold POSs	66	2.1E-05
M...O, Q	Hot POSs after refueling	6	2.5E-05

The fire frequencies for room types were then distributed among individual rooms using various weighting methods, including the Berry model [5]. Even though the activities in certain plant areas may vary significantly during the plant shutdown, in absence of good data the fire frequencies were distributed among individual rooms using the same parameters as in power operation fire PRA.

3.4 Plant Response Model

Most of the important fire compartments in shutdown fire PRA were also important in power operation fire PRA. Because of this, detailed analyses of fire impact on plant equipment were available for easy implementation in the shutdown fire PRA, even though the plant response and some parameters might differ from the power operation. No new fire simulations were made. However, after the initial assessment, some further refinements were done to remove conservatism related to shutdown PRA.

In Loviisa NPP, most safety related systems are divided into two redundancies, both with two 100 % capacity safety trains. The Technical Specifications state that one redundancy can be taken out of service for maintenance activities in the cold plant operation states. However, usually the maintenance activities are restricted to one safety train at a time and systems, the maintenance of which is not on the critical path, can even be fully operable at any given time. In the internal event PRA this is not very significant, since the difference in risk is small between two and three or four available safety trains. However, the power and instrumentation cabling of both safety trains of one redundancy often follow the same routes, and are therefore vulnerable to fires. In these cases the number of available safety trains after the fire can vary from zero (both safety trains under maintenance) to two (no safety trains under maintenance). Therefore, to remove excessive conservatism, maintenance unavailability was modeled according to plant maintenance schedules.

Further assumptions made to reduce work load and/or conservatism include:

- Manual operation of motor operated valves was assumed successful in case of cable failures
- Recovery of one 400 kV transformer was assumed possible in case of 110 kV grid failure
- Main feed water and emergency feed water systems were assumed inoperable due to lack of knowledge of cable routes and small risk impact
- Operation of residual heat removal systems from the switchgear rooms was assumed possible in case of automation failures

Although the hot POSs were modeled using power operation fire PRA analysis cases, 342 new analysis cases, each including 1 to 5 fire scenarios, were created for the shutdown fire PRA. Most of them are applicable to all cold POSs, and some also for the POS E.

4. RESULTS

The CDF due to shutdown fire is 4.2E-07/yr. which is about 3 % of the total shutdown CDF and under 2 % of the total CDF. The result is roughly as expected and similar results have been reported by other VVERs [6].

As shown in Fig. 1, fires inside the control building dominate the risk. Switchgear rooms and cable spreading rooms below them are the biggest contributors. I&C system cabinets and related cable spreading rooms are also significant. Fires inside the reactor building only amount to 1 % of the total shutdown fire risk.

Figure 1: CDF Contributions of Plant Areas

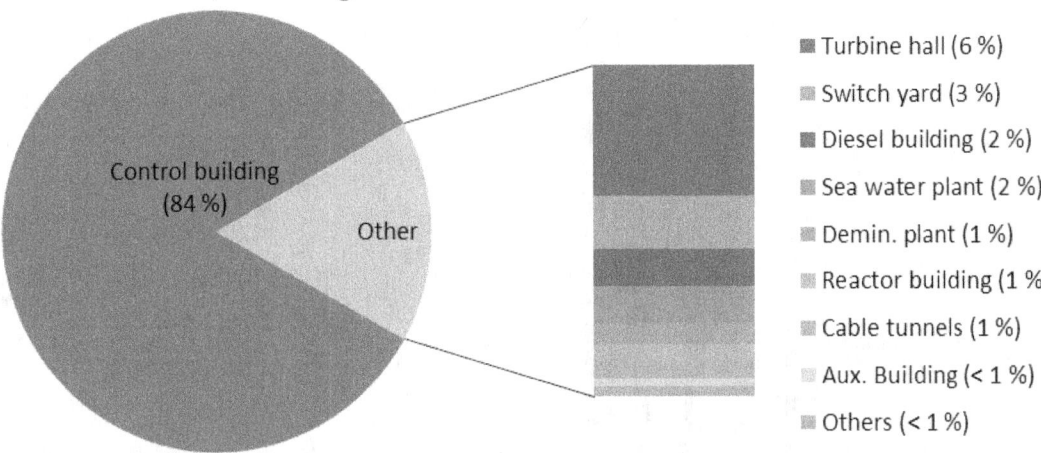

Contributions of individual IEs to the shutdown fire CDF is shown in Fig. 2. Most of the risk is related to IEs that cause loss of AC or DC power. In addition to LDCP and LOOP, many of the fire scenarios modeled as partial loss of service water are in fact caused by loss of AC supply to one or two trains.

Figure 2: CDF Contributions of Initiating Events

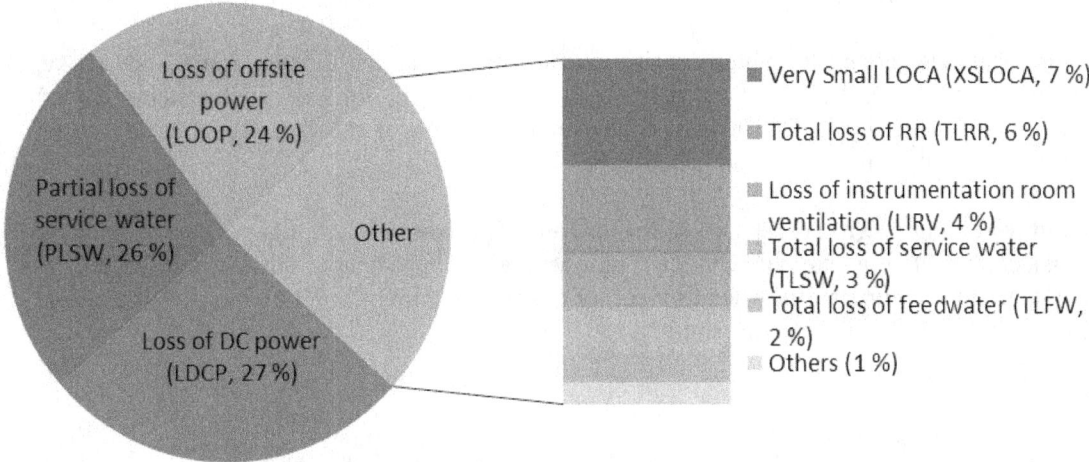

The distribution of CDF by POSs is shown in Fig. 3. The majority of the CDF is concentrated in the cold POSs, when a lot of systems are unavailable due to maintenance. This is where the importance of moving from conservative modeling towards best estimate is underlined. A considerable decrease in CDF has been achieved by taking into account that systems can be - and more likely than not are - available also when not required by the Technical Specifications. Although already normal plant practice, this verifies the importance of taking PRA information into account when planning the maintenance schedules.

Systems that are shared by both units and powered by either one also proved to be very important. This is due to many fire scenarios involving loss of power from one or both redundancies in one unit. E.g. the backup RHR system, originally built to mitigate the turbine building fire risk, is essential when coping with various loss of power events and the shutdown fire CDF would be about a hundred times higher without it. Operator performance in using the available and non-damaged systems is also important.

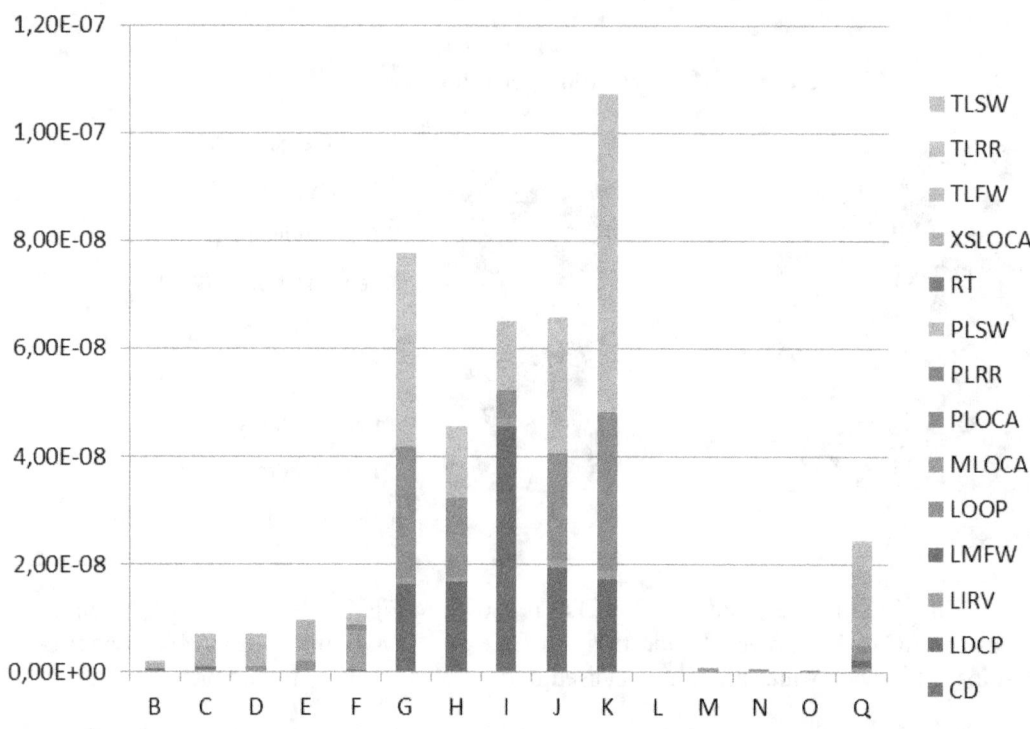

Figure 3: CDF Contributions of Plant Operating States

As the shutdown fire risk is relatively small compared to other shutdown risks, no new plant modification needs were identified.

The biggest uncertainties related to shutdown fire risk are related to the extent of fire and smoke damage, operation of the backup RHR system and the possibility of restoring systems under maintenance. Some uncertainty is also related to the quantification of the fire scenarios, as some scenarios are screened out by the cut-off limit.

During a forced repair shutdown, the relative contribution of fire risks is much higher. This is because many of the risks of a refueling and maintenance shutdown are absent. This can have a major effect on the allowed outage time optimization, when comparing the risk impact of continued power operation and a shutdown to repair failed safety equipment.

5. CONCLUSIONS

The shutdown fire risk is relatively small compared to other shutdown risks. However, the value of some modifications already implemented against fire risks and many good plant practices has been found to be even higher than expected.

With the completion of the shutdown fire PRA, the level 1 PRA model for Loviisa 1 is now comprehensive, including internal, weather and man-made external, seismic, flooding and fire events. This allows for greater confidence when using the PRA results for various applications, such as risk informed inspection of Technical Specifications and analyses of plant modifications.

References

[1]   M. Lehto et al. *"Fire risk analysis for Loviisa 1 during power operation"*, Proc. PSA'96, Park City, Utah, Sept. 29–Oct. 3, 1996.

[2]   DUKE Power Company. *"Fire Data for Loviisa 1 Fire-PSA"*, (1995).

[3]   *"OECD FIRE Database"*, Fire Incident Record Exchange project operated under OECD/NEA, (2008).

[4]   J. K. Vaurio and K. E. Jänkälä. *"Evaluation and comparison of estimation methods for failure rates and probabilities"*, Reliability Engineering and System Safety, 91, pp. 209–221, (2006)

[5]   D. L. Berry and E. E. Minor. *"Nuclear Power Plant Fire Protection - Fire Hazard Analysis (Subsystems Study Task 4) (NUREG/CR-0654)"*, U.S. NRC, 1979, Washington DC.

[6]   V. Vladimirova et al. *"Fire Risk Analysis of NPP "Kozloduy" Units 3&4"*, Proc. PSA2008, Knoxville, Tennessee, Sept. 7-11, 2008.

Modeling and Quantification of Team Performance in Human Reliability Analysis for Probabilistic Risk Assessment

Jeffrey C. Joe[a*] and Ronald L. Boring[a]

[a] Idaho National Laboratory, Idaho Falls, USA

Abstract: Probabilistic Risk Assessment (PRA) and Human Reliability Analysis (HRA) are important technical contributions to the United States (U.S.) Nuclear Regulatory Commission's (NRC) risk-informed and performance based approach to regulating U.S. commercial nuclear activities. Furthermore, all currently operating commercial nuclear power plants (NPPs) in the U.S. are required by federal regulation to be staffed with crews of operators. Yet, aspects of team performance are underspecified in most HRA methods that are widely used in the nuclear industry. Furthermore, there are a variety of "emergent" team cognition and teamwork errors (e.g., communication errors) that are 1) distinct from individual human errors, and 2) important to understand from a PRA perspective. The lack of robust models or quantification of team performance is an issue that affects the accuracy and validity of HRA methods and models, leading to significant uncertainty in estimating human error probabilities (HEPs). This paper describes research designed to model and quantify team dynamics and teamwork within NPP control room crews for risk informed applications, thereby improving the technical basis of HRA, which improves the risk-informed approach the NRC uses to regulate the U.S. commercial nuclear industry.

Keywords: HRA, PRA, Teams, Operating Crews.

1. INTRODUCTION

All currently operating commercial NPPs in the U.S. are required by the Code of Federal Regulations [10 CFR 50.54(m)] to be staffed with crews of operators. There are also a variety of "emergent" team cognition and teamwork errors (e.g., communication errors) that are 1) distinct from individual human errors, and 2) important to understand from a PRA perspective. That is, failures at the team level, such as Groupthink [1] and other failures in team cognition, are an emergent phenomenon (i.e., these errors cannot be made by an individual working alone; they can only occur when teams are working together), and can be significant contributors to plant risk. Team errors have been documented as contributing factors for a number of major industrial accidents. Both the Three Mile Island and Chernobyl NPP accidents had team errors (e.g., lack of team situation awareness, groupthink during problem solving and decision-making, and failure to do independent verifications) as causal factors. Similarly, Gladwell [2] attributes the crash of Korean Air flight 801 to the first officer and navigator being unwilling to challenge an error the captain made (i.e., the 'power distance' gap between the leader and subordinates was too large, leading to a break down in communication). Had these teams recognized how dysfunctional team dynamics can impact individual, team, and overall system performance, and implemented an approach to mitigate their effects, the severity of these accidents would have likely been attenuated, and could have possibly been avoided altogether.

Issues with teamwork, and especially accidents caused by failures in teamwork, have prompted a considerable amount of human factors research. One of the earliest efforts was Crew Resource Management [3], though a number of other researchers have since made significant contributions to the field [4, 5]. A number of additional researchers have also studied nuclear power plant operating crews and the challenges they encounter [6, 7, 8, 9, 10, 11], but none of this research has been done within the specific context of informing HRA and PRA. Some HRA methods do consider team cognition and teamwork [12, 13], but are not widely used, in part due to the fact that they are relatively new and complex methods. Most current HRA models focus on performance shaping factors (PSFs)

---

* Corresponding Author: Jeffrey.Joe@inl.gov

that affect the individual's cognition (e.g., how stress, procedures, and fitness for duty affect diagnosis and action, or alternatively, detection, sensemaking, decision-making, and action). Little effort has focused on team cognition and teamwork (i.e., the thinking and actions of the team), or the PSFs that affect team cognition and teamwork (i.e., quality of leadership, leadership style, awareness of what others on the team are thinking and doing, social pressure to conform, etc.). Given that crews operate commercial NPPs, HRAs examining individual cognition and errors may be, depending on what needs to be modeled and understood, at the wrong 'level of analysis'.

This 'level of analysis' issue can be further exacerbated by the fact that PRA models typically count the success or failure of the human at the team level. If one member of the crew fails at a task, but another member of the crew is able to recover, PRA counts this recovery as a successful human action. Thus, in the context of PRA for commercial NPPs, it may be more appropriate for HRA to model errors committed by humans at the team level. Furthermore, given that both 1) the maturity of conduct of operations in NPPs and 2) training guidelines provided by industry entities, such as the Institute for Nuclear Power Operations (INPO), have provided guidance on how operating crews should work together, there is team performance information available that could be included in HRA. Given the two issues, it is somewhat puzzling that HRA has focused on mitigating individual errors and failures, and seemingly ignored team level issues.

This discrepancy in the 'level of analysis' with respect to how HRA and PRA model human errors is illustrated in Figure 1. The fault tree on the left is a generic model of how human errors are modelled with the SPAR-H method [14]. The fault tree on the right is also based on SPAR-H, but models individual <u>and</u> team performance. This model of individual and team performance is at the same level of analysis that PRA models typically use to model human and/or team success or failure. However, most of the widely used HRA methods do not explicitly model human error within the context of working in teams.

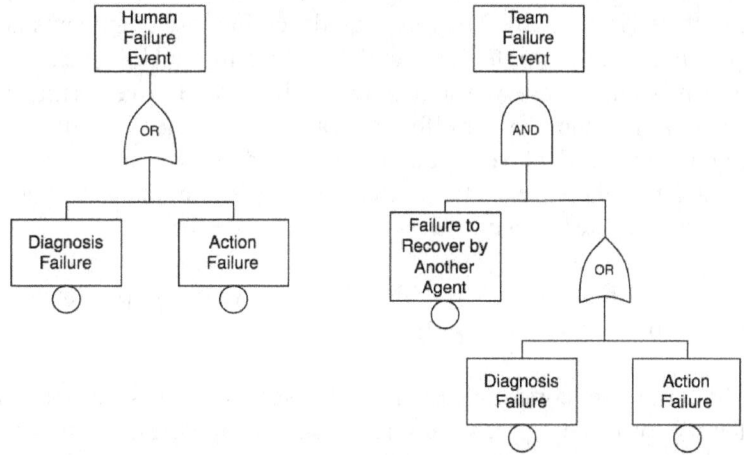

Figure 1. Fault trees showing how the logic changes when modeling individual <u>and</u> team level failures.

Examples of events that involve team performance issues relevant to PRA include:
1. Where an individual error (e.g., slip, lapse, mistake, etc.) manages to propagate through training and the rigorous crew conduct of operations, and ultimately challenges plant safety,
2. Where different members of the crew have all drawn the same erroneous conclusion from the available information, and then collectively commit an error that challenges plant safety,
3. Where different members of the crew have drawn different conclusions from the available information, fail to resolve the discrepancy (leading to outcomes such as incorrect diagnosis), and then collectively commit an error that challenges plant safety,
4. Where the crew is presented with a rare and complex event (e.g., severe accident) that they are initially unable to address because the required diagnoses and subsequent actions are beyond written procedural guidance.

Clearly, all of these examples are more appropriately modeled with the team failure fault tree than the individual human failure fault tree in Figure 1. Moreover, the subsequent success or failure of the crew's diagnoses and actions for these examples, and the safety consequences of their performance, will depend greatly on their ability to work as a team in order to effectively utilize their combined operating experience and training, among other things, to overcome the challenges described in these four examples. Given these issues, it is apparent to us that a technically defensible approach that is also methodologically straightforward is needed in HRA and PRA to address how emergent team performance and team errors contribute to risk.

2. RESEARCH ACTIVITIES

Analytical and experimental research is needed to establish the technical basis for modeling and quantifying team performance for inclusion in HRA and PRA. Ideally, the research activities need to be well-grounded in basic human psychology, compatible with current and future PRA frameworks, not be labor-intensive, be resistant to misuse and misapplication, and applicable to all NPP contexts involving teams (e.g., full-power, shutdown, normal operations, severe accidents, etc.). To achieve these ideals, two analytical tasks that could be pursued are 1) the feasibility of modeling and quantifying team performance using standard PRA tools and techniques, and 2) a review of widely used HRA methods, including [14, 15, 16, 17], to determine the suitability of including an HRA at the team performance level in these methods. Follow on experimental research tasks that make use of microworlds, or part-task simulation could then be conducted. These three tasks are described in more detail below.

For the modeling and quantification task, the feasibility of using standard fault tree and event tree approaches to model and quantify aspects of team performance, as identified through various sources of information about the conduct of operations in commercial NPPs, should be explored. For example, one source of information on nuclear conduct of operations is through the review of various documents from INPO, such as [18, 19, 20]. The INPO guidance in these documents is well grounded on NPP operational experience from multiple commercial utilities in the U.S. (i.e., "good practices"), and is based on research that is informed by basic human psychology. By overlaying and applying the event and fault tree logic onto the principles for effective team performance formulated by INPO, such as effective communication, independent verification, and other good practices for the conduct of nuclear operations, the research results from this task can effectively model team performance issues that are significant contributors to overall plant risk.

For example, one of INPO's "good practices" for NPP operators, which is also documented in [21], is to use three-way communication. According to [19]:

> The person originating the communication is the sender and is responsible for verifying that the receiver understands the message as intended. The receiver makes sure he or she understands what the sender is saying. First, the sender gets the attention of the receiver and clearly states the message. Second, the receiver repeats the message in a paraphrased form, which helps the sender know if the receiver understands the message. During this exchange, the receiver restates equipment-related information exactly as spoken by the sender. Third, the sender informs the receiver whether the message is properly understood, or corrects the receiver and restates the message. (pg. 17)

This approach to effective communication can be modeled using event and fault trees. Figure 2 shows the three-way communication approach modeled as an event tree, thereby showing the progression of various success and accident sequences. Note, however, that these event trees have been augmented to include more explicitly the team aspects of communication that are implied in three-way communication.

Three-way Communication

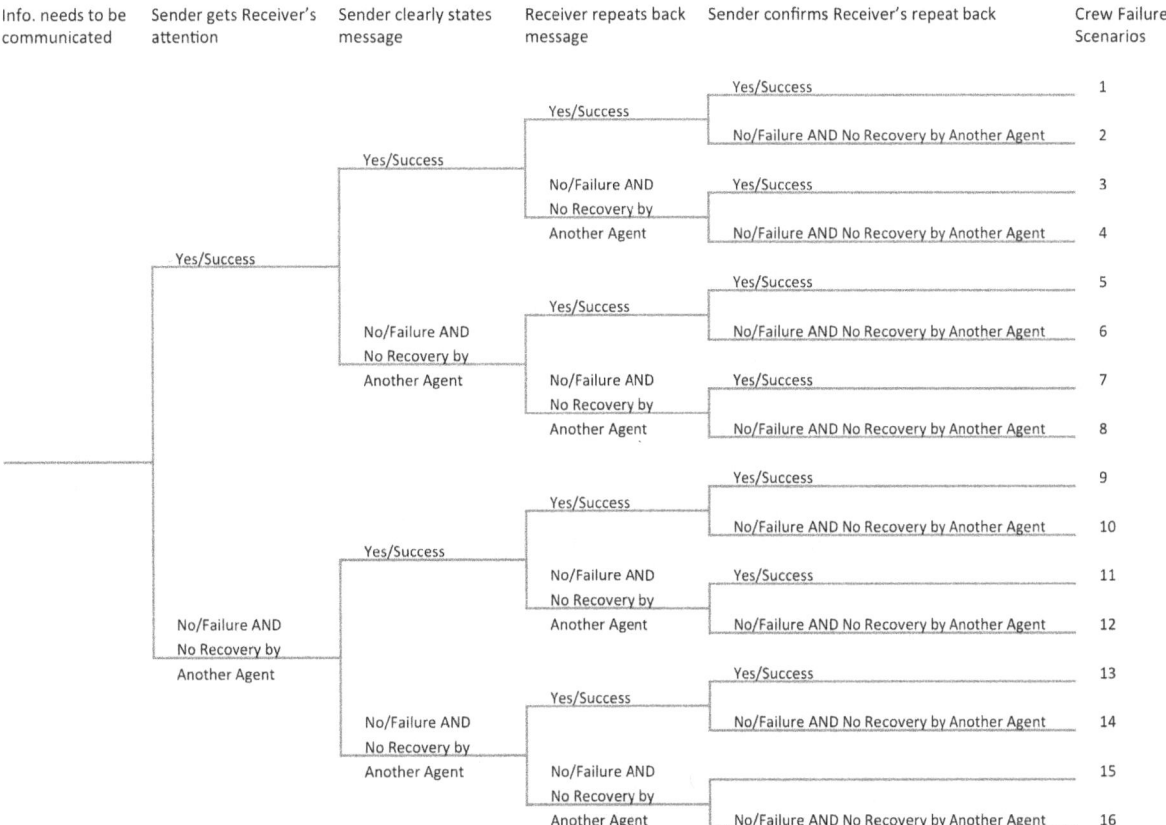

Info. needs to be communicated	Sender gets Receiver's attention	Sender clearly states message	Receiver repeats back message	Sender confirms Receiver's repeat back	Crew Failure Scenarios

Figure 2. Event tree for three-way communication

Figures 3a-3d shows example fault trees that model some of the credible events and PSFs, which contribute to three-way communication failures. Note that not all possible events and PSFs are depicted in these fault tree examples.

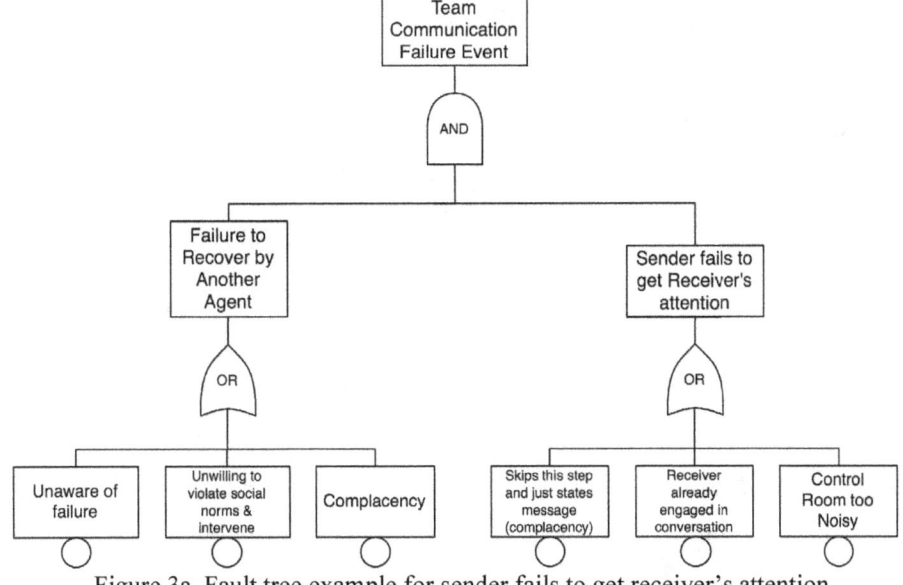

Figure 3a. Fault tree example for sender fails to get receiver's attention

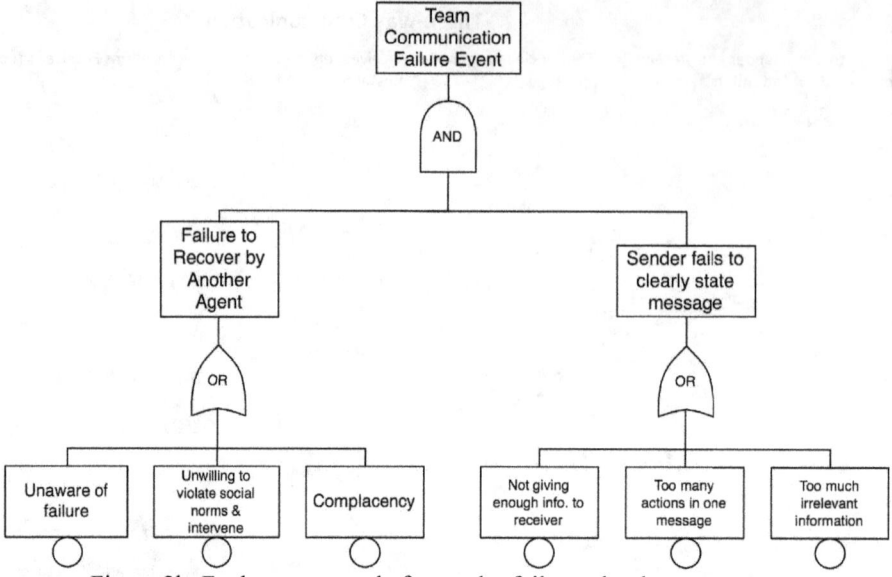

Figure 3b. Fault tree example for sender fails to clearly state message

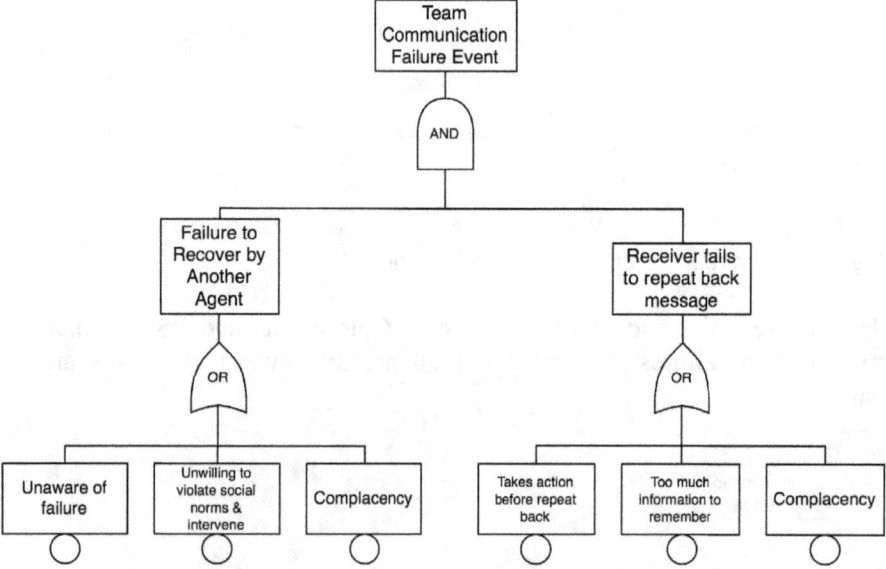

Figure 3c. Fault tree example for receiver fails repeat back

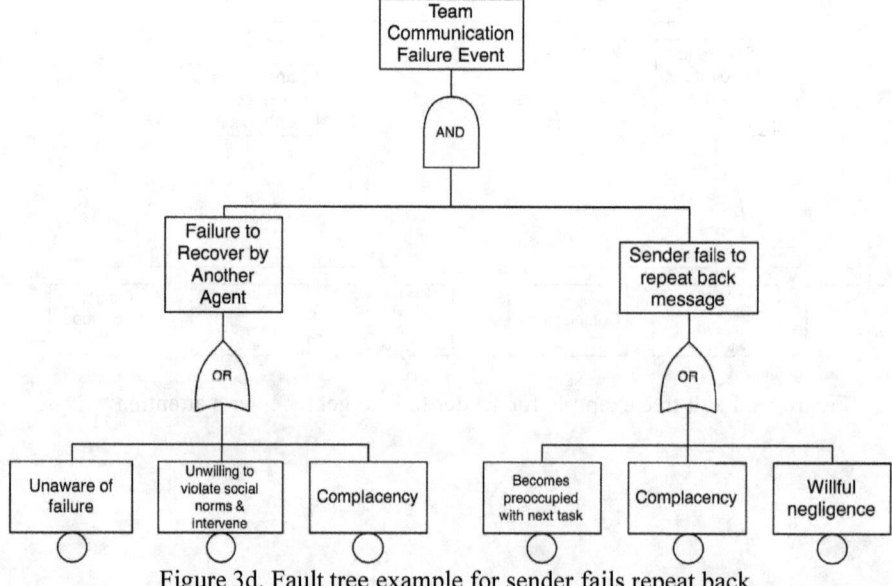

Figure 3d. Fault tree example for sender fails repeat back

Figures 2 and 3 show how it is feasible to model team performance in commercial NPPs in event and fault trees. Given that this is just one example from one information source (e.g., INPO), additional research to model other team aspects of commercial nuclear power's conduct of operations (e.g., independent verification and peer checking) that are obtained from relevant sources of information is needed in order to determine more completely whether team performance can be modeled using standard HRA and PRA tools. Other team errors (e.g., leadership/supervision errors), as described in [22], also need to be investigated for their risk significance across all operating modes in NPPs (e.g., full-power, shutdown, normal ops, severe accidents, etc.).

In order to assess how receptive existing HRA methods are to the inclusion of modeling and quantifying team performance, a literature review of those methods must be conducted. For those methods that can accommodate modeling and quantifying team performance, a technical basis for their inclusion needs to be established. In the event that no existing widely used HRA methods can accommodate team performance modeling and quantification, then work to develop a stand-alone team-level HRA method that will use 'best' practices [23], and tie in the most critical aspects of individual cognition that have a direct bearing on team cognition and teamwork should be performed. An example of such a stand-alone augmentation of existing HRA methods was recent work to address human reliability factors in fires [24]. While it is preferable to work within the framework provided by existing HRA methods, most HRA methods do not allow extension, and it is sometimes necessary to cover unique facets of human and team behavior under separate methods and aggregate them at the PRA level.

The experimental research task could make use of microworlds, or part-task simulation, to conduct proof-of-principle studies to validate the hypotheses that team performance issues are 1) emergent and different from individual performance issues, and 2) measurable contributors to risk. The results of this experimental research would help establish the technical basis that HRA methods can meaningfully model and quantify risk emanating from team performance issues. Microworlds have traditionally been used to assess individual operator performance, but research should expand on current microworld research platforms such as the one developed at the Idaho National Laboratory and University of Idaho [25] to assess group tasking and team performance. Early phase research could involve developing a crew framework representative of traditional crews (e.g., Reactor operator, Balance-of-plant/Turbine operator, Senior Reactor operator, Shift Technical Advisor), while later research would study crew performance using advanced digital instrumentation and control systems in traditional and nontraditional crew configurations, such as multi-modular reactor operations being proposed by small modular reactor (SMR) vendors. Previous research has suggested operators working on advanced digital workstations may tend to lose sight of the big picture and experience a breakdown in traditional three-way communication protocols (e.g., the keyhole effect documented in [26]). Human operators working as a team with automated agents is also a new concept for the operation of SMRs that will likely have some risk significance [27]. In short, the risk implications of a breakdown in teamwork in advanced control rooms are an important topic of research.

Additionally, as the NRC uses PRA and HRA to inform their risk-informed approach to regulating advanced NPPs, this research could leverage research conducted at places such as the Halden Reactor Project. Halden's simulator research to develop advanced control room instrumentation and control systems (e.g., large overview displays), which are designed to address team performance issues (e.g., team situation awareness) in a manner applicable to HRA and PRA, among other things, and their experience and subject matter expertise in running crews of NPP operators in simulators should provide important insights into team performance challenges in advanced control rooms.

In summary, more HRA related research is needed to address the effect of emergent team performance issues on overall plant risk. This research needs to focus on the modeling and quantification of team errors, which result from the interactions of individual crewmember errors, but also explicitly includes the possible recovery actions by other crewmembers or agents. Furthermore, by using fault and event trees to model and quantify team performance, and by using microworlds to validate the results, the subsequent team HRA method's products and results should meet the ideals described above, including compatibility with current and future PRA frameworks, and applicability to all NPP contexts involving teams (e.g., full-power, shutdown, normal ops, severe accidents, etc.).

3. CONCLUSION

Aspects of team performance are underspecified in most HRA methods that are widely used in the nuclear industry. The lack of robust models or quantification of team performance is an issue that affects the accuracy and validity of HRA methods and models, leading to significant uncertainty in estimating HEPs.

The objective of this research is to model and quantify team dynamics and teamwork (i.e., "team performance") within NPP control room crews for risk informed applications. We believe analytical and experimental research needs to be conducted to establish the technical basis for modeling and quantifying team performance for inclusion in HRA and PRA. Analytical work could begin with modeling and quantifying team performance. Experimental research could make use of microworlds, or part-task simulation, to conduct proof-of-principle studies to validate the hypotheses that team performance issues are 1) emergent and different from individual performance issues, and 2) measurable contributors to risk. In short, this research would establish the technical basis that HRA methods can meaningfully model and quantify risk emerging from team errors such that the epistemic uncertainty associated with their current under-specification would be minimized. The benefits of this research will be the improvement of the state-of-the-practice for HRA and PRA, which in turn supports an effective nuclear regulator that helps assure the safety and reliability of modern nuclear technologies.

Acknowledgements

INL is a multi-program laboratory operated by Battelle Energy Alliance LLC, for the United States Department of Energy under Contract DE-AC07-05ID14517. This work of authorship was prepared as an account of work sponsored by an agency of the United States Government. Neither the United States Government, nor any agency thereof, nor any of their employees makes any warranty, express or implied, or assumes any legal liability or responsibility for the accuracy, completeness, or usefulness of any information, apparatus, product, or process disclosed, or represents that its use would not infringe privately-owned rights. The United States Government retains, and the publisher, by accepting the article for publication, acknowledges that the United States Government retains a nonexclusive, paid-up, irrevocable, world-wide license to publish or reproduce the published form of this manuscript, or allow others to do so, for United States Government purposes. The views and opinions of authors expressed herein do not necessarily state or reflect those of the United States government or any agency thereof. The INL issued document number for this paper is: INL/CON-14-31339. The lead author also thanks Allison C. Joe for her help in preparing this article.

References

[1]   I. Janis, *"Groupthink: Psychological Studies of Policy Decisions and Fiascoes,"* Wadsworth Cengage Learning, 1982, Boston.

[2]   M. Gladwell, *"Outliers: The Story of Success,"* Brown and Company, 2008, New York.

[3]   B. Kanki, R. Helmreich, and J. Anca, *"Crew Resource Management,"* Academic Press/Elsevier, 2010, San Diego.

[4]   E. Salas and S. Fiore, *"Team Cognition: Understanding the Factors that Drive Process and Performance,"* American Psychological Association, 2004, Washington, DC.

[5]   C. Bowers, E. Salas, E., and F. Jentsch, *"Creating High-Tech Teams: Practical Guidance on Work Performance and Technology,"* American Psychological Association, 2005, Washington, DC.

[6]   J. Park, W. Jung, and J. Yang, *"Investigating the Effect of Communication Characteristics on Crew Performance Under the Simulated Emergency Condition of Nuclear Power Plants,"* Reliability Engineering and System Safety, 101, 1-13 (2012).

[7]   S. Kim, J. Park, and S, Byun, *"Crew Resource Management Training for Improving Team Performance of Operators in Korean Advanced Nuclear Power Plant,"* In IEEE International Conference on Industrial Engineering and Engineering Management (IEEE IEEM), 2055-2059, (2009).

[8]   Y. Chung, W. Yoon, and D. Min, *"A Model-Based Framework for the Analysis of Team Communication in Nuclear Power Plants,"* Reliability Engineering and System Safety, 94(6), 1030-1040, (2009).

[9]   P. Carvalho, M. Vidal, and E. de Carvalho, *"Nuclear Power Plant Communications in Normative and Actual Practice: A Field Study of Control Room Operators' Communications,"* Human Factors and Ergonomics in Manufacturing, 17(1), 43-78, (2007).

[10]   J. O'Hara and E. Roth, *"Operational Concepts, Teamwork, and Technology in Commercial Nuclear Power Stations,"* In C. Bowers, E. Salas, and F. Jentsch (Eds.) *Creating High-Tech Teams: Practical Guidance on Work Performance and Technology.* American Psychological Association, 2005, Washington, DC.

[11]   E. Roth, R. Mumaw, and P. Lewis, *"An Empirical Investigation of Operator Performance in Cognitively Demanding Simulated Emergencies,"* (NUREG/CR-6208). U.S. Nuclear Regulatory Commission, 1994, Washington, DC.

[12]   Y. J. Chang and A. Mosleh, *"Cognitive Modeling and Dynamic Probabilistic Simulation of Operating Crew Response to Complex System Accidents Part 1: Overview of the IDAC Model,"* Reliability Engineering and System Safety, 92, 997-1013, (2007).

[13]   A. Mosleh and M. Azarkhil, *"Dynamic Behavior of Operating Crew in Complex Systems: An Object-Based Modeling and Simulation Approach,"* University of Maryland Center for Risk and Reliability technical report, (2013).

[14]   D. Gertman, H. Blackman, J. Marble, J. Byers, and C. Smith, *"The SPAR-H Human Reliability Analysis Method,"* (NUREG/CR-6883). U.S. Nuclear Regulatory Commission, 2005, Washington, DC.

[15]   A. Swain and H. Guttman, *"Handbook of Human Reliability Analysis with Emphasis on Nuclear Power Plant Applications (THERP), Final Report,"* (NUREG/CR-1278). U.S. Nuclear Regulatory Commission, 1983, Washington, DC.

[16]   Electric Power Research Institute, *"An Approach to the Analysis of Operator Actions in Probabilistic Risk Assessment, Final Report,"* (EPRI TR 100259), 1992, Palo Alto, CA.

[17]   S. Cooper, J. Wreathall, C. Thompson, M. Drouin, and D. Bley, *"Knowledge-base for the New Human Reliability Analysis Method: A Technique for Human Error Analysis (ATHEANA),"* (BNL-NUREG-63256). Brookhaven National Lab, 1996, Upton, NY.

[18]   Institute of Nuclear Power Operations, *"Verification Practices: Process Description,"* (INPO AP-931), 2004, Atlanta, GA.

[19] Institute of Nuclear Power Operations, *"Human Performance Tools for Workers: General Practices for Anticipating, Preventing, and Catching Human Error During the Performance of Work,"* (INPO 06-002), 2006, Atlanta, GA.

[20] Institute of Nuclear Power Operations, *"Human Performance Reference Manual,"* (INPO 06-003), 2006, Atlanta, GA.

[21] U.S. Department of Energy, *"Human Performance Improvement Handbook: Human Performance Tools for Individuals, Work Teams and Management,"* (Vol. 2; DOE-HDBK-1028-2009), 2009, Washington, DC.

[22] A. Whaley, J. Xing, R. Boring, S. Hendrickson, J. Joe, and K. Le Blanc, *"Building a Psychological Foundation for Human Reliability Analysis,"* (NUREG-2114). U.S. Nuclear Regulatory Commission, 2014, Washington, DC.

[23] A. Kolaczkowski, J. Forester, E. Lois, and S. Cooper, *"Good Practices for Implementing Human Reliability Analysis (HRA),"* (NUREG-1792). U.S. Nuclear Regulatory Commission, 2005, Washington, DC.

[24] S. Lewis and S. Cooper, *"Methodology for Low Power/Shutdown Fire PRA,"* (NUREG-1921). U.S. Nuclear Regulatory Commission, 2012, Washington, DC.

[25] R. Boring, D. Kelly, C. Smidts, A. Mosleh, and B. Dyre, *"Microworlds, Simulators, and Simulation: Framework for a Benchmark of Human Reliability Data Sources,"* Joint Probabilistic Safety Assessment and Management and European Safety and Reliability Conference, 16B-Tu5-5, (2012).

[26] D. Woods, E. Roth, W. Stubler, and R. Mumaw, *"Navigating Through Large Display Networks in Dynamic Control Applications,"* Proceedings of the Human Factors and Ergonomics Society Annual Meeting, 34(4), 396-399, (1990).

[27] J. Joe, J. O'Hara, H. Medema, and J. Oxstrand, *"Identifying Requirements for Effective Human-Automation Teamwork,"* Proceedings of the 12th International Conference on Probabilistic Safety Assessment and Management (PSAM 12, Paper #371), (INL/CON-14-31340), 2014, Honolulu, HI.

Comparison of Task Loads between Usages of Computer-based Procedures in an Advanced Control Room

Yochan Kim[a*], Wondea Jung[a], and SeungHwan Kim[a]

[a]Korea Atomic Energy Research Institute, Daejeon, Republic of Korea

Abstract: With the development of a computer-based control room in an APR1400, the behaviors of operators in the control room have changed. To investigate the effects of the computerized instrument and control systems on workloads, the workloads of operators in an APR1400 who employ three different usages of a computer-based procedure were compared. The COCOA framework, a task-loading approach of workload evaluation, was employed to evaluate the workloads, and some statistical analyses were conducted to compare them. We performed a total of 22 experiments in a full scope simulator of an APR1400 under LOCA and SGTR scenarios, and obtained workload scores in cognitive, communicative, and operative dimensions. The results showed that the SS-centric usage requires many activities to the SSs, and the other usages require fewer activities to the SSs than the SS-centric usage. Based on the findings, we discussed whether the workloads between operators in an MCR can be adjusted by the CPS usages.

Keywords: Computer-based control room, Computer-based procedure, Workload, COCOA.

1. INTRODUCTION

With the development of a computer-based control room in an APR1400 (Advanced Pressurized Reactor-1400), the behaviors of operators in the control room have changed. Main human-system interfaces of a computer-based control room include an advanced alarm system, computer-based procedure (CBP) system, and graphic display system [1]. Because operators interact with the plant system using digital devices such as touch screens or mice, it is certain that the behaviors of the operators differ from the behaviors in conventional control rooms.

However, Kim et al. indicated that digitalized interfaces can also affect the cognitive tasks or activities of operators [2]. For example, shift supervisors (SSs) in a computer-based control room can directly notice or monitor plant information from workstation-based information systems or a large display panel; hence, the SSs can accurately understand the plant situation without a report from the board operators (BOs). Meanwhile, the new features of digitalized control rooms may demand new operative tasks, which have not yet been performed in conventional control rooms. For example, the control room of an APR1400, which is a computer-based control room, requires SSs to follow a CBP by clicking on every instruction. Because clicking and following the CBP is a new and additional task to SSs who manage the overall situation of a plant, it is possible for SSs to have a higher workload than other operators and SSs in conventional control rooms.

In this paper, we compared the workloads of operators in an APR1400 who work with three different usages of the CBP. The first usage is similar to the method used when operators in conventional CE-based plants usually follow an emergency operating procedure. Using the first usage, only the SS checks all instructions of the CBP line by line and directs actions to the other operators. The second usage appoints each BO as a manager of the selected steps. Instructions in each step of the procedure are then conducted and checked by the appointed BOs in the CBP. After a BO checks all instructions of the step, the SSs simply review the behaviors of the BO during the step and progress to the next step. The third usage requires an SS to simply check only the key steps in the procedure. The SS commits the progression of the procedure to the appointed BOs in the CBP and checks the BO's operations during the key steps. The workloads of the operators were compared by the COCOA (cognitive, communicative, and operative activity) framework [3]. The COCOA framework, a task-loading approach of workload evaluation, calculates the operator's workloads based on a task analysis or experiments. We conducted a total of 22 experiments in a full scope simulator of an APR1400 under LOCA (Loss of coolant accident) and SGTR (steam generator tube rupture) scenarios, and obtained the workload scores in cognitive, communicative, and operative dimensions.

The remainder of this paper is organized as follows. Section 2 introduces techniques including the COCOA framework to evaluate the operator workloads. Section 3 explains the experimental environment and CBP usages to compare the workloads in different CBP usages. Section 4 shows the results of the comparison. Section 5 discusses insights regarding the distribution or balance between the operator workloads.

2. RELATED WORK

The techniques used to evaluate a workload can be categorized by self-assessment or subjective rating scales, performance measures, psychophysiological measures, and task-loading measures [4, 5]. The subjective rating scales assess the feelings of the workload, effort, mood, or fatigue based on self-reported rankings or scales. Although various scales have been used, the following scales are primarily addressed from the literature: NASA-TLX (NASA task load index), SWAT (subjective workload assessment technique), MCH (modified cooper harper scale), ZEIS (sequential judgment scale), and so on. The performance measures evaluate how well a subject performs the given task using the performance time, error or success rate, or response latency. To analyze the margin of mental resources and elevate the diagnosticity of the measures, the subjects are sometimes demanded to perform additional tasks. Psychophysiological measures are used to calculate the physical responses of the body of a subject. Cardiac activity, brain activity, respiratory activity, speech measures, or eye activity are used to evaluate a workload. The task-loading measures are used to analytically evaluate the demands of given tasks using mathematical modeling, task analyses, simulation modeling, or expert opinions.

The COCOA framework was developed to sensitively evaluate the workloads of operators in digitalized systems and explicitly compare the workloads between operators who collaborate as a team [3]. This framework provides the taxonomy of activities conducted by operators in computer-based control rooms. Three dimensions of activities are defined in Table 1 below. The cognitive activities are defined using task definitions of the CORA method [6]. The communicative activities were obtained by selecting frequent activities from the speech-act code scheme [7]. In addition, the operative activities were established based on a task analysis of the operations in the control room of an APR1400.

Table 1: Operator activities of COCOA framework

Dimension	Activity	Description
Cognitive activity	COMPARE	Comparing two or more entities of system states.
	DIAGNOSE	Recognizing or determining the cause of the system states by signals or parameters.
	EVALUATE	Checking a system state with consideration of other system parameters or states
	EXECUTE	Performing a single prescribed action (ex. open, close, turn on, etc.)
	MAINTAIN	Sustaining a specific system state by executing or regulating systems.
	MONITOR	Continuously observing the system states or parameter.
	RECORD	Writing down or logging the system states or events.
	REGULATE	Changing the quantity, speed, or direction of the system parameters or states.
	VERIFY	Checking an entity of system states
	SCAN	Briefly reviewing specific system states by displays or other information
	PLAN	Formulating a path to achieve specific goals.
	IDENTIFY	Recognizing the overall state of a specific system.
Communicative activity	COMMAND	Ordering an operator to execute or regulate a system or component.
	COMMAND-ACK	Informing whether a listener understood a command.
	INQUIRY	Asking about system states or parameters.

	REPLY	Answering a question.
Operative activity	SWITCH_SCR	Altering a display screen to read or operate specific parameters or states.
	OPEN_CTRLPNL	Opening a control panel to operate a specific system.
	CLOSE_CTRLPNL	Closing a control panel that has been opened.
	CLICK_EXECUTE	Pushing a button to operate a single entity (ex. opening a valve)
	CLICK_REGU_UP DN	Pushing a direction button to regulate a system parameter or state (ex. opening a valve)
	CLICK_REGU_FA STUPDN	Pushing a double-direction button to regulate a system parameter or state (ex. opening a valve)
	CLICK_ENABLE D	Pushing a button to enable a control panel to be usable.
	CLICK_MANNUA L	Pushing a button to manually operate a single entity.
	CONF_STEP	Clicking a button confirming that all instructions of a step have been completed.
	CONF_SUBSTEP	Clicking a button confirming that an instruction of a step has been completed.
	ACK_ALARM	Checking and silencing a notifying alarm
	CONF_CHANN	Pushing a button to verify the channel of an entity to be operated.

In this study, the numbers of activities that operators conducted during the experiments were counted. First, which activities can be used was identified from the required procedures. Who conducted the activities were then analysed by audio-video records. To calculate the numbers of operative activities, all operative behaviours of crews were also tracked using the video records.

3. METHOD

3.1. Control Room of APR1400

The APR1400 is a CE-type nuclear power plant, which was developed based on the OPR1000 (Optimized Power Reactor-1000) design. Hence, there are similarities between the two plants. First, the crew in a main control room consists of five members: SS, RO (reactor operator), TO (turbine operator), EO (electric operator), and STA (shift technical assistant). Most responsibilities of each member between the two plants are also similar. When an emergency situation occurs, an SS generally follows the procedures and instructs the BOs including RO, TO, and EO to obtain information or take an action for coping with the situation. The BOs then inform the plant situations or execute the actions as the SS instructed. In this study, who follows the procedure, an issue related to CBP usages, can be changed in the experiment design, as described in section 3.2. An STA usually manages the critical safety functions of the plant. In addition, the contents of the emergency operating procedures of the APR1400 and OPR1000 are quite similar.

Because the APR1000 employs a computer-based control room, all operators use a workstation-based information system. Each operator obtains information from four personal displays, a large display panel, or advanced alarm system. One of the four displays is dedicated to an operating CBP system. All operators control the system status, parameters, or CBP system by clicking mice or touch screens.

The example of CBP screens can be seen in figure 1. When an operator who controls the CBP enters a step, the CBP shows instructions of the steps in the right part of it. After the CBP controller performs an instruction and clicks the instruction, the CBP then highlights the box with a black circle. If all instructions are checked and black circles are shown the instructions, the controller can click the 'complete' button, which is in the lower part of CBP. After the controller reviews what they did and clicks the complete button, the CBP shows the next step.

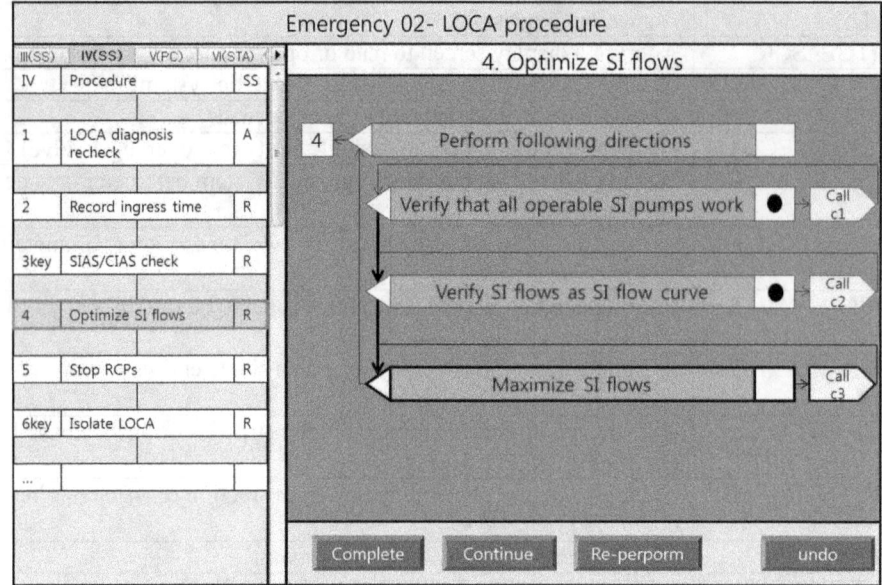

Figure 1: Snapshot of CBP system (conceptual image)

3.2. CBP Usages

Twenty-two experiments were conducted to compare the workloads of operators according to the CBP usages. Currently, the reserved operators of the APR1400, who are participants of these experiments, use the SS-centric usage, which has been used in many conventional control rooms. That is, only an SS manages the CBP including the clicking activities, while the BOs simply follow the SS's directions. The BO-SS-collaborative usage, which is newly proposed for this study, requires an SS to ingress a new step and entrust a BO with conducting the instructions of the step. The BO-centric usage lets the BOs manage most steps of the CBP, and an SS simply reconfirms the BOs work during key steps. Table 2 summarizes the difference between the CBP usages. It is noticeable that an SS can interrupt the behaviors of other operators and manage the CBP at any time, if necessary, even when a BO manages the step under BO-SS collaborative or BO centric usage. In addition, the BOs should announce what they did when they click an instruction of the CBP.

Table 2: Three CBP usages

Usages	Role of SS about CBP control	Role of BO about CBP control
SS-centric (current usage)	An SS checks and progresses all sentences of CBP.	BOs monitor CBP screens and follow the SS's directions.
BO-SS-collaborative (newly proposed)	An SS initially instructs to a BO to check a step. After the BO checks all instructions of the step, the SS review the CBP.	After the SS's initial announcement, the BO checks and performs the instructions written in the given step.
BO-centric (newly proposed)	When progressing to a key step, An SS reviews results come by the previously performed steps and instructs things to do until the next key step to BOs.	Until the next key step, the appointed BO verifies all checkpoints of CBP and performs the instructions of procedures.

3.3. Experimental Design

Experiments in a full-scope simulator were conducted under the following conditions.
- Independent variable: CBP usage
- Dependent variable: workload analyzed by COCOA method
- Participants: reserved operators of APR1400
  - Three teams in experiments for SS-centric usage
  - Five teams in experiments for BO-SS-collaborative
  - Three teams in experiments for BO-centric usage
- Scenario:

- LOCA (performing all steps of standard post trip action and diagnostic procedure and steps number 1 through 18 of LOCA procedure)
- SGTR (performing all steps of standard post trip action and diagnostic procedure and steps number 1 through 16 of SGTR procedure)

The quantity of operator activities for each CBP usage was compared with activities of other usage. In addition, a statistical analysis with a one-way ANOVA followed by Duncan's post hoc test (P < 0.05). The ANOVA and Duncan tests appraise which groups have different averages with others.

4. RESULTS

The activity frequencies of operators during both LOCA and SGTR are depicted in figures 2 and 3. The blue, red, and green bars indicate the average frequencies of SS-centric, BO-SS-collaborative, and BO-centric usages, respectively. The annotated terms, such as SD (significantly different) and NSD (not significantly different), indicate that the quantities of two bars that the terms indicate are statistically equal or not. The NSDs over tailed range marks imply all frequencies under the marks are not statistically different.

From the resulted values of the workloads, we obtained the following findings:
- With the SS-centric usage, the SSs conducted many cognitive, communicative, and operative activities than other operators.
- The BO-SS collaborative and BO-centric usages significantly reduced the SS's cognitive and operative activities than the SS-centric usage.
- The RO's activities during the BO-SS collaborative and BO-centric usages are larger than activities during the SS-centric usage.
- Significant differences between workloads caused by the BO-SS collaborative and BO-centric usages were not found.
- The TOs conducted more cognitive and communicative activities during the SGTR than the LOCA situation.
- The EOs generally conducted fewer activities than other operators and were not affected by the CBP usages.

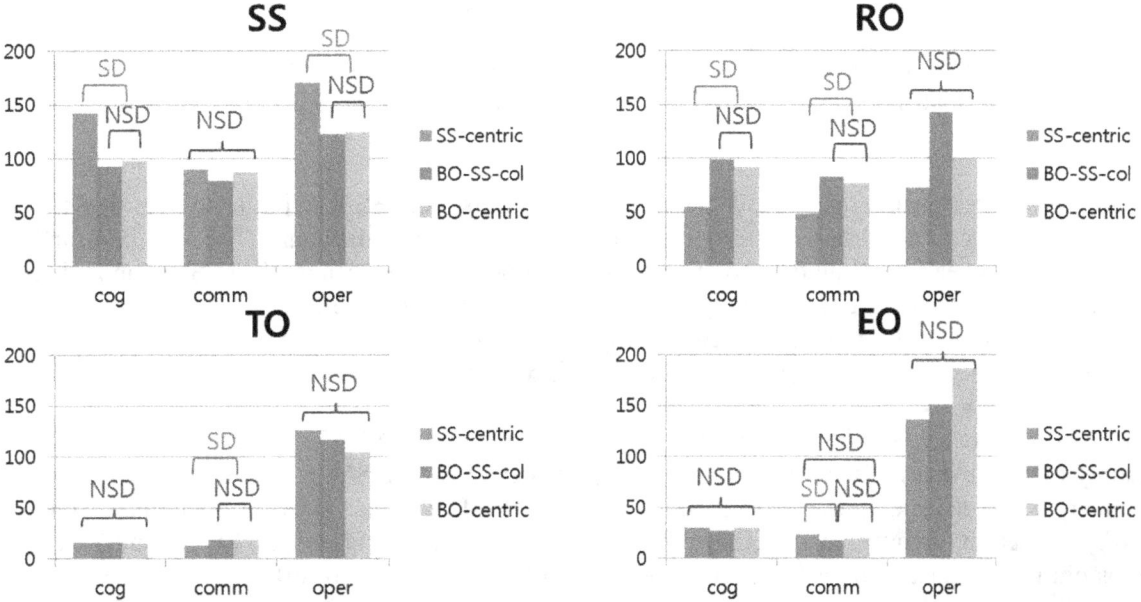

Figure 2: Difference between activity frequencies of CBP usages during LOCA scenario

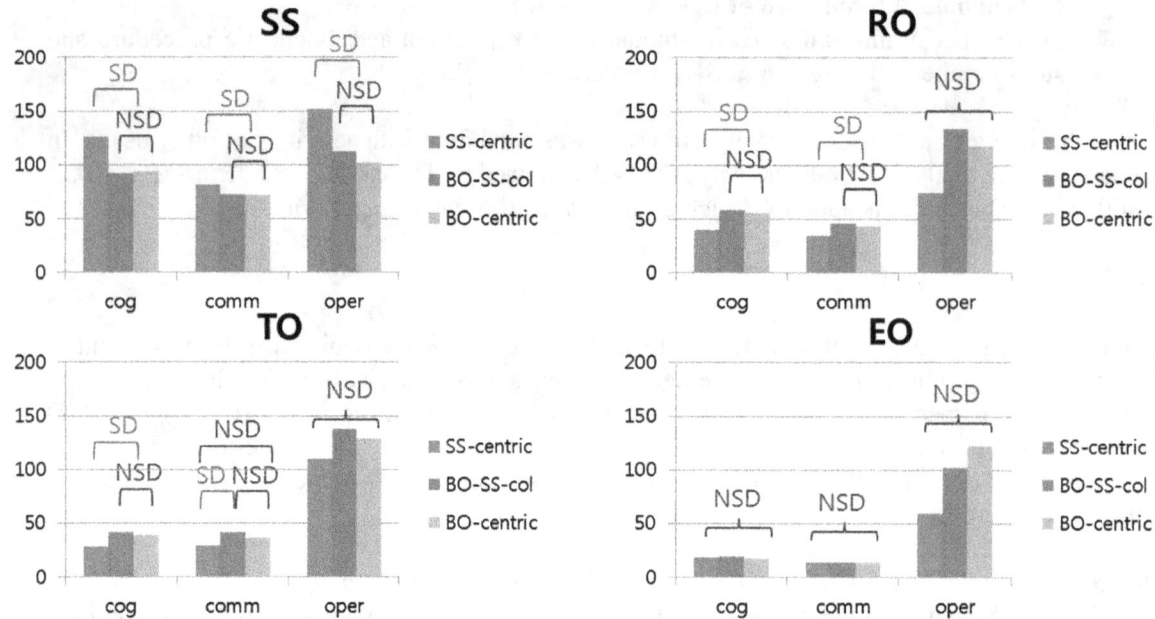

Figure 3: Difference between activity frequencies of CBP usages during SGTR scenario

5. DISCUSSION

It was found that the SS-centric usage requires many tasks of SSs. This is because controlling the CBP carries many operative activities as well as cognitive activities. It is not clear to insist which usage is a better or worse strategy than another, because this issue depends on the capability of the operators. It is noticeable that the COCOA framework evaluates workloads coming from given tasks rather than a lack of subjective knowledge, experience, or ability. However, it is also obvious that the SSs of SS-centric usage, which should incessantly identify the overall situations of the plant, have too many activities related to the CBP.

Thus, it is important that the other usages decreased the SS's activities. During the LOCA scenarios, the levels of SS activities were similar to the RO activities in the BO-SS collaborative and BO-centric usages. During SGTR scenarios, the RO and TO activities were similarly increased in both usages. These results imply that completely or partly entrusting BOs with CBP control transfers the task loads of SSs to BOs. The BO-SS collaborative and BO-centric usages enabled the SSs to manage the overall situations of the systems.

Although it seems reasonable that the BO-centric usage transfers more SS's activities to BOs than the BO-SS collaborative usage, the transferring effects were not much different. The ANOVA and Duncan tests also revealed that all types of activities of all operators between the BO-SS collaborative and BO-centric usages are not significantly dissimilar. This is because the SSs of BO-centric usage also needed to continuously monitor procedure progressions and the task loads of monitoring works were similar to the task loads during the BO-SS collaborative usage.

The differences between activity frequencies of TOs in LOCA and SGTR scenarios are probably affected by the differences of task characteristics of both scenarios. The SGTR requires more turbine- or secondary-loop-related activities than the LOCA situations. This reason provides a similar insight about why the activity frequencies of EOs were lower than for other operators. The LOCA and SGTR scenarios did not demand many activities of EOs. Station blackout or loss of offside power scenarios may show different results.

6. CONCLUSIONS

We evaluated the workloads of operators in a computer-based control room by the COCOA method, which was recently developed. Three types of CBP usages were defined and the effects of these usages on the workloads were investigated. The obtained results showed that the workloads between operators in a control room can be reassigned according to the CBP usages. Newly proposed ergonomic features can enhance the operator's efficiency or accuracy. However, in certain cases, these features can incur additional efforts or workloads of the operators. To prevent excessive workloads on specific operators, it is necessary to consider a reallocation of operative tasks, a customization of the interfaces, or system education. We believe that the results of these evaluations may provide an empirical basis of this consideration.

Acknowledgements

This work was supported by Nuclear Research & Development Program of the National Research Foundation of Korea grant, funded by the Korean government, Ministry of Science, ICT & Future Planning (Grant Code: 2012M2A8A4025991).

References

[1]   J. O'Hara, J. Higgins, J. Kramer. "*Advanced information systems design: technical basis and human factors review guidance*" (NUREG/CR-6633), Washington DC: US NRC, (2000).
[2]   Y. Kim, (2012). Y. Kim Y, J. Kim, S. Jang, W. Jung. "*Empirical investigation of communication characteristics under a computer-based procedure in an advanced control room*", J Nucl Sci Technol, 49 (10), pp. 988-998, (2012).
[3]   S. Kim, Y. Kim, W. Jung. "*Proposal on Framework for Measurement of Workload of Operators in Environment of Advanced Main Control Room*", In: ANS 2013 Winter Meeting and Nuclear Technology Expo, Washington, DC, (2013).
[4]   D. Embrey, C. Blackett, P. Marsden, J. Peachey. "*Development of a Human Cognitive Workload Assessment Tool*", Human Reliability Associates, (2006).
[5]   B. Cain "*A Review of the Mental Workload Literature*", Technical Report, RTO-TR-HFM-121-Part-II, (2007).
[6]   E. Hollnagel. "*A cognitive task analysis of the SGTR scenario*", Technical report, NKSRAK-1(96)R3, (1996).
[7]   S. Kim, J. Park, S. Han, H. Kim. "*Development of extended speech act coding scheme to observe communication characteristics of human operators of nuclear power plants under abnormal conditions*", J Loss Prevent Proc, 23, pp. 539-548, (2010).

Study on Analysis Method of Operator's Errors of Situation Awareness in Digitized Main Control Rooms of Nuclear Power Plants

Pengcheng Li[a], Li Zhang[a,b], Licao Dai[a], Jianjun Jiang[a], and Difan Luo[a]

[a] Human Factor Institute, University of south China, Hengyang, People's Republic of China
[b] Hunan Insitute of Technology, Hengyang, People's Republic of China

Abstract: Situation awareness (SA) is a key element that impacts operator's decision-making and performance in nuclear power plants (NPPs). The subsequent complex cognitive activities can not be correctly completed due to errors of situation awareness (ESA), which will lead to disastrous consequences. In order to investigate and analyze operator's situation awareness error in digitized main control room (DMCR) of the nuclear power plants, the model of ESA is established, the classification system of SAE is developed based on the built SAE model, and the method of ESA is also constructed on the basis of the observation of simulator and operator surveys. Finally, a case study is provided to illustrate the concrete application of the method. It provides a theoretical and practical support for the operator's SAE analysis in the digitized main control room of nuclear power plants

Keywords: Situational Awareness Error; Analytical Method; Digitized Main Control Room

1. INTRODUCTION

Situation awareness (SA), which is used within human factor research to explain to what extent operators of safety–critical and complex real systems know what is going on the system and the environment, is considered a prerequisite factor for effective decision making and performance[1]. In the accident disposition process of complex industrial systems, operators can not correctly complete the following complex cognitive activities which will bring disastrous consequences because of loss of situational awareness (LSA). For example, the Three Mile Island nuclear accident, the operator failed to understand the state of primary loop in Three Mile Island nuclear accident [2], and pilots lost the proper understanding of flight status and so on in the variety of aviation accidents[3]. Endsley [4] found that the cause of 88% of commercial aviation accidents caused by human error has some kind of connection with LSA. Jones and Endsley [5] pointed out that 69% of incident reports contain information gathering errors of SA of air traffic controllers by incident report analysis occurred in the field of air traffic control. Therefore, SA is a key element that impacts operator's decision-making and performance, and ultimately may lead to accidents.

With the improvement of level of system automation, operator's errors of situation awareness (ESA) become increasingly prominent. In this light, since the late 1980s, research on SA continues to gradually receive a considerable amount of attention from the high-risk field of civil aviation, air traffic control, nuclear power plants, hospital etc. It has become a hot research, and a number of theoretical model related to SA are established, in which the famous one is the SA model established by Endsley[6] based on information processing method. The model divides SA into three levels: Perception of the elements in the current environment (Perception), comprehension of the current situation (Comprehension) , and projection of future status (Projection), and the factors affecting SA are identified, which includes individual factors and system / task factors (see Figure 1). Endsley's SA model seems generic and comprehensive, as it is based on general cognitive processes, and provides a broad theoretical analysis framework for many application areas, but it doesn't in detail describe the cognitive processes and influencing factors of SA. For example, the identification of causes of system malfunction, it should belong to SA. Therefore, it is not conducive to the classification and investigation of ESA, and it is also difficult to make specific preventive measures to prevent ESA.

In addition, with the rapid development of computer, control, and information technology, the instrumentation and control (I&C) system of nuclear power plants (NPPs) is transformed from traditional analog control to digital control, the man-machine interface (MMI) in control room is transformed from the traditional monitor and control board to the computer-based workstation. In this respect, the role of operator is changed from the past "manipulator" to "monitor and manager", and the operating environment in advanced digital main control room (MCR) is very different from the

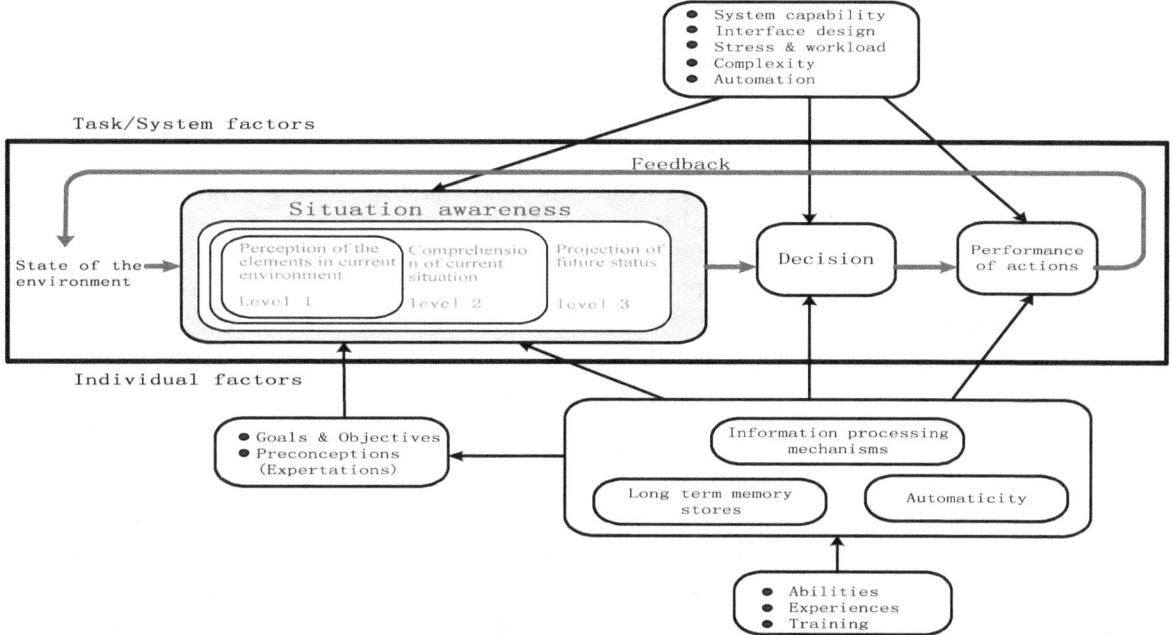

Figure 1 Endsley's Three-level model of situational awareness

environment in traditional analog control room. Digital human-system interface (HSI) has changed the operating context including information display (a huge amount of information with limited display), procedures (computerized procedures), control (soft control), task (interface management tasks) and so on, which may bring new human factor problems, especially operator's SA problems [7]. For example, the control panels in conventional control room are spatially fixed, and information displays on panels are visual, it is useful to understand the status of the entire system of NPPs for the operators, but the positions of information showed by modern computer-based display are not fixed, the relationships among information are divided, information displayed on screen is more abstract and upper information, and computer-generated information displays are not limited by physical space, the amount of displayed information is huge etc., but information that can be directly observed is limited through computer screen, a lot of dynamic information is hidden. To obtain plant state information, operators must implement so-called "interface management tasks" such as navigating, configuring and arranging etc.,[8]. The "secondary tasks" related above will increase operators' cognitive load, consume their attention resources and generate a "keyhole effect "[9]etc. , thus affect operator 's SA, which makes operator out-of-the-loop [10]. Therefore, SA issues may be more prominent in digital main control room of NPPs, digitization of HSI has changed the operator's SA cognitive mechanisms, so the traditional theoretical models of SA are difficult to meet the current requirements of ESA analysis. Therefore, we attempt to establish a framework of ESA analysis to guide ESA analysis.

2. OPERATOR'S SA MODEL IN DIGITAL MAIN CONTROL ROOMS

2.1. Operator's SA in Digital Main Control Rooms

With the development of technology and improvement of automation level, operator's major tasks represent as cognitive tasks in complex social-technical systems such as NPPs, including: (1) monitoring and detection, (2) situation assessment, (3) response planning, and (4)response implementation [11]. Endsley[6] views SA as "the perception of the elements in the environment within a volume of time and space, the comprehension of their meaning, and the projection of their status in the near future". According to Endsley's definition, situation awareness process includes a series of cognitive activities (including at least monitoring and detection, situation assessment), these errors related to the cognitive activities are part of ESA. In this light, we think that SA is "operators actively try to construct a coherent, logical explanation to understand the unit/system state and what is going on by various information processing on the basis of the collected parameter information in digital NPPs", this series of cognitive processes and assessment results are called SA.

Operator's behavioral characteristics of SA are identified by field observations of simulator training and operator interviews in digital main control room of NPPs. In digital control system of NPPs, many of complex tasks are replaced by automation (e.g. whether high head safety injection operate? whether loss of feedwater? Whether at least one steam generator with activity< high measurement limit? etc.) due to increased automation level, which are diagnosed and determined by operator support system. In the information collection processes of SA, when an abnormal event occurs, one or more alarms will appear, operator will examine the occurred alarm and its reasons to take appropriate measures to deal with it. Information that operator needs to monitor is provided by computer-based display screen, the information that requires operator monitoring is more dispersed. For example, when operator needs to collect the failed components to determine the failure state of a system, operator needs to monitor the status of multiple parameters (e.g. error signal input to controller, requirement information of components, and actual state of components, etc.). In general, in the stage of monitoring and detection, the operator's task is mainly to gather information, including single information and more information. The operator's cognitive activity is monitoring/detection by "seeing/listening", "information searching or location", "information recognition" and "information verify" for individual information, and their activities are information filtering, screening, etc. for more information, which are combined into a cognitive function, that is "multiple information gathering". Further, in case of an accident, since the state-oriented procedure (SOP) is used in digital control room of NPPs, operator 's tasks need to implement are more simple tasks in general, mainly including information comparison (such as temperature , pressure comparison), simple judgment (such as acknowledge the alarms occur? At least one stream generation is isolated?), a simple calculation (the pressure difference is greater than 10 Bar.g between two SG?) etc. Therefore, the assessment for these simple tasks is called "information comparison", i.e., the parameters identified in the monitoring process compare with the parameters procedures required, and to confirm the state of components. Furthermore, the combination of several simple tasks can identify the status of system components (such as to determine whether the RCP (French acronym, namely reactor coolant system) pump shuts down, the determination made by operator needs to judge a series of standards), the comparison result of single task/parameter is obtained through comparison between actual data and requirement data, then the larger component or subsystem state is identified by information integration (including understanding and reasoning), the series of cognitive processes may be relate to cognitive functions which involve "information comparing", "information integrating", and "state explanation". For diagnosis of more complex incidents (although the DOS (Document of orientation and stabilization) procedure does not need operator to determine what kind of accident occurs, but the judgment of six key parameters and their combinations are also complex), it needs to identify the state of system according to the combination of state of various subsystems, the process also relates to the cognitive functions related above. Similarly, for different fault, failure and incident or accident, their causes need to be identified, it is conducive to the selection and development of response plans. Furthermore, for novel situation and other, the current state of system needs to be assessed and the next state needs to be predicted by deeper reasoning according to their own knowledge and experiences, as well as limited data. Similarly, the severity of accident, the future development trend of state of component, subsystem and system need to make projection. Therefore, the cognitive functions of operator in situation assessment process also involve "cause identification", and "state projection". Therefore, the model of SA is built on the basis of information processing theory and cognitive task analysis as shown in Figure 2.

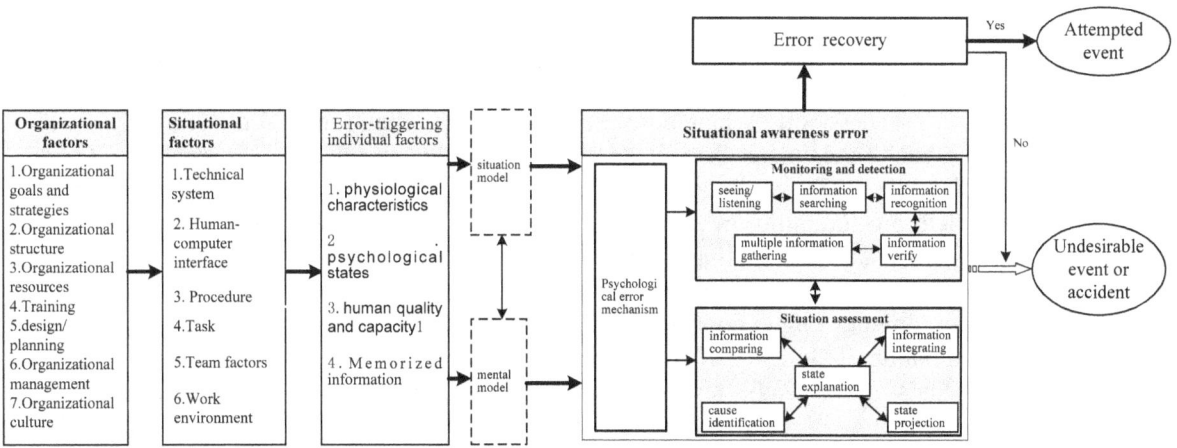

Figure 2 Operator's SA analysis model

2.2. Influencing Factors of Operator's SA in Digital Main Control Rooms

Reason [12] writes: "Human error is a consequence, not a cause. Errors are shaped and provoked by the upstream workplace and organizational factors." ESA is no exception. As far as influencing factors of human errors be concerned, according to Want's viewpoint [13]: the antecedents triggering human errors (or unsafe actions) are psychological precursors, such as human motives, plans, expectations, attention, and the way of reasoning, etc., the influencing factors causing psychological precursors are environmental conditions, which is called as latent failures. These latent failures in principle are under management and control of organization, however, when the organization's management and control fails, it will cause an accident. Therefore, in order to identify the root causes of organization causing ESA and prevent it from the source of ESA. The model of influencing factors of SA is established based on the established organization-oriented "structure-behavior" model [14] and context analysis of digital main control room of NPPs, it is shown in Figure 2.

The plant state is influenced by human behavior, human behavior is influenced by the states of the mind and body as well as the plant state, and the mental state is influenced by both the plant state and human behavior[15]. Operator's SA is not only affected by external dynamic contextual factors but also by internal factors (such as personnel quality and capability), and there are complex interactive relationships between the internal/individual factors and situational factors as well as organizational factors (such as training), we can think that the individual factors are the direct factors causing ESA because operator's physical, psychological and other factors (the deficiencies of inherent characteristics such as knowledge) are influenced by external contextual factors, when they are in imbalance state, which will result in human cognition and action errors. Although it is difficult to change person's own conditions to improve person's cognition and action reliability, we can control the states of situational and organizational factors to improve person's cognitive or behavioral reliability because the qualities of individual factors are mainly affected by situational factors (such as poor man - machine interface) and organizational factors (such as safety culture), so we can improve person's cognitive or behavioral reliability to reduce ESA by controlling the qualities of situation situational and organizational factors, and it will reduce the action errors (active errors). Therefore, the established influencing factor model includes organizational factors, situational factors and individual factors. The ESA is influenced by individual factors, and the individual factors are influenced by situational factors, as well as situational factors are influenced by organizational factors, but do not rule out the existence of jumping affecting between factors. For example, the individual factors are directly affected by organizational factors (such as the improper plan of work will lead to continuous work, thus it will cause operator's fatigue).

The cognitive field, situation assessment, involves two related concepts: the situation model and the mental model [16]. Situation model is operator's understanding of the specific current situation, and it is constantly updated as new information is received. The mental model is built up through formal education, system-specific training, and operational experience, and it is stored in the brain and relatively fixed, and it is represented in the knowledge bases of long-term memory (LTM). The situational awareness process is operators use their general knowledge to interpret the information they observe and understand its implications to construct a situation model, the current state of component/system/plant is understood by matching the situation model and mental model. If the both models are correct, and each other matches well, then the reliability of situation assessment will be very high. If the situation model (for example, there is not enough information is provided) or mental model (for example, there are limitation in knowledge) has some flaws, or incorrect matching, then ESA will occur. The operator's understanding of outside state of system (situation model) is impacted mainly by external contextual factors, including "system factors", "computer-based human-system interface factors", "environment of main control room", "team factors", "task factors" and "procedure factors" and so on. For example, the level of automation of systems may affect operator 's situational awareness, Kaber and Endsley [17] indicates operator can not promptly and accurately obtain some important information, which will lead to wrong cognition and wrong manipulation, eventually causing serious consequences due to increased levels of automation, it is known as so-called "human out-of-the-loop performance". Out of the loop performance problems are characterized by a decreased ability of the human operator to intervene in system control loops and assume manual control when needed in overseeing automated systems. In addition, the mental model is influenced by training and education including ways of training, training programs, training tools, required resources allocation of training, special education support, supervision of training process, evaluation of training effectiveness, quality assurance of training and own knowledge and practical experience etc. Furthermore, situation model and mental model, and matching process between them are also affected by the inherent limitations of individual information processing including attention (such as attention tunneling), memory (such as memory capacity limitations), expectations(something expected to see) , goals (goal-driven information searching) and ways of information processing and strategies (mode matching, story building process, meta-cognition, etc.) [17, 18].

3. THE CLASSIFICATION FRAMEWORK OF ESA ANALYSIS

A classification scheme, as an ordered set of categories, is necessary both to define the date that should be recorded and to describe the details of an event, and it is basis for ensuring consistency of results of event analysis.

3.1. Classification of ESA and Psychological Error Mechanism

ESA can be used error modes of SA to describe, and it is closely related to failures of human cognitive function, and can be expressed as cognitive function failures and failure modes. According to the definition of SA related above, SA processes majorly include the cognitive processes of monitoring and situation assessment. The cognitive functions include "seeing/listening", "information searching or location", "information recognition", "information verify" for individual information and "multiple information gathering" in monitoring process. Furthermore, the cognitive functions involve "information comparing", "information integrating", and "state explanation", "cause identification", and "state projection" in situation assessment process. In the digital main control room of NPPs, operators need to "compare" the factual information and procedure required information to identify whether the parameter is abnormal, in which operators are prone to make errors such as comparing error, no comparing or delaying comparing. Similarly, for other cognitive function in SA, there are also other failure types of cognitive function, we use keywords such as "none", "late", "wrong" and "loss" etc. to describe. The specific categories of ESA are shown in Table 1.

Operator's cognitive function failure has its corresponding psychological error mechanisms (PEMs). PEMs describe the psychological nature of the cognitive function within each cognitive domain/process, the cognitive biases that are known to affect performance. If we can find PEMs corresponding to different cognitive failure modes, then it is useful for error reduction and mitigation. However, they may require significant understanding of psychological aspects of an error, which may not always be obtainable from incident reports, and the existing psychological analysis tools are not sufficient support a deeper understanding of psychological mechanisms of error. The PEMs for cognitive domain of monitoring and situation assessment are identified on the basis of the results of previous studies [19-21], and they are listed in Table 1.

Table 1: The classification of ESA

Cognitive processes	Cognitive activities	error modes	Specific errors (relevant keywords)	Psychological error mechanism
monitoring/ detection	C1:Seeing/Listening	Seeing/Listening error	-None,late,wrong, loss	Expectation bias; perceptual confusion; distraction / preoccupation; task overload; perceptual tunnelling; spatial confusion; low vigilance, attention detection not in time, visual fatigue; frequency preference; keyhole effect, etc.
	C2:Information searching or location	Information searching or location error	-None,late,wrong	
	C3:Information recognition	Information recognition error	None,late,wrong	
	C4:Information verify	Information verify error	None,late,wrong	
	C5:Multiple information gathering	Multiple information gathering error	-Omission, Irrelevant, Insufficient, Redundant	
Situation Assessment	Information comparing	Information comparing error	-None,late,wrong	Lack of knowledge; no considering side effects; integration failure; false assumption; misinterpret; cognitive fixation (halo effect); similarity interference; memory block; memory capacity overload; loss of positions; keyhole effect, etc.
	Information integrating	Information integrating error	-None,late,wrong	
	State explanation	State explanation error	-None,late,wrong,loss	
	Cause identification	Cause identification error	-None,late,wrong	
	State projection	State projection error	-None,late,wrong	

3.2. Classification of Influencing Factors of SA

With reference to the classification of influencing factors of SA, which have been studied by some researchers[22-25], but their researches are lack of hierarchical, systematic and comprehensive, such as their classifications of influencing factors are not specific and detail enough to describe the characteristics of influencing factors, some studies only focus on individual factors effects on SA, and there is no considering the root causes of organization etc. Therefore, the detailed classification of organizational factors, situational factors and individual factors is built based on collected references including the first generation techniques such as THERP (Swain and Guttmann, 1983), HCR(Hannaman et al., 1985), SLIM (Embrey, 1984), HEART (Williams,1992), the second generation techniques such as CREAM (Hollnagel, 1998), ATHEANA (Cooper et al., 1996), CAHR (Sträter, 2000) and third generation HRA techniques such as OPSIM (Dang, 1996), IDAC (Chang, 2007) are collected, and also consider the classification of performance shaping factors (PSFs) in the CSNI classification (Hollnagel, 1998), SPAR-H (Gertman et al., 2005) and the HRA good practices (Kolacakowski et al., 2005), and combine with the study results of classification of organizational factors from the previous studies (Li et al., 2009)[14]. The classification tries to follow five principles of classification: (1)concrete, (2)assessable and measurable, (3) non-repetitive and non-cross; (4)consistency, and (5)comprehensive. The specific classification of organizational factors may be a process, such as planning formulation, task allocation, may indicate a state, such as lack of goals, the number of personnel, may indicate the certain property of upper-layer factors, such as the style of training, etc., as shown in Table 2. It should be noted that the above classification system basically considered the influencing factors of human activities, not just for the influencing factors of SA, but we think that the classification of influencing factors is also applicable for the analysis of ESA.

Table 2: The classification of individual, situational and organizational factors

Influencing factors	Subclass	Specific factors
Individual factors	Psychological state	Cognitive modes and tendencies: alertness, attention to current task, attention to surrounding environment, cognitive bias, Stress：frustration, conflict, pressure, uncertainty. Strains and feelings: time-constraint load, task-related load, non-task related load, passive information, confidence in performance. Perception and appraisal: perceived severity of consequence associated with current diagnosis/decision, perceived criticality of system condition, perceived familiy with situation, perceived system confirmatory/contradictory responses, perception of alarms(quantity, intensity, importance),perceived decision responsibility, perceived complexity of strategy, perceived task complexity, perception of problem-solving resource , awareness of role/responsibility, done quickly psychology, habit psychology. Intrinsic characteristics: motivation (desire, demand), attitude, morale, character and personality, self confidence, problem solving style
	Physiological state	Suddenness of onset, pain or discomfort, fatigue, hunger or thirst, physical movement constriction, lack of physical exercise, disruption of circadian rhythm, sensory loss, individual size / body condition
	Memorized information	None or incorrect of Recall perceptual information, none or incorrect of memory of previous execution action, none or incorrect of memory of current execution action (diagnosis, action and results), none or incorrect of memory of prospective execution action sequence, none or incorrect of memory of the stored information(procedural and declarative knowledge)
	Quality and capability	Knowledge, experience, skills / capacity, social roles, and level of moral
Situational factors	System	Degree of automation, the complexity of system, redundancy of system, system reliability, software reliability, compatibility and coupling degree of system configuration, inspection and test of output of system , system feedback, response speed/delay of system, number and speed of information presentation, information interference, number of simultaneous goals, required judgment beyond level of skill and experience, time stress/available time determined by system design
	Human - computer interface	Monitor and controller reliability, structure relationship of screens, range of display, display precision, display information cognizability, display information understandability, accessibility of control equipment , operability / availability of control equipment, accuracy of controlled location, requirement of special tools, complexity of interface management tasks，information display format, amount of information displayed, consistency of information in different displays, cognizability of software control icon, location of soft control icon on display , ease of operation of soft controller , type of soft control, data input and state feedback of controller, ease of discrimination of alarm, easy of search of alarm, keyhole effect
	Task	Perceptual requirements, motor requirements(speed, strength, precision), anticipatory requirements, interpretation of task, task complexity, narrowness of task, task frequency and repetitiveness of tasks, task criticality, required long-term and short-term memory, calculation requirements, results feedback (knowledge of results), task type (dynamic VS step-by-step activities), task novelty, task speed requirements, high jeopardy risk, threats (failure, loss of job), nature of task (monotonous, degrading or meaningless work)
	procedure	Format or type of procedure, logic structure, forms of presentation, function / availability / validity of procedure, complexity, level of detail, accuracy requirement to activity, adequacy and integrity of description of warnings and cautions, activity criterion, understandability of procedure , interpretation margin, number of steps, required time for completion, clarity of instruction and terminology, level of standardization in use of terminology, decision making criterion, number of logical conditions(branches), number of simultaneous tasks
	Environment	Physical access, temperature, humidity, air quality, radiation, lighting, Color, noise , vibration, degree of general cleanliness, G-force extremes, atmospheric pressure extremes, oxygen insufficiency, external interference / distraction events
	Team factors	Team structure and level of personnel allocation, type of team communication, quality and validity of communication, team cohesion, team leadership, team cooperation and coordination, dynamic characteristics of team, role and responsibility of team
Organizational factors	Organizational Goals and strategies	Organizational objective (safety, performance): lack of objective, integrity objective system, consistency of objective system, priority of objective system, objective specificity, contradiction between current objective and long-term objective, Strategy: organizational policies/systems, formulation of high level plans, work methods/strategies, centralization of organizational decision-making, priority of management, problem identification and solution, determination of organizational structure, responsibilities, authority, Organization/workshop practices
	Organizational structure	Number of staff, control range, number of organizational level, location of decision / authority , roles and responsibilities, authorization, communication paths
	Organizational resources	Information resources: superior instruction, information of analysis method, information of process, manual. of activities, methods, tools. Material resources: equipment, tools, parts, materials Human resources: employee selection, performance evaluation, reward / incentive. Economic resources: available funds .Time resources: effective time, available time. Other resources: such as space resources
	Organizational Management	Organization: task allocation, personnel allocation, resource allocation, time arrangement, shift organization, work preparation, staff placement. Management: level of management such as human resources management .Control: supervision, control (such as quality control), verification and evaluation. Leadership: Leadership. Coordination: cooperation and coordination
	Education / Training	Way of training, training programs, training tools, required resources allocation of training, special education support, supervision of the training process, evaluation of training effectiveness, quality assurance of training.
	Organizational culture	Organizational climate: organizational cohesion, organizational knowledge, organizational learning, information sharing, sense of belonging of employees, group identity. Safety culture: tradeoff between safety and economy, safety standards and rules, safety attitudes, safety practices, safety measures, experience feedback, violations, documentation.
	Organizational plan / design	Strategic planning, safety planning, objective design, system design, work process design, programming/procedure design, work design.

3.3. Classification of recovery of ESA

Human errors are reduced by "error suppression" approach in the past, but a lot of practice shows that it is difficult to completely eliminate human errors. Another approach that may be a sensible way of combating the human error problem in high-risk technologies is error detection and correction. Error recovery includes three processes[26]: (1) error detection; (2) error explanation; and (3) error correction. There may exist in a variety of failures in error recovery process. Therefore, the classification of error recovery failure is determined according to the process of error recovery, as shown in Table 3.

Table 3: The classification of failures of error recovery

Error recovery type	Error recovery sub-type		description
Error detection	Self-examination	Detect mismatch between expectations and outcomes	Difficulties in perceiving or attending to actual outcomes and setting-up or remembering expectations about effects can result in failures of detection due to job factors such as poor interface design, high workload etc.
		Compare effects of equipment failures& self-produced errors	Biased attitudes and responsibility of explaining away errors can impede detection, the undesired outcomes can be easily attributed to equipment rather than one's own performance.
		Detect mismatch between plans and executed actions	Action-based detection takes place by a perception of some aspect of the erroneous action either auditorially, visually, or proprioceptively.
		Detect mismatch between intentions and plans	In the conceptual or planning stages, operator doesn't recognize wrong intentions(i.e. higher-level goals), or doesn't recognize the formulated plan is not suitable for achieving the goals etc.
	External examination	System feedback failure	System does not provide error externalization functions or poor externalization functions
		Error detection failure by supervisor	External personnel's monitoring, evaluation and communication each other etc. are not sufficient
Error explanation	Locate error in the interpretation of the situation		To establish error corrective plans, operator need to try to identify or explain the causes of the error, in which it may be produce explanation error due to available time and knowledge etc.
	Locate error in goals or plans		The errors in goals or plans are not identified because of limited time available and so on.
	Locate error in the specification of the task sequence		The errors in task sequence are not identified because of limited time available and so on.
Error correction	Re-assess situation		Situation reassess errors may occur during the re-evaluation of the state
	Develop corrective plan		After reassessment of the situation, the corrective plan errors may occur in developing corrective plans due to less time available ect.
	Execute corrective plan		The corrective plan is not successfully implemented

4. ANALYSIS METHOD OF ESA

The retrospective analysis method of ESA consists of the following five steps, as shown in Figure 3.

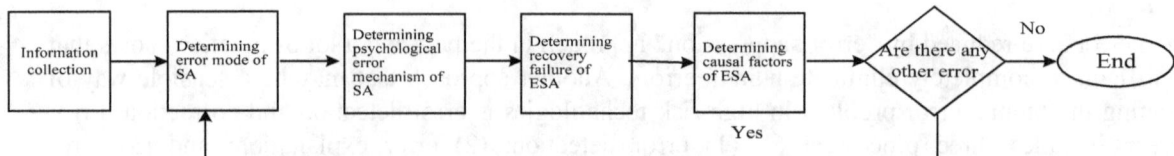

Figure 3 The analysis procedure of ESA

(1) Information collection. Using structured information collection method, which includes task analysis, goal-means analysis, cognitive function analysis and contextual analysis. Task steps, type, structure are determined by task analysis, and personnel, system, components and their relationships etc. are determined. Goal-means analysis focus on tasks or subtasks itself, the task goals and required methods and resources achieving task goals etc. are determined by Goal-means analysis; Cognitive function analysis is the determining of required cognitive functions related to SA to complete a given task, the classification of cognitive function of SA can be used to cognitive function analysis. Contextual analysis is a process of collecting and integrating the macroscopic contextual information on the given task. It includes the determining of task scenario, system/subsystem or components state, work environment and organization and management information etc.

(2) Determining error mode of SA. The most likely SA error mode characterizing human ESA id determined according to the detailed classification of ESA and the collected information, and we can also find a reasonable ESA by reasoning and argumentation if ESA is not very obvious.

(3) Determining psychological error mechanism of SA. The psychological error mechanism of SA is determined by reasoning according to the identified ESA and characteristics of errors recovery failure. The identification of psychological error mechanism of SA helps to understand the specific task context how to affect human cognition and behavior, and helps to find the causal factors triggering ESA and prevention of human error.

(4) Determining recovery failure of ESA. The possible recovery failure of ESA is identified through checking one by one according to the identified error mode of SA, and the classification of error recovery failure.

(5) Determining the causal factors of ESA. Based on analysis related above and investigation and verification of contextual environment, the reasons can be found in the corresponding classification of influencing factors, and can date back to the root cause of organization. If there are other ESA in an incident, a similar analysis will be continued until all of ESA are completed.

5. AN EXAMPLE

A case, namely "the high-high signal of water level of 2nd steam generator (SG) of 4th unit of Ling Ao NPP superimposing the signal of P7 causes shutdown of reactor", is used to demonstrate the proposed ESA analysis method.

5.1. Event Summarization

May 21, 2011, the 4th unit of Ling Ao NPP falls back to hot shutdown due to the water quality in secondary circuit. May 22, the unit goes critical after restoring water quality in secondary circuit, the reactor power rises to 8.0% Pn, and the feed-regulating valves and rotation speed of main feed pumps of three steam generators are in manual control. 17:53, the team leader of unit finds "L4ARE001MP" (steam generator feed water and vapor pressure difference) is low (1.2bar, far less than the required value 4.2bar), so he requires the operator to adjust the speed of feed pumps to rise pressure difference in order to control rotation speed control as soon as possible, then the feed pumps are put into operation in automatic mode. The signal of high-high water level (P14 signal) of the 2nd steam generator emerges when the operator raises rotation speed of feed water pumps, and at the same time, the reactor power fluctuate over 10% Pn, the signal of P10/P7 occurs. The reactor automatically shutdowns because of the signal of high-high water level superimposing the signal of P7. Then the operators implement DOS procedure to the unit. 18:57, the unit is in stable state and exits the DOS procedure.

5.2. Information collection

The critical path of event is identified and built by information collecting and analyzing as shown in Figure 4, we can see the incident mainly includes two human errors from Figure 4, that is, the operator of primary loop can not correctly adjust the opening degree of small valve of ARE (Feedwater flow control system), and the operator of secondary loop adjusts the rotation speed of APA102 pump (Motor-driven feedwater pump system) too fast. Then task analysis is carried out for a specific human error to determine the required cognitive functions and cognitive errors for completing the task. Furthermore, contextual environment analysis is carried out to determine the cause of errors. We take "the operator of primary loop can not correctly adjust the opening degree of small valve of ARE" as an example to demonstrate information collection. The following is collected basic information:(1) Task——the manual adjustment of three small flow control valve of ARE; (2)Goals and means—— Controlling water level of three SG control, which needs basic manual operation; (3)Required cognitive function—— detection (detect water level changes), explanation (state of water changes), decision (adjust to what extent of water level) and execution (implement of adjusting response); (4)Contextual environment—— SG water slowly declines, the man-machine interface related to valve, which waits for verification of results.

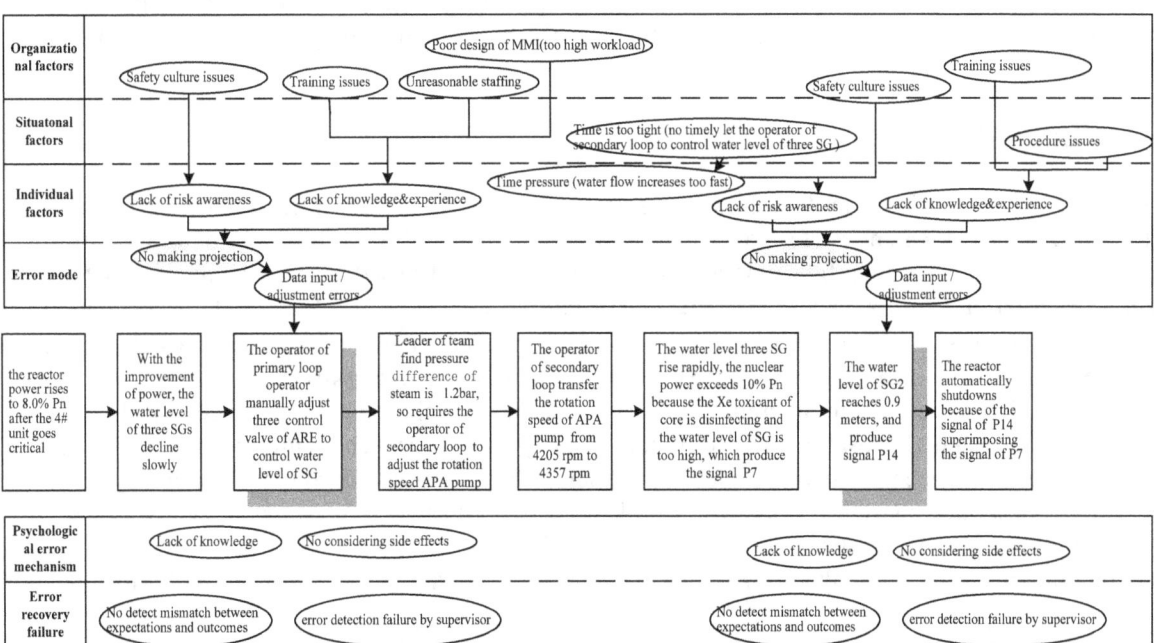

Figure 4 The event and cause factors analysis of the application case related above

5.3. Determining action and cognitive error modes

With reference to "the operator of primary loop can not correctly adjust the opening degree of small valve of ARE", the action error mode is apparently "adjustment error". The cognitive cause causing action error is "operator does not make projection", that is operator does not predict whether this adjustment produce an adverse effects on the system when the opening degree of the valves is adjusted to 80%.

5.4. Determining psychological error mechanism

The psychological error mechanism obviously is "lack of knowledge" because the operator does not assess the rationality of opening degree, and he does not consider the risk of 80% opening degree of the valves, its psychological error mechanism can be thought as "no considering side effects" by analysis.

5.5. Determining error recovery failure

According to the classification of error recovery failure and the identified human error modes, we can obtain the error recovery failure "detect mismatch between expectations and outcomes" corresponding to the error "the operator of primary loop can not correctly adjust the opening degree of small valve of ARE ". Similarly, another error recovery failure is identified by contextual analysis, which is "error detection failure by supervisor" belonging to classification of "External examination", because the director of team does not diction the operator's error.

5.6. Determining causal factors of human error

According to the operator's cognitive and action error modes and contextual environment analysis, the individual factors causing human error are "lack of knowledge and experience" and "lack of risk awareness". The causes of "lack of knowledge and experience" are attribute to "poor training", which belong to organizational factors, and "poor man-machine interface" which belong to situational factors, because the secondary loop operator's task load is too large, which may be due to poor man-machine interface, so secondary loop operator task load is too large , so that the primary loop operator is arranged to do the secondary loop operator's work which does not belong to him and does not know well. The cause giving raise to "lack of risk awareness" is due to "poor safety culture". In addition, the investigation results of another human error can be obtained by the same analysis procedure as shown in Figure 4.

6. CONCLUSION

SA issues are more prominent in digital control systems of NPPs, and it is a key element impacting operator's decision-making and performance. The analysis method of ESA is also an important part of human error analysis. The cognitive functions of SA are identified based on task analysis and field simulator observation in this paper, and the classification system of ESA is built according to cognitive function of SA. ESA are placed in a more all-sided contextual environment to establish the model of SA from organizational perspective, and the model-based classification system used to analysis of ESA is provided from inherent cognitive mode, the psychological error mechanism, error recovery failure to external causal factors etc. It combines "person approach" with "system approach", and the classification system is more detailed and concrete than the traditional classification system., and provides a structured analytical framework to identify the root cause of organization causing ESA. Therefore, the established analysis method of ESA will provide theoretical basis and practical application guidance for ESA analysis, prevention and reduction in digital control system of NPPs.

Acknowledgements

The financial support by Natural Science Foundation of China (71371070, 71071051, 71301069) and Scientific Research Program supported by Lingdong Nuclear Power Plant (KR70543) and Innovation Ability Construction Projects based on the new Industry- Academy- Research Cooperation of Hunan Province（2012GK4101）is gratefully acknowledged.. We would like to express gratitude to the staff in two Chinese nuclear power plants (Qinshan Phase 3 and Lingdong) for the interviews and investigations which facilitate the research included in this paper. The anonymous reviewers and the editor of this paper are also gratefully acknowledged for their constructive comments and suggestions..

References

[1] S.W. Lee, J. Park, A.R Kim., P.H Seong. "*Measuring situation awareness of operation teams in NPPs using a verbal protocol analysis*", Annals of Nuclear Energy, 43, pp.167-175, (2012).

[2] X.H. He, X.R. Huang, "*Human reliability analysis in industrial systems: Principles, Methods*

and *Applications*", Tsinghua University Press, 2007, Beijing.

[3] R. Woodhouse, R.A.Woodhouse. "*Navigation errors in relation to controlled flight into terrain (CFIT) accidents*", In: Proceedings of the 8th International Symposium on Aviation Psychology, Columbus, OH, 1995.

[4] M.R. Endsley. "*A taxonomy of situation awareness errors*", In: "Western European Association of Aviation Psychology 21st Conference", Dublin, Ireland, 1994.

[5] D.G. Jones, M.R Endsley. "*Investigation of situation awareness errors*", In: Proceedings of the 8th International Symposium on Aviation Psychology. Columbus, OH , 1995.

[6] M. R. Endsley."*Toward a theory of situation awareness in dynamic systems*",Human Factors,37 pp.32-64,(1995).

[7] J.M. O'Hara, J.C. Higgins, W. Brown. "*Identification and evaluation of human factors issues associated with emerging nuclear plant technology*", Nuclear Engineering and Technology, 41, pp. 225-236, (2009).

[8] J.M. O'Hara, W.S. Brown, P. M. Lewis, et al. "*The effects of interface management tasks on crew performance and safety in complex, computer-based systems: detailed analysis*", Washington D.C: U.S. NRC, 2002.

[9] P.H. Seong. "*Reliability and risk issues in large scale safety-critical digital control sytems*", Springer,2009, New York.

[10] D.B. Kaber, C.M. Perry, N. Segall, et al. "*Situation awareness implications of adaptive automation for information processing in an air traffic control-related task*", International Journal of Industrial Ergonomics, 36, pp. 447-462,(2006).

[11] S.J. Lee, M.C. Kim, P.H. Seong. "*An analytical approach to quantitative effect estimation of operation advisory system based on human cognitive process using the Bayesian belief network*", Reliability Engineering and System Safety, 93, pp.567-577,(2008).

[12] J. Reason. "*Managing the risks of organizational accidents*", Ashgate Pub Ltd, 1997, Aldershot.

[13] P. G. D. Want VD, "*Tripod incident analysis methodology*", In: J. van Steen (Ed.), Safety performance measurement. Rugby,UK: Institution of Chemical Engineers, 1996. 99–106

[14] P.C. Li, G. H. Chen, L.C. Dai, L. Zhang. "*A fuzzy Bayesian network approach to improve the quantification of organizational influences in HRA frameworks*", Safety Science, 50 , pp. 1569-1583, (2012).

[15] Y. Jin, Y. Yamashita, H. Nishitani. "*Human modeling and simulation for plant operations*", Computers and Chemical Engineering, 28, pp.1967-1980,(2004).

[16] M.O. John, C.G. James, K. Joel. "*Advanced information systems design: technical basis and human factors review guidance*", Washington, DC: U.S. NRC, 2002.

[17] D.B. Kaber, M.R. Endsley. "*The effects of level of automation and adaptive automation on human performance, situation awareness and workload in a dynamic control task*", Theoretical Issues in Ergonomics Science, 5, pp.113-153,(2004).

[18] M.R. Endsley, B. Bolte, D.G. Jones. "*Designing for situation awareness: an approach to user-centered design*", Taylor & Francis, 2003, London.

[19] J. Reason, "*Human Error*", Cambridge University Press, 1990, New York.

[20] S. Shorrkck, B. Kirwan. "*Development and application of a human error identification tool for air traffic control*", Applied Ergonomics, 33, pp.319-336,(2002).

[21] B. Kirwan. "*Human error identification techniques for risk assessment of high risk systems-Part 2: towards a frame work approach*", Applied Ergonomics, 29 , pp. 299-318, (1998).

[22] D.G. Jones, M.R. Endsley. "*Sources of situation awareness errors in aviation*", Aviation, Space, and Environmental Medicine, 67 , pp. 507–512, (1996).

[23] A. Sneddon, K. Mearns, R. Flin. "*Stress, fatigue, situation awareness and safety in offshore drilling crews*", Safety Science, 56, pp. 80–88, (2013).

[24] M.R. Endsley, D.B. Kaber. "*Level of automation effects on performance, situation awareness and workload in a dynamic control task*", Ergonomics, 42, pp.462-492,(1999).

[25] S.K. Kim, S.M. Suh, G.S. Jang, et al. "*Empirical research on an ecological interface design for improving situation awareness of operators in an advanced control room*", Nuclear Engineering and Design, 253,pp.226–237, (2012).

[26] T. Kontogiannis. "*User strategies in recovering from errors in man-machine systems*", safety Science, 32, pp.49-68,(1999).

Study on Human Errors in DCS of a Nuclear Power Plant

DAI Licao[a] ZHANG Li[b] LI Pengcheng[a] HU Hong[b] ZOU Yanhua[b]

[a]Human Factor Institute, University of South China, Hengyang, P.R.China
[b]Hunan Institute of Technology, Hengyang, P.R.China China

Abstract: More and more main control rooms in advanced nuclear power plants (NPP) use computer-based displays and controls, which are called digital control systems (DCS). DCS changes some technological aspects in a NPP control room, including information display systems, alarm systems, controllers and components and computer-based procedure systems. These changes on man-machine interface (MMI) alter the ways of operators acquiring information and controlling the system and thus give rise to new human error issue. In order to investigate the impact of the new MMI on human reliability, the researchers conducted a study in a reference plant with DCS. The practical operation data as well as the experimental data were acquired to study the causes, effects and recovery factors of the new human errors. The research makes an effort on providing a foundation for human error prevention in a DCS and human reliability analysis.

Keywords: digital control system; human error; human factor events; human reliability analysis

1. INTRODUCTION

Human factor is a major contributor to the safety of nuclear power plant (NPP). The operators conduct monitoring and control tasks in the main control room of a nuclear power plant. In a digital main control room, the man-machine interface differs greatly from that of a traditional one. Digital control system (DCS) changes some technological aspects in a NPP control room, including information display systems, alarm systems, controllers, components, computer-based procedure systems and computerized operator support systems. An analog-based conventional control room uses hard-wired displays, such as alarm tiles, indicators, lights, gauges and scales and hardwired controls, such as switches, buttons, knobs and handles. In emergency, operators walk around the panels to acquire information and manage the events on the panels by following paper procedures. In a DCS, information is displayed on video display units (VDU) and large display system (LDS) and controls are made by mouse and keyboard. The digitalization of main control room gives rise to many changes in terms of human factor. The human factor analysts would identify the human errors thus caused and try to understand the reasons and the mechanism of recovery. The final purpose is to propose a plan to reduce these errors to improve the safety of the nuclear power plant.

The empirical study on human errors should be based on the identification and categorization of external forms of human errors. The basic external form is the "task unit" of human behavior, for instance, "clicking a mouse" or "switching between screens" in a DCS. Higher level of "task unit" includes conducting a series of actions or a set of actions for the purpose of realizing the particular function of a system. For example, in case of post-accident management, operators fail to isolate the ruptured SG.

According to Rasmussan[1], human behavior is categorized into three types, the skill-based , rule-based and knowledge-based. The behavior of low consciousness involves low level task unit which is always skill-based and rule-based. Higher level of knowledge-based human behavior involves more complicated diagnosis process. Operators' action is divided into two phases, "forming the intention" and "action execution". The errors of intention are mistakes and those of action execution are slip and lapse. Slip is the failure of attention while lapse is the failure of memory[2].

In order to investigate the possible new errors and their mechanism in a DCS, the research group collected the human factor event reports on the spot and simultaneously did experiments. Researchers analyzed these reports and made a summary in the following sections.

2. ANALYSIS ON HUMAN FACTOR EVENTS

The research group collected human factor events by designing a record system for operators reporting their errors by themselves. These reports collected errors on the basis of "task unit". More than 400 human factor event reports were collected in the reference nuclear power plant with a DCS control room for a time period of 18 months. The events analyzed in this paper are mainly featured by DCS man-machine interface covering errors of action execution, monitoring and poor design of man-machine interfaces.

2.1 Errors of action execution

New DCS man-machine interfaces change the cognition of operators and thus change the action execution mode. Errors of action execution are the major external form of human errors in connection with DCS ,e.g. the observable phenomenon reflected from within human cognition. These errors fall into four categories, those of time, action, target and sequence. 29 events were recorded in which 20 occurred in the routine operating period accounting for 69% of the total. The others occurred when operators were re-trained on full-size simulators.

The reports show that the majority of the errors occurs on manipulating the control objects. It mainly includes two types of errors, one is target error and another is action error. The former is the operators acting on wrong control objects and 15 events are related to this. Most of these cases are in the configuring of screen displays, e.g. secondary task in a DCS when operators opened a wrong picture. Some cases involve operators controlling on wrong objects within one picture. Action error means that operators omit a step in the procedure or act in a wrong sequence.

As for the tasks when errors happened, most errors (4 of 29) happened when operators conducted isolation and when operators reset the state of the equipments (4 of 29) on the operation displayed on computers.

As for the causes of these errors, the failure of operators' attention and memory is a major contributor. Long-time work and poor design of man-machine interfaces might cause the operators lose their attention and memory. In one case, operators were not familiar with the picture and made VVP steam pressure waved up and down.

As for the consequences of the events, five of them were considered having impacts upon systems, including total loss of L3KIC, the stoppage of L8DVN, unexpected initiation of L3RIS002PO, L4RRA safety valve moved six times and L3RIS003VP not in automation position when it was started.

As for the recovery of the errors, 13 of the 29 cases were recovered instantly and most of them (8cases) were recovered by operators themselves, two were recovered by other people or supervisors and three of them were recovered by alarms or annunciators. The other recovery facts could not be identified in reports.

The errors of action execution fall into two categories, error of omission (EOO) and error of commission (EOC)[3]. EOO comes from lapse, e.g. the failure of memory making operators forget the whole task or a part of the task. EOC has two sources, one is the mistake in forming the intention to execute the task, and another is that the task is not executed in a correct manner due to the distraction of attention. Analysis found that in the recorded human factor events most of them were EOCs. For five of the recorded human factor events, operators seemed to feel difficult to act on controllers andd gadgets. Interview and observation showed that operators were difficult to control or to feel the feedback on the computer. After digitalization, because of the diversified controlling objects displayed on computers combined with poor interface design, some simple skill-based actions, for example, pressing a button, might require high consciousness and cause serious consequences. In other words, the low consciousness task in traditional analog control room could possibly transfer into behaviour requiring high-consciousness.

2.2 Errors of Monitoring

While nuclear power plant becomes more and more automatic, the major routine task of operators in a DCS control room becomes monitoring on the screens. 18 human factor events in regard to monitoring were recorded in which 11 happened in main control room and 7 on the spot. The events could be classified into two categories. The first category was that when operators finished operations on the equipments, they failed to track the state of the equipments or the systems. Eight human events were involved, for example, failing to monitor the water-level gauge when conducting water make-up; not efficiently monitoring the plant state in nitrogen purging and when adjusting L4RRI B flow rate and L4RRI020VN was turned off and not replaced to a normal state, etc.. The second was that operators failed to detect the abnormal conditions when equipment or system went wrong. There were ten such kind of human factor events. For example, operators didn't detect abnormalities after taking over from the preceding shift.

As for the causes of the errors, time pressure may be the main cause. It seems that the operators were prone to lose monitoring on the state of equipments or system at shift handover. The second reason is related to work plan. Eight of the recorded human factor events were in the period of overhaul when operators were involved in a lot of work. So the critical parameters were not efficiently monitored. The third reason may be related to poor design of DCS man-machine interface.

As for the recovery, alarms, annunciators, the third person or other organization barriers are strong recovery factors. In monitoring process, no self-recovery event was recorded.

All monitoring errors are EOOs. Operators paid no attention on or omitted information due to the failure of memory or attention.

2.3 Errors in relation to man-machine interface

Operators do their routine work (computer-based monitoring and control) on man-machine interface in a DCS. The man-machine interface has great impacts upon the human performance and reliability which needs a careful consideration in human reliability analysis.

The recorded events show that all the problems on man-machine interface are because of poor design of pictures and control gadgets, such as the relationship between control system components not very clear, the inappropriate controller positions, improper unlock and interlock, poor matching of controllers with system response and poor design of input fields and formats etc..

Nearly all the simple tasks on computer screens are skill-based. The computer screen is mono-dimensioned and has a narrow display scope. The information that activates the action of operators is easy to be ambiguous to cause slips and lapses. They both involve errors in performing well-practiced actions with low consciousness.

Though in the reference plant there were many human errors possibly having some kind of relationship with the design of the man-machine interface, the operators detected and recovered from most of the errors by themselves. There existed a strong self recovery factor.

3. Conclusion and summary

From the collected data we could find that the computer-based behavior in a DCS changed the way of acquiring information and control input. This change makes a different allocation of attention and memory resources of the operators from that in a traditional analog control room[4-7]. Therefore it may produce new human errors or human errors may appear in new forms. DCS makes the following changes.

Firstly, DCS may change the operators' cognitive workload. In DCS, operators' cognitive workload differs greatly form that in a traditional control room. This difference comes from the physical changes of the man-machine interface[8-12]. An analog-based conventional MCR uses hard-wired displays and controls while in a DCS operators sitting before a set of computers monitoring and controlling. This changes how operators percept, process and feed back the information and thus changes operators' cognitive workload.

DCS changes how people percept the information. In a conventional analog control room, the control panels with a multi-dimensioned physical structure display information directly. This directly-displayed information seems to be more compatible with the indications of the actual state of the plant. In DCS, most of the information is hidden behind the screens and operators need to make some efforts on secondary task to get these information. More cognitive load is needed.

DCS also changes the way how operators process the information. The information is processed in human brain by working memory, the allocation of attention and the extraction from the long-term memory. In a conventional control room, the majority of the input of the phonological loop is visual information, such as the light of annunciators, the form of the control knobs and the position of the gauges etc. and the auditory information (alarms or sounds). The visual information is processed on visuospatial scratchpad of working memory and decoded into phonological repetition and stored[6]. In DCS, the semantic representation of symbols, texts and charts is directly processed by phonological loop. In addition, the efficiency of working memory heavily depends on its capacity. In a conventional control room, the panels and controllers possess a multi-dimensional form with a certain physical construction. Through training operators easily form distinctive chunks when processing information. But the semantic representation on computers appears to be more ambiguous. This may give rise to different stability of working memory and thus produce influence on the cognitive workload.

DCS also changes the way human allocates the resource of attention. In a conventional control room, the entire information stimulus is at a fixed position on panels. The information is visually salient. Furthermore, data-driven information could be more easily detected and operators could access them more conveniently. In DCS, the attention seems to be model driven while in a conventional one it seems to be a combination of model-driven and data-driven data.

Secondly, DCS may change the role of individual crew member. DCS changes the behavior of a single operator and therefore the role of an operator in a crew. In a traditional control room, a sole operator is just one part of a crew cognition. When accident happens, all the crew members are sitting before the panels and share the information timely. They discuss, negotiate, exchange information and make judgment. In DCS, RO1 and RO2 are respectively a complete cognition unit. The independent work of each operator causes new risks. One risk is that the errors of the operator are not easily detected by other people and thus produce impacts upon the crew reliability. Another risk is that the operator may sink into following the procedure when accident happens and may lose situation awareness. In the reference plant, experiments on full-size simulator show that RO1 & RO2 have low situation awareness and this is supplemented by another crew member US (unit supervisor). The situation awareness of the crew heavily depends on the communication between crew members.

The third significant change may take place in the course of following procedures. In a traditional control room, operators use paper procedures while in DCS operators use computer-based procedure. The way of following procedures are different, including how to get the necessary information, the way of supervising and the monitoring of system feedback. Nevertheless, the ways by which the procedures are presented make the execution of procedures different. Furthermore, the paper procedure is convenient for operators to track what have been done and predict what will happen. In DCS, if you want to realize these, you need to do many secondary management tasks. Therefore it will possibly increase the opportunities of errors.

From the collected reports and observations upon simulator training we know that DCS makes difference. It changes the cognitive workload of operators and the role of each member in a control room. It also changes the way of following procedures and may possibly produce new errors. Further research is still needed on the above-mentioned issues and thus to improve the human reliability in a DCS.

Acknowledgements

The financial support by Natural Science Foundation of China (71371070, 71071051, 71301069) and Scientific Research Program supported by Lingdong Nuclear Power Plant (KR70543) , Innovation Ability Construction Projects based on the new Industry- Academy- Research Cooperation of Hunan Province（2012GK4101）and Construct Program of the Key Discipline (1201) in Hunan Province (Management Science and Engineering) are gratefully acknowledged.

References

[1] J. Rasmussen. *"Information processing and human-machine interaction: An approach to cognitive engineering"*. North Holland , 1986,New York.

[2] J. Rasmussan. *"Skills, rules and knowledge: signals, signs and symbols, and other distinctions in human performance models"*, IEEE transactions on systems, Man & Cybernetics, 13,pp.257-266, (1983)

[3] A.D. Swain, Guttmann H E. *"Handbook of Human Reliability Analysis with Emphasis on Nuclear Power Plant Application"* . NUREG/CR-1278, Sandia National Laboratories, Washington, DC.,(1983).

[4] J. Rasmussen, & K.J. Vicente. *"Coping with human errors through system design: Implications for ecological interface design"*. International Journal of Man-machine Studies,31,pp.517-534,(1989).

[5]P.M.Sanderson. *"Cognitive work analysis. In J.M.Carroll(Ed.) HCI models, theories, and frameworks: Toward a multi-disciplinary science"* (pp.225-264). San Francisco: Morgan Kaufmann.

[6] A. D.Baddeley, & Hitch, G. J. *"Working memory"*. The psychology of learning and motivation (Vol. VIII, pp. 47 – 89)". New York: Academic Press. (1974).

[7] J.A. Spurgin. *"Human Reliability Assessment"*, CRC Press, Taylor & Francis Group, Boca Raton, FL 33487-2742 (2010).

[8] N.J.Cooke. *"Varieties of knowledge elicitation techniques"*. International Journal of Human-computer Studies,41,pp.151-173,(1994).

[9] K.A.Ericsson & H.A. Simon. *"Protocol analysis: Verbal reports as data"*. Cambridge, MA: MIT Press (1985).

[10]E.Hollnagel,& D.D.Woods. *"Cognitive systems engineering: New wine in new bottles. International Journal of Man-machine Studies"*,18, pp.583-600,(1983).

[11] J.Rasmussen, & K.J.Vicente, *"Coping with human errors through system design: Implications for ecological interface design"*. International Journal of Man-machine Studies,31,pp.517-534,(1989).

[12]P.M.Sanderson. *"Cognitive work analysis". In J.M.Carroll(Ed.) HCI models, theories, and frameworks: Toward a multi-disciplinary science (pp.225-264). San Francisco: Morgan Kaufmann (2003).*

Experience feedback from Fukushima towards Human Reliability Analysis for level 2 Probabilistic Safety Assessments

V. Fauchille, H. Bonneville, J.Y. Maguer

Institut de Radioprotection et de Sûreté Nucléaire
BP. 17
92262 Fontenay-aux-Roses Cedex
FRANCE

Abstract: In the years 2000, the IRSN developed its first level 2 Probabilistic Safety Assessment (PSA) for the 900 MWe French PWRs. It was an ambitious project and one of the important tasks was to build a Human Reliability Analysis (HRA) model able to model the human actions to be implemented after the core melted. These actions are performed by operators in the main control room or by field operators outside but most of the decisions are taken, on the basis of the Severe Accident Management Guide (SAMG), by the crisis organization.

A Human and Organizational Reliability Analysis in Accident Management (HORAAM) model is born from this enterprise. It is based on the "Decision Tree method". HORAAM has been developed from the observation of the nuclear crisis exercises that are regularly practiced in France. Several influence factors which particularly affect human and organizational reliability in such a situation were identified.

Currently HORAAM is used at IRSN but it has never been compared to the experience feedback of a real accident. After the Fukushima accident, IRSN conducted a study to confront HORAAM with the difficulties encountered to implement actions after the core meltdown. The purpose of this article is to present the main conclusions drawn from this study.

Keywords: PSA / HRA / FUKUSHIMA / SAMG / NUCLEAR CRISIS

1. INTRODUCTION

In the years 2000, IRSN developed its first level 2 Probabilistic Safety Assessment (PSA) model for the French 900 MWe PWRs. It was an ambitious project and one of the important tasks was to build a Human Reliability Analysis (HRA) model for the actions of the Severe Accident Management Guide (SAMG). Human and Organizational Reliability Analysis in Accident Management (HORAAM) model was developed for this purpose.

On the damaged unit, after the core melted, actions are implemented by the crew but they result from decisions that may be taken by the nuclear crisis organisation. The experience feedback from operators applying Emergency Operating Procedures (EOPs) on simulators, widely used for level 1 HRA models, couldn't be reused to design HORAAM. First indeed, actions are oriented to the single goal of reducing the release consequences but the level 2 HRA model had also to include the information exchange and the decision processes which are crucial mechanisms for the crisis center activities in such a context. To validate their assumptions, the HORAAM developers have used feedback analyses of the nuclear crisis exercises that are regularly practiced in France.

HORAAM has then been used at IRSN to produce HRA data for level 2 PSAs for the French PWRs but was of course never compared with a real accident. In 2012, IRSN conducted a study to confront HORAAM with difficulties encountered during the Fukushima-Daiichi accidents to implement actions after the core meltdown. This paper presents the main conclusions of this study.

2. PRESENTATION OF HORAAM

2.1. The decision tree method

HORAAM is a "decision tree (DT) model". The DT methodology relies on the hypothesis that the failure probability of a human action can be evaluated through a limited number of factors. These factors are called "Influence Factors" (IFs) and they illustrate the context of a human action. They are chosen in order to take into account in the best possible way the factors that have an effect on the implementation of the action and particularly the related difficulties: IFs must be indeed numerous enough to take into account the main aspects of the action but not too numerous in order to limit the complexity of the event tree. A theoretical example of a DT is presented on Table 1 below. Its goal is to evaluate the failure probability of an action implemented in the main control room to cope with an initiating event through four IFs.

The IFs are ranked from the most influent to the less influent. The number of modalities (or branches) is adjusted to reflect the relative influence of the IFs. If possible, IFs take only two modalities to limit the complexity of the model. Next step once the structure of the DT is finalized is to quantify the probabilities associated to each branch of the tree. Several quantification technics can be used but all are a combination of "anchor probabilities" (statistics from observation for few branches) and "partial probabilities" (expert judgment).

Thus the individual effect of each IF is integrated to yield the corresponding probabilities and to account for dependencies between IFs. Different combinations of IFs can lead to greatly different Human Error Probabilities (HEPs).

Table 1: Typical Decision Tree with 4 Influence Factors

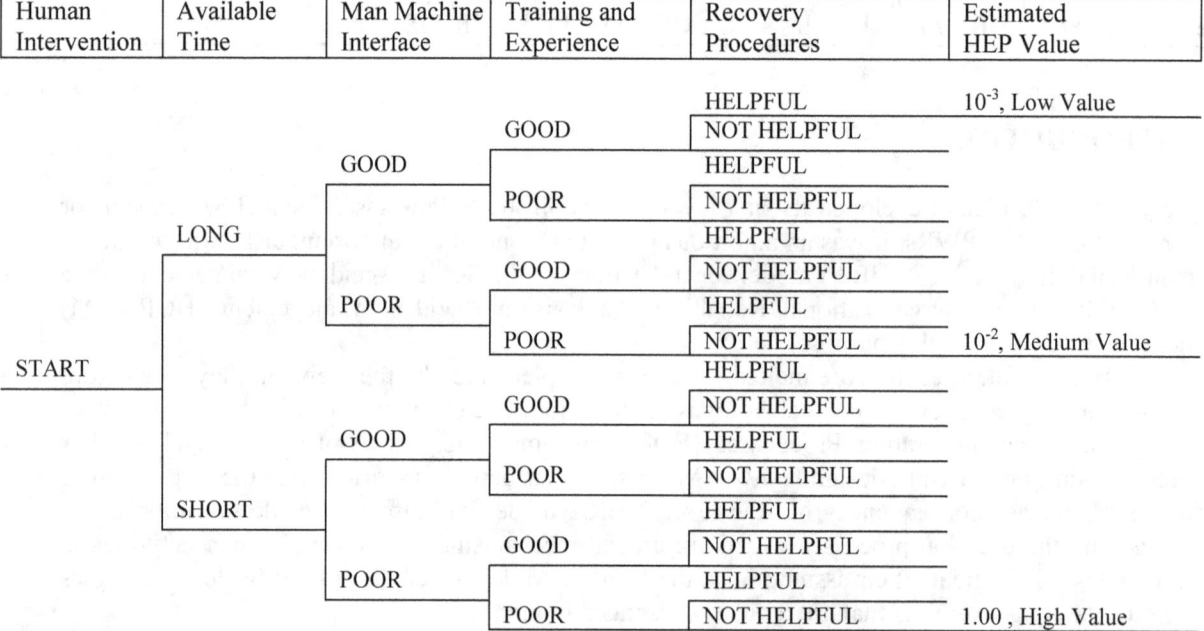

Human Intervention	Available Time	Man Machine Interface	Training and Experience	Recovery Procedures	Estimated HEP Value
			GOOD	HELPFUL	10^{-3}, Low Value
				NOT HELPFUL	
		GOOD	POOR	HELPFUL	
				NOT HELPFUL	
	LONG		GOOD	HELPFUL	
				NOT HELPFUL	
		POOR	POOR	HELPFUL	
				NOT HELPFUL	10^{-2}, Medium Value
START			GOOD	HELPFUL	
				NOT HELPFUL	
		GOOD	POOR	HELPFUL	
				NOT HELPFUL	
	SHORT		GOOD	HELPFUL	
				NOT HELPFUL	
		POOR	POOR	HELPFUL	
				NOT HELPFUL	1.00 , High Value

2.2. The DT of HORAAM

The IFs of HORAAM were selected by the developers after crisis exercises. Thereafter they conducted interviews of several crisis experts. These interviews were used, on one hand to validate the pre-selected IFs and to rank them in function of their decreasing influence (cf. 2.2.1), and on the other hand to quantify the HORAAM DT (cf. 2.2.2).

In order to elaborate this model, data were collected from different crisis centers (IRSN, headquarters of EDF, plants). Different series of reactors and different scenarios were observed too.

2.2.1 The influence factors of HORAAM

From the observation of crisis exercises, developers identified several influence factors which particularly affect human and organizational reliability in a situation of core meltdown. Finally, the DT of HORAAM is constituted of the following seven IFs, each of them having 2 or 3 modalities. They are ranked in descending order of influence in table 2.

Table 2: The 7 IFs of HORAAM

	Influence factors	Description
1	Time for decision	The time necessary to obtain, check and process information and make a decision about the required action. This influence factor has three modalities "short" "medium" or "long".
2	Information and measurement means	This IF refers to the quality, reliability and efficiency of all measurements and information available in the control room and means of transmitting them to crisis teams. This influence factor has two modalities "satisfactory" or "unsatisfactory".
3	Decision difficulty	This IF refers to the difficulty in taking the right decision. This influence factor has three modalities "easy" "medium" or "difficult".
4	Difficulty for the operator	The difficulty of the action (quality of the procedures, experience and knowledge in the control room or in the plant) is evaluated independently of work conditions. This influence factor has two modalities "easy" or "difficult".
5	Difficulty induced by environmental conditions	This IF takes into account the on-site conditions in which the actions decided upon, have to be performed (radioactivity, temperature, smoke, gas, exiguity...). This IF has two modalities "normal" or "difficult".
6	Scenario difficulty	This IF refers to the difficulty of the global context of the current accident scenario in which a decision must be made. This influence factor has two modalities "easy" or "difficult".
7	Degree of involvement of the crisis organization	Local crisis organization on the plant site or the whole national crisis organization. This influence factor has three modalities "not involved", "local crisis team involved" or "local and national crisis teams involved".

2.2.2 Quantification of the DT

For the quantification of the DT of HORAAM, developers turned to expert judgement through interviews of the crisis experts. Even if each IF has only two or three modalities, the DT of HORAAM has about one hundred branches. This is obviously a too large number to obtain a statistical result for each branch. Another classical technique of quantification would be asking to the crisis experts to give a mark to the influence of the seven IFs and then converting these marks into probabilities but this way of doing is not correct because of dependencies between IFs.

Therefore, developers turned to an alternative method where IFs were considered in triplets or doublets. Several triplets which were identically identified by most of the different experts were used as anchor values of the DT. So, the triplets that were identified to lead undoubtedly to the success of actions were associated to a failure probability of 10^{-4} (the lower probability of the DT), those which were identified to lead undoubtedly to the failure of actions were associated to a failure probability of 1. Between these branches a non-quantified area of the DT remained. In order to quantify the remaining branches, the developers used favorable and unfavorable doublets (with a sufficient agreement between the experts), associated with the marks given by the experts.

3. QUICK OVERVIEW OF THE FUKUSHIMA-DAIICHI NUCLEAR ACCIDENT

The Tohoku district-off the Pacific Ocean earthquake occurred at 14:46 on March 11, 2011. The three units of the Fukushima-Daiichi nuclear plant which were operating were automatically shut down. Due to the earthquake, the external power supplies failed but emergency diesel generators started to carry on the cores cooling. However, one hour later, a tsunami flooded the site and degraded most of the diesel generators, most of the switchboards and most of the batteries. On unit 3 only, batteries remained available. In the main control room of unit 1 and unit 2, light, reactor parameter measurements and active cooling systems were lost.

On unit 1 a semi-passive system did not operate for complex reasons linked to the closing of some valves. The reactor was no longer cooled and the core began to melt down about two hours after the tsunami. Water injection in the vessel began only at 6:00 on March 12 through fire-trucks, probably too late to prevent vessel failure.

On unit 2, just before the loss of all AC/DC power source, the shift team manually started a steam-driven cooling system (Reactor Core Isolation Cooling: RCIC) that operated for around 70 hours without any other operation of the operators (except for the shifting of water sources from the condenser to the torus). Due to the failure of electrical supplies and the loss of all measurements on the reactor, the operators were unable to check the status of the RCIC (flow rate) for hours. RCIC failed on the 14th of March and water injection were resumed (through fire-trucks) only after more than 7 hours. No water injection for more than 7 hours resulted in a core melting (of which the extent is still difficult to precise).

For unit 3, the situation was slightly different because batteries remained available. However the safety systems RCIC and HPCI finally failed on the 13th of March. Delays in opening the safety valves in order to decrease the vessel pressure to be able to inject water through fire-trucks caused a core melting. As for unit 2, it is still difficult to evaluate its extent.

From the early moment of the tsunami arrival, working conditions of the operators were very degraded on the site. These elements are analyzed in chapter n°4 thereafter.

4. EXPERIENCE FEEDBACK FROM FUKUSHIMA

The first step of the study carried out by IRSN consisted in determining the scope of the analysis. It was decided to concentrate on the first few days that followed the tsunami. Although differences in design exist, it was admitted that the ultimate actions consisting in water injection and depressurization could be compared in the case of a similar accident in a PWR. Just below are presented the context factors that had a large impact on the implementation of actions at Fukushima. In chapter n°5 "Evaluation of HORAAM" these context factors will be confronted to the IFs of HORAAM.

This study, performed in 2012, was based on public information on the Fukushima accident available at this time (mainly reference [1]) and IRSN understanding. Some misunderstandings are obviously possible.

4.1. Unavailability of the Safety Parameter Display System (SPDS)

The Fukushima nuclear power station was equipped with the Safety Parameter Display System (SPDS) for collecting the main data from the different units, as the state of the valves and the pumps, the thermohydraulic parameters... Usually SPDS is available in the main control room, but also in the local crisis center and at the headquarters of TEPCO. Thereby, all actors involved in the crisis share actual data to assess the situation.

But, after the tsunami, because of the total loss of electrical power, SPDS was unavailable. Information from the main control room was delivered to the other actors of the crisis by phone. Information sharing was slow and some pieces of information were lost.

For example, the Local Crisis Team (LCT) was informed late about the manual stop of the HPCI of unit 3. Moreover operators could not open a relief valve to decrease the pressure in order to start low pressure injection and, as a result, the core melted few hours later. Had the LCT been aware of the situation, they might have taken actions to avoid the core melting.

4.2. Communication between the main control room and the field operators

Ordinarily, the crew in the main control room communicates with the field operators by mobile phones. But at Fukushima, the mobile phones were unavailable which had the effect of slowing down the safety actions implementation. To get information, the crew always had to wait for the operators return from the field. Moreover, facing unexpected situations, field operators always had to come back to the main control room to receive instruction from the crew. The complex actions could take hours, as for the first venting of the containment of the unit 1 (four hours were needed).

4.3. No data from sensors – erroneous data

Rapidly after the tsunami, on units 1 and 2, batteries were lost and no data about the reactors status were available. This has affected the operation of both reactors and operators spent a lot of energy trying to retrieve data. But these obvious consequences were not the only ones. The main consequence was that, ignoring the real status of each reactor, the crisis organization had difficulties to prioritize actions.

4.4. Darkness in MCR and outside

Because of the station blackout part of the work was performed in complete darkness either in the MCR or outside the MCR. Operators used flashlights or even their mobile phones. It made it difficult sometimes to locate a valve or to find a way through passageways or rooms in some areas of the plant.

4.5. High dose rates

Dose rates grew up over time. It is the first cause of failure for human actions. Operators had to use full-face masks and charcoal filters and to wear level B or C clothes or coveralls even inside the building of the units 1 and 2, which slowed down all the operations.
Moreover, in order to implement actions outside the MCR, teams were sent in turns to the work site. Despite this, several actions couldn't be implemented due to the high level of radiation.
It has to be noticed too that the fire trucks drivers from NANMEI Company stopped working claiming their right of withdrawal: dose rates were beyond the scope of their contract.

4.6. Seismic aftershocks

During the few days that followed the tsunami, seismic aftershocks were numerous. They contributed to slow down the implementation of actions outside the MCR. On the way to their work, they could stop field operators who had to turn back to the Seismic Isolation Building.
Site Superintendent Yoshida knew that the seismic aftershocks endangered the field operators' life and it was a reason for him to limit actions outside the MCR.

4.7. Flooding and debris

Following the tsunami, buildings were flooded. Many components were unavailable and in addition, due to the remaining level of water, local actions were impossible in several rooms. Even the checking of the RCIC operation on unit 2 was not possible because the level of water was higher than the boots of the field operator who was sent there. When they opened the door, water gushed out of the room and they could not go in.

Another consequence of the tsunami was large areas of the site littered with debris. It took five hours to drive to unit 1 a fire truck which was parked close to unit 5. Debris were removed by hand.

4.8. Hydrogen explosions

Hydrogen explosions occurred three times (on units 1, 2 and 4). The first one occurred on unit 1 on the 12th of March (the day following the tsunami): five workers were injured in the explosion. Each time the explosion strongly disturbed the actions outside the MCR. Some workers rescued the injured people from the site and the others were evacuated to the Seismic Isolation Building. Investigations were carried out and any recovery work could not start again before the area was deemed safe. On reactor 3, water injection was indeed interrupted for almost six hours after the hydrogen explosion. Moreover hydrogen explosions have disrupted the work in progress.

5. EVALUATION OF HORAAM

Influence factors are ranked in the decision tree by decreasing order of importance. The first step of the study was the assessment of the IFs: the choice and the ranking. The confrontation of HORAAM to experience feedback from Fukushima shows that both were good. It confirms that the first two IFs "Decision period" and "Information and measurement means" are essential. If these IFs are disabled the mission will probably fail. In the PSA meaning, the action is considered "not implemented" or "implemented too late to avoid a degradation" (the core meltdown or the failure of a component).

5.1. Validation of the first four IFS of HORAAM

The first four IFs are validated without modification:

- IF n°1 "Time for decision"

The IF "Time for decision" is the first IF in HORAAM, so this IF is considered as contributing the most to the success or the failure of a mission. Experience feedback from Fukushima confirms this importance but some considerations can be added. The time before core degradation (few hours) was short for the reactor 1 and it was not sufficient for a correct diagnosis of the reactor status, taking into account the instrumentation unavailability. For the reactor 2 and 3, the time before core degradation was longer (70 hours and 40 hours), but the time available from decision to action implementation was not sufficient. Considering the degraded situation, it took a long time to implement all the actions, even the simplest ones. A short time available in degraded situation obviously leads to failure. One lesson may be that the notion of "short time available for HRA" is longer in a degraded situation, as it was the case at Fukushima, than the one in "normal" environmental conditions (around ten hours in degraded situation compared to around one hour in "normal" environmental conditions).

- IF n°2 "Information and measurement means"

Just after the tsunami, reactor 1 and 2 have lost all Direct Current (DC) power supplies. As a result the operators were unable to monitor plant parameters including the reactor water level. One direct consequence was that they were unable to apply the EOPs because there was no procedure taking into account the events where all Alternative Current (AC) and DC power sources would be lost. Moreover, SPDS was unavailable, even for unit 3, so the crisis teams had to try to understand the situation at each unit based on information obtained by phones. Site Superintendent thought it would be impossible to take any action necessary to control the nuclear plants without the plant parameters, especially those for the reactor water level and pressure. He put a priority to restore equipment necessary to measure the main reactors parameters to the detriment of other actions.

- IF n°3 "Decision difficulty"

Considering the lack of electrical power supplies and the batteries depletion, pumps or valves had to be operated locally, directly on the component. Each of these actions endangered the health of the field operators through dose rate, seismic aftershocks and hydrogen explosions. Site Superintendent who coordinated the operations was also responsible for the radiological releases outside the plant and the protection of the health of the operators in the plant. His decision was a compromise between the plant operation and the protection of the health of the field operators.

- IF n°4 "Difficulty for operators"

The IF n°4 "Difficulty for operators" is designed to sort actions well known by operators (normal operation or well-trained operations on simulators) from actions that operators implement rarely as unusual pipes alignments.
At Fukushima, after the loss of the safety injection means (IC, RCIC, HPCI), operators had use mobile water injection means and they had to face difficulties:
- workers from NANMEI Company who were used to driving and operating the fire trucks stopped working because of dose rates ;
- TEPCO operators were not skilled to operate the fire trucks for water injection into the core of the reactors ;
- no procedures were available, especially for connecting the fire hoses of the fire engines to the fire pumps discharge ports outside the turbine building.

At Fukushima, the long duration to put into operation this alternative water injection demonstrates, if necessary, the importance of IF n°4.

5.2. Comments about IFS n°5 to n°7

Based on the analysis of the Fukushima accident, IFS n°5 to n°7 don't appear to be inappropriate but comments are needed:

- IF n°5 "Difficulties induced by the environmental conditions"

The Fukushima accident shows (cf. 4.4 to 4.8) how bad environmental conditions can slow down the implementation of actions. Moreover, several operations were cancelled because of the high level of dose rate. The relevance of IF "Difficulties induced by the environmental conditions" is not to be questioned but on the contrary, it appears that this IF has a too low influence on the resulting probabilities. Therefore, HORAAM should be updated to reassess IF n°5.

- IF n°6 "Difficulty of the scenario"

As it was the case for IF "Difficulties induced by the environmental conditions", the relevance of IF n°6 "Difficulty of the scenario" is not to be questioned. Surely the situation at Fukushima induced by the external hazard of an earthquake and a tsunami, the total loss of electrical power, the loss of the heat sink and the meltdown of the core of tree units is a difficult scenario. All these elements induced difficulties to implement any action. However, it appears that difficulties induced by the type of scenario are already included in the other parameters of HORAAM. The crisis experts who were interviewed for the development of HORAAM had highlighted the interest of a particular IF "to take into account multiple failures" but finally the process of quantification (based on the interviews of the same experts) attributed a low influence for this IF. So, after this study, the suppression of IF n°6 is recommended.

- IF n°7 "Degree of involvement of the crisis organization"

The IF n°7 "Degree of involvement of the crisis organization" is correlated to the attendance of crisis teams. Four hours after the occurrence of an initiating event, the French crisis organisation is supposed to be available and its influence no longer changes. In HORAAM, IF n°7 concerns only medium and long durations. The failure probability of a human action decreases slightly when the local crisis team is available and it decreases a little more when the national crisis organisation is available. Of course the experience feedback from Fukushima does not question the relevance of an IF dedicated to the crisis organization but it is important to note that the crisis organisation has to manage simultaneously many priorities when several units of a nuclear site are damaged. HORAAM should take into account this lesson. Therefore, IF n°7 "Degree of involvement of the crisis organization" would become "availability of the crisis organization for the unit management" with tree modalities "1: Only the crew is available" or "2: the local crisis team is available" or "3: the national and the local crisis team are available together". Yet, for each action, the modality of IF n°7 would be fixed depending on the state of achievement of other actions, either on the same unit or on other disabled units. For the latter case, HORAAM could be used for site PSAs.

6. CONCLUSION

The purpose of the IRSN study was to confront HORAAM, its HRA model for level two PSAs, with the difficulties encountered at Fukushima to implement actions after the core meltdown. At Fukushima, most of the data needed to operate the units were unavailable, communication means were disabled, the system to share information between all the crisis actors was unavailable, hours after hours dose rate increased, part of the work was performed in a complete darkness, the ground was littered with debris, the basements of the buildings were flooded, often works in process were interrupted by seismic shocks, and worse, hydrogen explosions endangered the life of field operators.

With a DT HRA model, as HORAAM, the IFs assess the context of a human action in order to derive its failure probability. The choice and the ranking of the IFs are essential. The study confirmed that the first four IFs: "Decision period", "Information and measurement means", "Decision difficulty" and "Difficulty for operators", were well selected. If these IFs are poorly rated, the human action will probably fail.

However, the study concluded that the last three IFs called for comments. These IFs are not indeed inappropriate but improvements need to be made. IF "Difficulty of the scenario" is relevant but it is redundant with several other IFs and it can easily be suppressed. IF "Difficulties induced by the environmental conditions" is essential. It must be kept and its influence on the quantitative results should be increased. Lastly, IF "Degree of involvement of the crisis organization" which only relies on the presence of the crisis organization should be adapted to take into account the extent of the crisis organization operationality (depending on the number of simultaneous tasks to be performed simultaneously). This evolution would also result in extending the scope of HORAAM to the assessment of several units together (site PSAs).

References

[1] Investigation Committee on the Accident at Fukushima Nuclear Power Stations of Tokyo Electric Power Company, Interim Report, December 26, 2011

[2] F. Ménage, A. Vogel, B. Chaumont, "Using a decision tree to estimate human error probabilities in a level 2 PSA: the HORAAM method", PSA 99

[3] G. Baumont, F. Ménage, J.R. Schneiter, A. Spurgin, A. Vogel, "Quantifying human and organizational factors in accident management using decision trees: the HORAAM method", Original Research Article - Reliability Engineering & System Safety - Volume 70, Issue 2, Pages 113-124 (November 2000)

[4] V. Fauchille , L. Esteller, E. Raimond, N. Rahni, "Application of the Human and Organizational Reliability Analysis in Accident Management (HORAAM) method for the updating of the IRSN level 2 PSA model", PSAM 9

[5] M. Villermain, E. Raimond, K. Chevalier, N. Rahni, B. Laurent, "Method for examination of accidental sequences with multiple containment failure modes in the French 900 MWe PWR level 2 PSA", PSAM 9

The Impacts of Supervisor Attributes and Supervision-Related Policies on Safety and Environmental Outcomes and Reporting Behavior

Christopher J. Jablonowski[a], John J. Tolle[b]
[a] Shell Exploration and Production Company, Houston, TX, U.S.A.
[b] Value Discovery LLC, Houston, TX, U.S.A.

Abstract: This paper specifies detection-controlled regression models to investigate the drivers of health, safety, and environmental (HSE) performance and reporting behavior. The analysis confirms some results from previous research and also tests new hypotheses, with emphasis on supervision-related practices and policies. Most of the results are general and thus applicable to other regions, to other operators, and very likely to other industrial sectors. The results can be used to drive decisions regarding operating practices and HSE management system policy.

Keywords: Detection-controlled Estimation, Reporting, Safety, Supervision.

1. INTRODUCTION

Health, safety, and environmental (HSE) managers are responsible for analyzing HSE performance and continuous improvement. Quantitative analysis poses special challenges because there is no theoretical basis for assumptions regarding the functional form of HSE incident phenomena, incident data is often unbalanced (few incidents), data is not collected (typically) in cases when there are no incidents, and incidents are not always reported. However, these challenges can be met with common-sense assumptions, improved data collection strategies, and advanced modeling methods.

This work specifies detection-controlled regression models similar to those employed in previous research to investigate the drivers of HSE performance and reporting behavior in oil and gas drilling [16,17]. The analysis confirms some results from previous research and also tests new hypotheses, with emphasis on supervision-related practices and policies. Most of the results presented here are general and thus applicable to other regions, to other operators, and very likely to other industrial sectors. The results can be used to drive policy decisions regarding operating practices and HSE management system policy by providing a basis to allocate resources to those policies with the largest benefit-cost ratios.

1.1. Imperfect Reporting of Safety Incidents

As described in the literature, underreporting of incidents in the workplace occurs across many sectors [26,20,25,31,27,28]. There are various reasons for underreporting, some are intentional (evasion) while others are unintentional (ignorance). Also, underreporting can occur at any level (worker, supervisor, manager). The purpose of this study is to investigate incidence and underreporting behavior in oil and gas drilling. It is acknowledged that the prospect also exists for overreporting, for example, fraudulent reports of incidents that did not occur made as an attempt to obtain a financial gain from the employer or insurance provider, but in this study it is assumed that there is no overreporting.

There is an emerging literature on the subject of incomplete detection based on the seminal work of Feinstein [10,11]. As Feinstein predicted, his model of detection controlled estimation (DCE) could be applied in various contexts. Studies have been completed in tax compliance, environmental compliance, health diagnosis, political science, and safety in oil and gas drilling [8,5,13,39,4,19,33,16,17]. The present study adds to the empirical foundation in oil and gas drilling by re-testing previous hypotheses with improved (more granular) independent variables, and by testing new hypotheses.

1.2. Implications of Imperfect Reporting

Imperfect reporting distorts the observations of incident data. A simple example demonstrates the impacts of underreporting, assuming that no fraud occurs (modified from [17]). Consider 100 hypothetical safety outcomes in Table 1. The columns represent whether or not an incident occurred, while the rows represent whether or not the incident was reported. In this unobservable "truth" case, the underreporting is evident. In practice, however, the underreported incidents are counted with the actual non-incidents. Thus, the analyst observes the data as depicted in Table 2.

Table 1. True Incident Data

		Incident Occurred?	
		Yes	No
Incident Reported?	Yes	10	--
	No	4	86

Table 2. Observed Incident Data

		Incident Occurred?	
		Yes	No
Incident Reported?	Yes	10	0
	No	0	90

Depending on the levels of imperfect reporting, the implications can be severe. The true frequency of an incident, $P(I)$, is equal to 14/100, while the analyst computes a value of 10/100. Of course the conditional probabilities are also affected. The data in Table 1 provides the *reporting rate*, defined as the conditional probability, $P(Report|Incident) = P(R|I)$. Here, this value equals 10/14, not 1 as indicated in Table 2. The complement of the reporting rate is the *underreporting rate*, $P(No\ Report|Incident) = P(NR|I)$. It is clear that in the presence of imperfect reporting, use of the data in Table 2 will distort any qualitative or quantitative analysis. If the imperfect reporting can be modeled explicitly, then more accurate assessments can be made of the true incident phenomena. Also, the analyst will learn about factors that affect the reporting rate.

2. REGRESSION MODELS OF IMPERFECT REPORTING

It is assumed that incidents are reported as the product of two independent and sequential events. First, an incident (or set of incidents) either occurs or does not occur. Second, an incident (or set of incidents) either is reported or not reported. This assumption facilitates the mathematical treatment and discussion. Two models are specified for this study.

2.1. Model of Perfect Reporting (No Underreporting, No Overreporting)

This model is specified and estimated to establish a base case for comparison, and reflects conventional practice in regression analysis of HSE incidents [11,14,15,6,34,35,22,7,21,40,18]. That is, this model estimates the case as depicted in Table 2.

Observations were collected from nine drilling rigs over a ~22 month period in 2011-2102 from one of Shell's (operator's) onshore development assets in the U.S. The unit of observation is defined as one well. Data is collected for each well i on each rig r in the study period. There are $r = 1 \ldots R$ rigs and $i = 1 \ldots N_r$ wells on each rig, $\boldsymbol{x}_{ri}$ is a $1xh$ vector of independent variables for well ri believed to affect incidence, and $\boldsymbol{\beta}$ is defined as a $hx1$ vector of coefficients to be estimated. This model specifies the incidence function as Poisson where $ln(\mu_{ri}) = \boldsymbol{x}_{ri}\boldsymbol{\beta}$. The probability for observation y_{ri} is represented as shown in Equation (1) with the resulting log-likelihood equation shown in Equation (2). Note that the marginal effect of a variable on the dependent variable, $\partial y/\partial x_h$, is equal to $\beta_h \bar{y}$.

$$Pr(Y_{ri} = y_{ri}) = \pi(y_{ri}; \mu_{ri}) = e^{-\mu_{ri}} \frac{\mu_{ri}^{y_{ri}}}{y_{ri}!} \tag{1}$$

$$L = \sum_{r=1}^{R} \sum_{i=1}^{N_r} -e^{\boldsymbol{x}_{ri}\boldsymbol{\beta}} + y_{ri}\boldsymbol{x}_{ri}\boldsymbol{\beta} - ln(y_{ri}!) \tag{2}$$

2.2. Model of Imperfect Reporting (Underreporting, No Overreporting)

This model is specified and estimated to investigate the impacts of underreporting. One set of independent variables are specified for an incidence function, while another set of independent variables are specified for a reporting function. This model requires a key assumption regarding the reporting process. When more than one incident occurs, there are three potential outcomes for reporting. One outcome is that all of the incidents are reported, a second outcome is that none of the incidents are reported, and a third outcome is that there is partial under-reporting and a subset of incidents is reported. In the derivation below, it is assumed that for each observation of the dependent variable, incidents are either all reported or all not reported, simplifying the computations.

If one allows for the possibility of partial reporting, the implications are severe. The number of conditional reporting probabilities that need to be estimated grows significantly, even when reasonable simplifying assumptions are made. In addition, the number of terms on the right hand side of the regression is in theory, infinite. For example, to compute the probability of observing one reported incident, the analyst would have to consider all potential values of incidence. The analyst could constrain this number to limit the scope of the computation, but the selection of the cutoff point would be arbitrary. For these reasons, the case of partial under-reporting is not specified here.

The incidence function is specified again as Poisson, and the reporting function is specified as a binary probit model (see [17] for a description of the probit model). $\boldsymbol{z}_{ri}$ is a $1xj$ vector of independent variables for well ri believed to affect reporting, and $\boldsymbol{\delta}$ is defined as a $jx1$ vector of coefficients to be estimated. The probability that observation y_{ri} on the dependent variable takes on a value greater than zero is shown in Equation (3) with the resulting log-likelihood function for all non-zero observations, m, shown in Equation (4).

$$Pr(Y_{ri} = y_{ri}) = \pi(y_{ri}; \mu_{ri})\Phi(\boldsymbol{z}_{ri}\boldsymbol{\delta}) \tag{3}$$

$$L_m = \sum_{r=1}^{R} \sum_{i=1}^{N_r} -e^{\boldsymbol{x}_{ri}\boldsymbol{\beta}} + y_{ri}\boldsymbol{x}_{ri}\boldsymbol{\beta} - ln(y_{ri}!) + ln\big(\Phi(\boldsymbol{z}_{ri}\boldsymbol{\delta})\big) \tag{4}$$

The probability that observation y_{ri} on the dependent variable takes on a value equal to zero is the sum of the probability that no incident occurred plus the probability that an incident occurred but was not reported, and this is shown in Equation (5) with the resulting log-likelihood function for all zero observations, n-m, shown in Equation (6).

$$Pr(Y_{ri} = y_{ri}) = \pi(y_{ri}; \mu_{ri}) + \big(1 - \pi(y_{ri}; \mu_{ri})\big)\big(1 - \Phi(\boldsymbol{z}_{ri}\boldsymbol{\delta})\big) = 1 - \Phi(\boldsymbol{z}_{ri}\boldsymbol{\delta}) + \pi(y_{ri}; \mu_{ri})\Phi(\boldsymbol{z}_{ri}\boldsymbol{\delta}) \tag{5}$$

$$L_{n-m} = \sum_{r=1}^{R} \sum_{i=1}^{N_r} ln\left(1 - \Phi(\mathbf{z}_{ri}\boldsymbol{\delta}) + e^{-e^{\mathbf{x}_{ri}\boldsymbol{\beta}}}\Phi(\mathbf{z}_{ri}\boldsymbol{\delta})\right) \tag{6}$$

The log-likelihood for the sample is $L = L_m + L_{n-m}$ and is maximized numerically. The asymptotic covariance matrix is estimated by evaluating the negative inverse of the Hessian at the maximum likelihood estimates. The identification conditions for this family of models are derived and explored in [10].

2.3. Dependent Variable

Incident data was collected for several categories of incidents: loss of primary containment, fires, near misses, property loss and damage, unsafe acts and conditions, and injuries and illnesses. For each well *ri*, these events were summed to create the primary dependent variable, and the events were not weighted in any way. While analysis can be performed on these data individually, the authors' experience suggests that it is more appropriate to view the collection of incidents (and potential incidents) as an overall *index* of HSE performance. A second reason for aggregating the data in this way is that unaggregated data is often too unbalanced to yield reliable statistical results, that is, the dependent variable does not exhibit sufficient variability (away from 0). Figure 1 provides the distribution of the dependent variable.

2.4. Independent Variables and Hypotheses

There are various taxonomies for organizing risk factors (see [1] for example). In this study, the independent variables are grouped into five categories. This structure is intended to help organize the analysis and discussion herein. For each variable, the hypothesis regarding the directional impact (sign) of the variable on incidence and reporting is stated, and whether or not the variable is expected to be statistically significant (at the 95% confidence level). Many of these expectations are based on results reported in [17], referred to in this section as "previous research."

2.4.1. Work Type

Variables in this category describe the attributes of the work being performed.
- PadSwitch: This binary variable takes a value of 1 if the well drilled on a different pad than the rig's previous well and 0 otherwise. This variable will test the hypothesis that the first well after rig-up increases the likelihood of incidents (e.g. shakedown issues). The expectation is that the sign of this variable will be positive and significant in the incidence function and insignificant in the reporting function.
- Gap: This variable is defined as the count of days between the start of the well and the end of the previous well. This variable will test the hypothesis that longer gaps between drilling operations disrupt established practices and policies and thus increase the likelihood of incidents. The expectation is that the sign of this variable will be positive and significant in the incidence function and insignificant in the reporting function. It is also recognized that there will be some correlation between this variable and the PadSwitch variable.
- WellType: This binary variable takes a value of 1 for development wells and 0 for all other wells. Previous research indicated that differences between well types in engineering design, operations, and site attributes increase the likelihood of incidents on development wells. The expectation is that the sign of this variable will be positive and significant in the incidence function and insignificant in the reporting function.
- DrillingDays: This variable is defined as the count of days from the start of the well to the end of the well. It is intended to control for exposure time and thus the expectation is that the sign of this variable will be positive and significant in the incidence function and insignificant in the reporting function.

Figure 1. Distribution of Dependent Variable

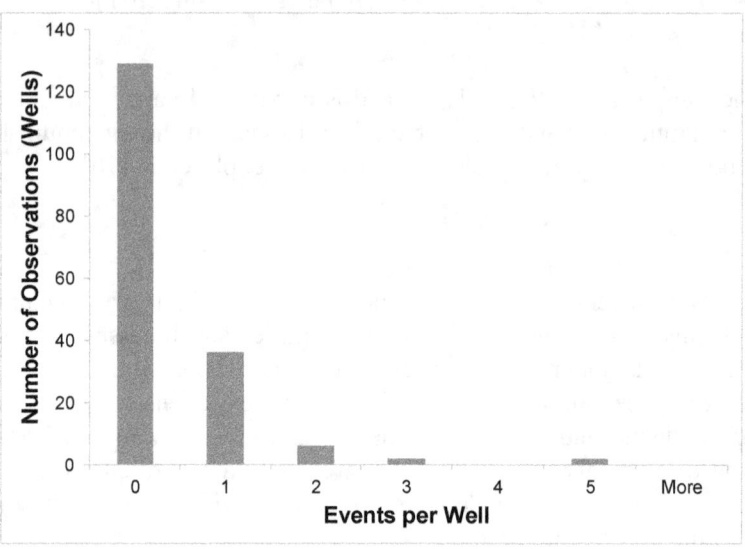

- Non-ProductiveTime: This variable is defined as the percent of total days on the well spent on non-productive activities (e.g. rig or other equipment breakdown, other unplanned events). This variable will test the hypothesis that such disruptions in normal drilling operations and switching between activities, often under increased time pressure, increase the likelihood of incidents. The expectation is that the sign of this variable will be positive and significant in the incidence function and insignificant in the reporting function.

Because the rigs were all drilling similar types of wells using similar procedures, other variables such as well design and underbalanced or managed pressure drilling could not be tested.

2.4.2. Equipment and Work Site

Variables in this category describe the rig and the location where the work is being performed.

- Rig#: Binary variables are defined for each of the nine rigs in the data set. These variables model differences in performance not captured by other variables. The expectation is that these variables will be insignificant in the incidence and reporting functions once other variables are controlled for (e.g. supervision), consistent with previous research.
- WeatherQuarter: Environmental conditions such as extreme heat or cold, or heavy rains or snows, may affect incidence. This variable is defined as a binary variable representing the calendar quarter in which the well was drilled. There are no expectations on the signs of these variables or their significance in the incident function. They are not expected to be significant in the reporting function.

The rigs were all from the same drilling contractor and were outfitted in a similar way (i.e. similar levels of automation), and the drilling sites all shared the same geography and degree of site remoteness, thus none of these variables could be tested.

2.4.3. Supervision

Variables in this category capture attributes of the onsite supervisor(s). The quality of supervision is an important factor in driving compliance with procedures [36]. How many are there? Is there well-to-well consistency? Are they owner employees or contractors? These types of questions have been investigated in several of the aforementioned studies. The results for these types of variables can be used to adjust policies on allocation of supervisory resources.

- Foreman-DaysForeman#: For each of the 59 foremen who worked within the study period, this variable is defined as the sum of days that the foreman worked on the well ("Foreman-Days"). These variables are intended to model the marginal impact of each supervisor on safety incidence. Based on previous studies, the expectation is that some supervisors will be positive (negative) and significant in the incident function, and possibly in the reporting function. An alternate version of this variable was defined as the sum of days worked on the well divided by the DrillingDays. The difference in results was negligible.

- Foreman-DaysTeam#: Some foreman worked on many of the same wells as other foreman, and of 44 of these foremen "teams" were identified. For each of these teams, this variable is defined as the sum of Foreman-Days for each foreman divided by the DrillingDays. Recent research suggests that team characteristics and communication norms can affect safety outcomes [3,24]. As is the case with the individual foreman variables, it is possible that some teams will be positive (negative) and significant in the incident function, and possibly in the reporting function.

- ForemanCount: This variable is defined as the number of different foremen who worked on the well. When more foremen are assigned to a well, it may lead to mixed messages and an increase in incidence, although previous research did not support this hypothesis. In contrast, it is also possible that having more different foremen work on a well, even if at different times, provides additional perspectives, sharing, and reinforcement of policies and best practices that decrease incidence. The expectation is that this variable will either be insignificant, or negative and significant, in the incidence function. The expectation is that for the same reasons this variable will be positive and significant in the reporting function, consistent with previous research. An interesting question is whether this supervisory *diversity* variable or the one described next that describes supervisory *concentration* will better explain incidence and reporting. Previous research did not test any hypotheses regarding concentration.

- Foreman-DaysPerRig-Day: This variable is defined as the sum of individual Foreman-Days divided by DrillingDays. For example, if Foreman#1 worked 10 days and Foreman #2 worked 15 days on a 15 day well, then this variable would equal (10+15)/15. It is intended to measure the concentration of supervision. It is possible that a larger concentration of supervision leads to better oversight and a decrease in incidence. However, it is also possible that large concentration can lead to confusion regarding who is in charge, mixed messages, and an increase in incidence. The *mixed messages* hypothesis here is different than that described in the previous variable definition because in this case the variable measures whether the messages occur at the same time. Therefore, the expectation on this variable in the incidence and reporting function is uncertain.

- ForemanConsistency: This variable is defined as the sum of individual Foreman-Days for those foremen who also worked on the previous well, divided by the sum of all Foremen-Days. Thus, a large value indicates that much of the supervision on the current well is the same as the previous well (more consistent). Previous research indicated that consistency increased the likelihood of incidents, supporting the hypothesis that the benefits of new perspectives and sharing of best practices across locations (rigs) outweighed the benefits of consistency. However, the variable as it is defined here is more refined and should produce a more reliable test of this hypothesis. Previous research used a simple binary variable to indicate whether the foremen were *exactly the same as* the previous well; for example, if only 3 out of 4 foremen were carried over from previous well, it would have been classified as not consistent. There is no expectation on the sign of this variable in the incidence function. The expectation is that the variable will be insignificant in the reporting function.

Three variables are defined to investigate the impact of foreman employment status. Previous research did not indicate any difference between operator and contractor foremen, but one hypothesis for this result was the fact that some of the contractor foreman were former operator employees and as such would reflect operator norms more so than a "pure" contractor. Foremen in each of these categories have different experiences and face different incentives, and the purpose is to test whether the degree of contractor supervision has an impact on safety incidence and/or reporting. This variable is tested in both the incidence and reporting functions, and there are no hypotheses regarding the sign of this variable.

- ForemanPureOperator: This variable is defined as the sum of Foremen-Days worked on the well by foremen who are full-time operator employees.
- ForemanPureOperatorOrFormerOperator: This variable is defined as the sum of Foremen-Days worked on the well by foremen who are full-time operator employees or a former full-time operator employees.
- ForemanPureContractor: This variable is defined as the sum of Foremen-Days worked on the well by foremen who are contractors and who have never worked for the operator.

While not a part of this study, future investigations could examine specific measures of supervisor training and competence, test for regional differences to put the spotlight on higher levels of management as suggested by others [12], and other supervisor policy options.

2.4.4. Safety Management System and Policy (SMS)

The importance of safety management systems (SMS) is self-evident. All of the rigs in this study were governed by the same SMS (e.g. inspection protocols), thus there is limited opportunity to investigate elements of the SMS. The variables in this category represent some tactics and policies deployed by the operator.
- Interventions: This variable counts the number of safety interventions made by workers and supervisors. For example, one worker may notice an unsafe act being committed by a fellow worker and intervene to stop the activity. Most companies have a mode for reporting such events, and unsafe conditions, as part of their SMS. It is commonly believed that these kinds of behaviors bolster the safety climate and improve safety performance [29], and previous research supported this hypothesis. However, it is also possible that large numbers of such interventions could be a sign of a poor safety climate. The expectation is that this variable (or percentage changes in lagged values) will be significant and negative in the incidence function. That is, larger values or percentage upticks are indicative of high levels of awareness and a good safety climate. This variable is expected to be insignificant in the reporting function. Variations of lagged specifications can also be considered as precursor candidates [32].
- WellCountOnRig: This variable is defined as the cumulative well count on each rig up to and including the current well. It is intended to capture the effect of experience on incidents and reporting. The expectation is that this variable will be negative and significant in the incidence function, and positive and significant in the reporting function. This reflects the expectation that as time passes, the operator's SMS and culture becomes more well-established, and that this improves safety and reporting performance. Previous research was inconclusive on this point. It is an important hypothesis because if indicated to be true, it may affect procurement strategy when picking up and dropping rigs.

In addition to the operator's SMS, the drilling contractor's SMS also affects safety performance and variables can be defined in a likewise fashion. As explained above, there was insufficient variability in the drilling contractor in the study period so this was not included. Future studies could include assessments of other safety policies like financial incentives for the drilling contractor [23].

2.4.5. Worker Attributes

Variables in this category provide information about the individual worker. This can be basic demographic information like age, experience in industry, and experience with the operator. The age and experience of workers are potential risk factors. One source reports that the majority of incidents involve workers with less than 5 years of experience, and that almost half of all incidents involve workers with less than 1 year on the job [2]. This category can also include measures of training and competence, and would speak to issues of training effectiveness [30,37]. Workers can also be described by levels of stress, fatigue, and workload [38]. While some of this data is collected on workers who suffer an injury, it is generally not collected for instances when no incidents occur (and this data is needed for a regression analysis). Because of this lack of data, and resource constraints which prevent its collection ex post, analysis of individual worker attributes was not included in this study.

3. REGRESSION ANALYSIS AND DISCUSSION

The model of perfect reporting was estimated first to identify *probable* drivers of incidence and/or reporting. That is, when one observes a statistically significant variable in this model, it is not discernible whether the effect is attributable to incidence or reporting behavior. However, it is a sign that the variable is probably important in one or both of the functions and careful attention is warranted in the model of imperfect reporting. When a variable does not indicate as significant in the model of perfect reporting, one cannot ignore the variable in the model of imperfect reporting. That is, it is possible that the incidence and reporting behaviors "cancel out" and thus are not observed in the model of perfect reporting.

Sample regressions for the Work Type variables are provided in Tables 3 and 4. These results are from the conventional Poisson model as specified in Equation (2); the detection-controlled models did not suggest any reporting-related effects. The results for the Work Type variables are as follows:

- The PadSwitch (Table 3) and Gap (Table 4) variables are both significant in the incidence function and have the expected signs. This result suggests that the first well after rig-up on a different pad increases the likelihood of incidents. Longer gaps between drilling operations also increase the likelihood of incidents. Because of the correlation between these variables, their effects cannot be measured precisely when both are included in the same regression.

- There is some evidence that indicates that the WellType variable is positive and significant in the incidence function, consistent with previous research. However, the effect was not consistently observed across the variety of specifications that were investigated, and in many cases it is statistically insignificant and excluded.

- DrillingDays is consistently significant (or weakly significant) as expected and is retained as a control variable for exposure time.

  - Non-ProductiveTime is insignificant in the incidence function, contrary to expectations. Disruptions in normal drilling operations, switching between activities, and increased time pressure do not appear to increase the likelihood of incidents.

Table 3. Work Type Variables (a)

Variable	Coefficient Estimate	z-statistic
PadSwitch	0.6928	2.0500
WellType	1.0680	2.5200
DrillingDays	0.0430	2.4200
Non-ProductiveTime	-2.4319	-1.0000
Constant	-2.6603	-4.7000

Table 4. Work Type Variables (b)

Variable	Coefficient Estimate	z-statistic
Gap	0.0371	1.7800
WellType	0.9025	2.2000
DrillingDays	0.0524	3.0400
Non-ProductiveTime	-2.3264	-0.9500
Constant	-2.7203	-4.7000

A sample regression for the Equipment and Work Site variables is provided in Table 5. The regression includes the significant Work Type variables and two Weather variables. This result is from the conventional Poisson model as specified in Equation (2); the detection-controlled model did not suggest any reporting-related effects. The results for the Equipment and Work Site variables are as follows:

- None of the Rig# binary variables are significant in the incidence or reporting function after controlling for Foreman-Days. In regressions where Foreman-Days is excluded, some of the Rig# variables are individually significant, but the explanatory power of the models is small. When significant Foreman-Days variables are included with the Rig# variables, the Rig# variables cease to be significant and the explanatory power of the models improves ~four-fold, suggesting that the true drivers of incidents and reporting are the foremen, not the rig. Based on this result, it was concluded that in regressions that contain Rig# and no Foreman-Days variables, that the Rig# variables are merely (weak) proxies for the Foreman-Days variables because of correlations between the two variables, i.e. some foremen worked repeatedly on the same rig. The same structure and result was observed in previous research.

- There is some evidence that indicates that two of the WeatherQuarter variables are significant in the incidence function. More incidents appear to occur in the hot summer months, and fewer in the fall. While heat-related factors make sense in terms of their ability to affect HSE performance, we have no hypothesis for the apparent decrease of incidents in the fall. However, the effects was not consistently observed across the variety of specifications that were investigated, and in many cases they are both statistically insignificant and excluded.

Comparing Tables 3, 4, and 5 is instructive because they demonstrate how coefficient estimates and significance tests change when alternate specifications are estimated. This is typical for all regression-based analysis, and it puts extra demands on the analyst to explain inconsistent and/or awkward results. In this case, the results are consistent across specifications.

Table 5. Equipment and Work Site Variables

Variable	Coefficient Estimate	z-statistic
Gap	0.0454	2.2000
WellType	1.1884	2.8300
DrillingDays	0.0688	3.6200
3Q (summer)	0.7101	2.4100
4Q (fall)	-0.8378	-1.9800
Constant	-3.5131	-5.3500

The results for the Supervision and SMS variables are based on the detection-controlled model as specified in Equations (4) and (6) because the analysis suggests some reporting-related effects for these variables. In most cases the regressions include the previously identified significant Work Type and Equipment and Work Site variables, although the coefficients are not measured precisely in all regressions, especially for regressions with larger numbers of independent variables. In some cases this is explained by correlations between variables, otherwise, we attribute the fluctuations between specifications to the complexity of the incident- and report-generating processes (e.g. excluded variables, joint effects, etc.), and to the numerical complexity of the detection-controlled model (for example, in some specifications the model does not converge). The results for the Supervision variables are as follows:

- Several of the individual foremen (Foreman-DaysForeman#) are significant in the incidence function, supporting the hypothesis that some foremen have a unique impact on safety performance. There is some evidence to support differential reporting behavior as shown in Table 6. In this example result, Foreman#12 appears to have more incidents relative to other foremen, and also appears to be less likely to report an incident once it occurs. But the results for individual foremen are somewhat sensitive to the specification, i.e. the choice and number of coefficients being estimated. Also, many foremen have small

Foreman-Days totals, making it difficult to estimate their individual impacts with precision. Interpreted in their totality, the results do not suggest significant differential reporting behavior between foremen. Both results are consistent with previous research.

Table 6. Foremen Variables

Variable	Coefficient Estimate	z-statistic
Incidence Function		
Gap	0.0247	0.9500
WellType	0.8643	1.9100
DrillingDays	0.0673	3.1800
3Q (summer)	0.4326	1.3300
4Q (fall)	-0.8528	-1.9100
Foreman#12	0.3268	4.8000
Foreman#43	0.0601	1.9200
Constant	-3.1622	-4.4600
Reporting Function		
Foreman#12	-0.3777	-2.5400
Foreman#35	-0.1488	-0.9400
Constant	1.8768	2.2300

- The foreman team variables (Foreman-DaysTeam#) were defined to test hypotheses about team impacts on safety performance. In some cases, the foreman variables and the team variables are correlated, and additional analysis was required to understand the structure of the relationship. Based on this analysis, it is clear that significant foreman teams (in incidence and/or reporting) are driven by the significance of their constituents; i.e. all but one significant teams have at least one individually significant constituent. In the one case where a significant team was comprised of two individually insignificant constituents, there is an apparent synergy between the two foremen that improved their joint performance. In summary, there does not appear to be a strong team impact on incidence or reporting, rather, it is the individuals who drive these outcomes. These variables were retained and used in cases where individual foremen variables were highly correlated.

- The evidence for the ForemanCount variable indicates that when more foremen work on a well there is a decrease in the likelihood of incidents and an increase in reporting, consistent with expectations. This result indicates that when additional foremen work on a well, the diversity of supervision serves to reinforce policies and best practices rather than to introduce uncertainty from mixed messages.

- The Foreman-DaysPerRig-Day variable was insignificant in the incidence and reporting functions. This result indicates that larger supervisory concentration does not yield a measurable impact, positive or negative, on incidence or reporting. It is important to note however that this variable is somewhat tightly clustered around a value of 2, potentially affecting the precision of the estimate. Also, when the value falls below this value it is not by much, therefore no conclusions can be drawn about the impact of concentrations less than 2.

- As described above, previous research indicated that foreman consistency increased the likelihood of incidents. This was a somewhat controversial result and it was desired to revisit this question in the present study. The previous research used a simple binary variable to indicate whether the foremen were *exactly the same as* the previous well. The ForemanConsistency variable defined above is more refined and provides a more definitive test of this hypothesis. The results using the new variable definition indicate that consistency is significant in the incidence function, and that more well-to-well consistency decreases the likelihood of incidents. There is no impact on reporting.

- The analysis of the three variables that describe foreman employment status indicate that employment status does not significantly affect incidence or reporting. This result was consistent whether individual foreman variables were included or not included in the regressions, reducing the risk that correlations between the two sets of variables was affecting the result.
- The Interventions variable is significant and positive in the incidence function, contrary to expectations. This result suggests that higher levels of intervention can be interpreted here as an indicator of a breakdown in the safety climate. By itself, this information is not very useful because each rig develops different norms regarding the level of reporting. Percentage changes in Interventions was also investigated but were not found to be significant. The Interventions variable is not significant in the reporting function.
- The WellCountOnRig variable was defined to test the hypothesis that as additional wells are drilled, the operator's SMS and culture becomes more well-established, and that this improves safety and reporting performance. The variable is significant and negative in the incidence function, consistent with expectations. There is weak evidence that the variable is positive and significant in the reporting function.

An auxiliary regression was specified to investigate the relationship between Interventions and the WellCountOnRig variable. A strong negative relationship was discovered; Interventions decrease the longer a rig is in the fleet. A typical comprehensive regression is shown in Table 7.

Table 7. Comprehensive Regression (example)

Variable	Coefficient Estimate	z-statistic
Incidence Function		
Gap	0.0554	1.9900
DrillingDays	0.0397	1.4700
Foreman#6	0.1016	1.8800
Foreman#35	0.2761	2.1700
Foreman#43	0.1461	2.7100
Foreman#45	0.1432	2.9000
Foreman#55	0.1940	3.2000
Foreman#56	-0.0610	-1.2500
ForemanTeam7/16	-2.8114	-1.4500
ForemanCount	-0.4660	-2.1500
ForemanConsistency	-1.4864	-1.7800
WellCountOnRig	-0.1037	-2.8600
Constant	1.2499	1.1000
Reporting Function		
ForemanCount	2.9462	2.0000
Foreman#12	0.7016	0.8800
Foreman#35	-0.3663	-1.3800
WellCountOnRig	0.2174	1.2200
Constant	-10.9078	-1.9900

Table 8 contains a summary of the expectations and findings for each independent variable. The first column lists the variable name, the second column summarizes the expectations, and the third column summarizes the findings.

Table 8. Summary of Individual Variable Expectations and Results

Variable	Incidence Function EXPECTATION	Incidence Function RESULT
PadSwitch	+	+
Gap	+	+
WellType (dev=1)	+	+ (weak)
DrillingDays	+	+
Non-ProductiveTime	+	insignificant
Rig#	insignificant	insignificant
WeatherQuarter	?	mixed (weak)
Foreman-DaysForeman#	mixed	mixed
Foreman-DaysTeam#	mixed	insignificant
ForemanCount	insignificant or -	-
Foreman-DaysPerRig-Day	?	insignificant
ForemanConsistency	?	-
ForemanPureOperator	?	insignificant
ForemanPureOperatorOrFormerOperator	?	insignificant
ForemanPureContractor	?	insignificant
Interventions	-	+
WellCountOnRig	-	-
Variable	Reporting Function EXPECTATION	Reporting Function RESULT
PadSwitch	Insignificant	insignificant
Gap	insignificant	insignificant
WellType	insignificant	insignificant
DrillingDays	insignificant	insignificant
Non-ProductiveTime	insignificant	insignificant
Rig#	insignificant	insignificant
Weather	insignificant	insignificant
Foreman-DaysForeman#	mixed	mixed (weak)
Foreman-DaysTeam#	mixed	insignificant
ForemanCount	+	+
Foreman-DaysPerRig-Day	?	insignificant
ForemanConsistency	insignificant	insignificant
ForemanPureOperator	?	insignificant
ForemanPureOperatorOrFormerOperator	?	insignificant
ForemanPureContractor	?	insignificant
Interventions	Insignificant	insignificant
WellCountOnRig	+	+ (weak)

As reported in Table 8, there are only a few variables that affect reporting. Using the regression model reported in Table 7, it is possible to compute the probability of a false negative for each zero observation, *P(Incident|No Report)*, or *P(I|NR)*, using the notation from the Introduction. The result provides a general indication of whether imperfect reporting is a significant problem. The probability is defined in Equation (7) using the notation from the Introduction

$$P(I|NR) = \frac{P(I)P(NR|I)}{P(NR)} = \frac{\left(1 - \pi(y_{ri}; \mu_{ri})\right)\left(1 - \Phi(z_{ri}\delta)\right)}{\pi(y_{ri}; \mu_{ri}) + \left(1 - \pi(y_{ri}; \mu_{ri})\right)\left(1 - \Phi(z_{ri}\delta)\right)}$$

$$= 1 - \frac{\pi(y_{ri}; \mu_{ri})}{1 - \Phi(z_{ri}\delta) + \pi(y_{ri}; \mu_{ri})\Phi(z_{ri}\delta)}$$

(7)

The average probability for all zero observations is 7%, suggesting that imperfect reporting is not a significant problem in this asset. This result suggests that future analysis probably can be completed without the more complex detection-controlled models without introducing significant bias. However, for definitive results, the detection-controlled estimates are always recommended.

4. CONCLUSION AND RECOMMENDATIONS

The results of this analysis are largely consistent with previous research, strengthening the case for action on specific points [17]. Some of the results are specific to the operator or the asset, but most are general and thus applicable to other regions, to other operators, and very likely to other industrial sectors. Based on these results, the following actions are recommended for all drilling operations:
- Refresh the focus on safety after all rig moves between drilling pads and extended delays between wells by organizing a formal HSE engagement event at the rig site before the next well starts.
- Identify specific differences in exploration and development wells that have the potential to cause differences in incident rates, and engage the engineering and operations staff to mitigate these risks.
- Institute a refresher course for all foremen and the workforce prior to the summer season to emphasize heat-related hazards.
- Engage/interview foremen who were identified as more or less likely to have incidents to ascertain the potential drivers of these performance differences.
- Assign additional foremen to each well to provide additional perspectives, sharing, and reinforcement of policies and best practices. Currently, about four foremen work on any one well. Increasing the number of different foremen who work on a well can be accomplished by splitting hitches between wells, or by having more foremen on location at the same time.
- Maintain some supervisory consistency on each rig by assigning one or more of the foremen on the previous well to the current well. In cases where this is not possible, organize a formal HSE engagement event at the rig site before the next well starts.
- Continue the current contractor HSE on-boarding process which appears to be successful in ensuring equivalent incident and reporting performance with the operator's full-time staff.
- Retain rigs for longer terms to firmly establish the SMS and reporting norms, i.e. a "first in, last out" model.

A final note of caution is needed regarding the use of this kind of information. First, when a relationship between incidents and a variable is identified and an intervention plan or policy is enacted to reduce risk, then over time the relationship between incidents and the variable will degrade and ultimately be eliminated if the intervention plan or policy is effective. For example, if it is recognized that switching pads increases the likelihood of incidents, and a new policy is effectively implemented to refresh the safety focus in such cases, then switching pads will not be identified as a risk if the analysis is repeated in the future. But this should not be interpreted as evidence that switching pads no longer increases the likelihood of incidents, rather it should be interpreted as evidence that the intervention policy is working. Also, this same phenomenon makes it difficult to identify risk factors that are already being mitigated by some policy. In this case, the lack of statistical evidence would not be a sufficient reason to alter or cancel an existing mitigation policy that is otherwise believed to be working.

5. ACKNOWLEDGEMENTS

The authors thank our colleagues who contributed technical expertise and data: Cody Buyer, Eryn Clark, Jonathan Dallaire, Cesar Gongora, James Goodwyne, Kevin Hoffman, Lake Johnson, Gregory Knott, R.J. Mendoza, Jose Mota, Brendan O'Shea, Donald Patton, Bobby Ramos, and Linda Randle. All conclusions, errors, and omissions remain the sole responsibility of the authors.

6. REFERENCES

[1]   Altabbakh, Hanan, Murray, Susan, Grantham, Katie, Damle, Siddharth. 2013. Variations in Risk Management Models: A Comparative Study of the Space Shuttle Challenger Disaster. Engineering Management Journal, 25 (2):13-24.

[2]   Are We Losing Ground in Rig Safety? Some Operators Say 'Yes." 1998. *Drilling Contractor*, May/June: 18.

[3]   Azarkhil, Mandana, Mosleh, Ali. 2012. Modeling and Simulation of the Impact of Team Characteristics on Crew Performance. Probabilistic Safety Assessment and Management Conference, Helsinki, Finland, 25-29 June 2012.

[4]   Bradford, W.D., Kleit, A.N., Krousel-Wood, M.A., Re, R.N. 2001. Testing Efficacy with Detection Controlled Estimation: An Application to Telemedicine. *Health Economics*, 10: 553-564.

[5]   Brehm, J., Hamilton, J.T. 1996. Noncompliance in Environmental Reporting: Are Violators Ignorant, or Evasive, of the Law?" *American Journal of Political Science*, 40 (2): 444-477.

[6]   Chunlin, H., Chengyu, F. 1999. Evaluating Effects of Culture and Language on Safety. *J. Pet Tech*, April.

[7]   Conchie, S., Donald, I. 2006. The Role of Distrust in Offshore Safety Performance. *Risk Analysis*, 26 (5): 1151-1159.

[8]   Erard, B. 1997. Self-selection with Measurement Errors: A Microeconometric Analysis of the Decision to Seek Tax Assistance and Its Implications for Tax Compliance. *Journal of Econometrics*, 81: 319-356.

[9]   Feinstein, J. 1989. The Safety Regulation of U.S. Nuclear Power Plants: Violations, Inspections, and Abnormal Occurrences. *Journal of Political Economy*, 97 (1): 115-154.

[10]   Feinstein, J. 1990. Detection Controlled Estimation. *Journal of Law and Economics*, 33: 233-276.

[11]   Fleming, M., Flin, R., Mearns, K., Gordon, R. 1996. The Offshore Supervisor's Role in Safety Management: Law Enforcer or Risk Manager. Paper SPE 35906 presented at the Third International Conference on Health, Safety, and Environment in Oil and Gas Exploration and Production, New Orleans, LA, USA, 9-12 June.

[12]   Fruhen, L.S., Mearns, K.J., Kirwan, B., Flin, R. 2012. Skills and Traits as Contributors to Senior Managerial Safety Commitment. Probabilistic Safety Assessment and Management Conference, Helsinki, Finland, 25-29 June 2012.

[13]   Helland, E. 1998. The Enforcement of Pollution Control Laws: Inspections, Violations, and Self-Reporting. *The Review of Economics and Statistics*, 80 (1): 141-153.

[14]   Iledare, O., Pulsipher, A., Dismukes, D., Mesyanzhinov, D. 1997. Oil Spills, Workplace Safety and Firm Size: Evidence from the U.S. Gulf of Mexico OCS. *The Energy Journal*, 18 (4): 73-89.

[15]   Iledare, O., Pulsipher, A., Dismukes, D., Mesyanzhinov, D. 1998. Safety and Environmental Performance Measures in Offshore E&P Operations: Empirical Indicators for Benchmarking. Paper SPE 49153 presented at the SPE Annual Technical Conference and Exhibition, New Orleans, LA, USA, 27-30 September.

[16] Jablonowski, C. 2007. Employing Detection Controlled Models in Health and Environmental Risk Assessment: A Case in Offshore Oil Drilling. *Journal of Human and Ecological Risk Assessment*, 13 (5): 986-1013.

[17] Jablonowski, C. 2011(a). Statistical Analysis of HSE Incidents and the Implications of Imperfect Reporting. *SPE Drilling & Completions*, 26 (2): 278-286.

[18] Jablonowski, C. 2011(b). Using Regression Analysis to Relate Safety and Environmental Outcomes to Incidence Factors. *SPE Projects, Facilities & Construction*, March: 33-40.

[19] Kleit, A.N., Ruiz, J.F. 2003. False Positive Mammograms and Detection Controlled Estimation. *Health Services Research*, 38 (4):1207-1228.

[20] Leigh, J.P., Marcin, J.P., Miller, T.R. 2004. An Estimate of the U.S. Government's Undercount of Nonfatal Occupational Injuries (Review). *Journal of Occupational and Environmental Medicine*, 46 (1): 10-18.

[21] Malallah, S. 2009. Leadership Influence in Safety Change Effort. Paper IPTC 13816 presented at the International Petroleum Technology Conference, Doha, Qatar, 7-9 December.

[22] Mearns, K., Whitaker, S., Flin, R. 2001. Benchmarking Safety Climate in Hazardous Environments: A Longitudinal, Inter-organizational Approach. *Risk Analysis*, 21 (4): 771-786.

[23] Osmundsen, Petter, Toft, Anders, Dragvik, Kjell Agnar. 2006. Design of Drilling Contracts-Economic Incentives and Safety Issues. Energy Policy, 34: 2324-2329.

[24] Park, Jinkyun, Jung, Wondea, Yang, Joon-Eon. 2012. The Use of SNA Metrics to Investigate the Relationship Between the Characterisitcs of Crew Communications with the Associated Crew Performance. Probabilistic Safety Assessment and Management Conference, Helsinki, Finland, 25-29 June 2012.

[25] Phimister, J.R., Bier, V.M., Kunreuther, H.C. (eds.). 2004. Accident Precursor Analysis and Management: Reducing Technological Risk Through Diligence. Washington, D.C.: The National Academies Press.

[26] Pransky, Glenn, Snyder, Terry, Dembe, Allard, Himmelstein, Jay. 1999. Under-Reporting of Work-Related Disorders in the Workplace: A Case Study and Review of the Literature. Ergonomics, 42 (1): 171-182.

[27] Probst, T.M., Brubaker, T.L., Barsotti, A. 2008. Organizational Injury Rate Underreporting: The Moderating Effect of Organizational Safety Climate. *Journal of Applied Psychology*, 93 (5): 1147-1154.

[28] Probst, Tahira M., Estrada, Armando X. 2010. Accident Under-Reporting Among Employees: Testing the Moderating Influence of Psychological Safety Climate and Supervisor Engorcement of Safety Practices. *Accident Analysis and Prevention*, 42: 1438-1444.

[29] Pullwitt, Tanja, Mearns, Kathryn. 2008. Safety Interventions-A Status Challenge for Managers and Supervisors. Paper SPE 111801 presented at the SPE International Conference on Health, Safety, and Environment in Oil and Gas Exploration and Production, Nice, France, 15-17 April.

[30] Ramirez, Alberto E., Prasad, Dharmesh. 2000. Safety Training Effectiveness. Paper SPE 61251 presented at the SPE International Conference on Health, Safety, and the Environment in Oil and Gas Exploration and Production, Stavanger, Norway, 26-28 June.

[31] Rosenman, K.D., Kalush, A., Reilly, M.J., Gardiner, J.C., Reeves, M., Luo, Z. 2006. How Much Work-Related Injury and Illness is Missed by the Current National Surveillance System? *Journal of Occupational and Environmental Medicine*, 48 (4):357-365.

[32] Saltmarsh, Elizabeth A., Saleh, Joseph H., Mavris, Dimitri N. 2012. Accident Precursors: Critical Review, Conceptual Framework, and Failure Mechanisms. Probabilistic Safety Assessment and Management Conference, Helsinki, Finland, 25-29 June 2012.

[33] Scholz, J.T., Wang, C.L. 2006. Cooptation or Transformation? Local Policy Networks and Federal Regulatory Enforcement. *American Journal of Political Science*, 50 (1): 81-97.

[34] Shultz, J. 1999. The Risk of Accidents and Spills at Offshore Production Platforms: A Statistical Analysis of Risk Factors and the Development of Predictive Models. Doctoral Dissertation, Carnegie Mellon University.

[35] Schultz, J., Fischbeck, P. 1999. Predicting Risks Associated with Offshore Production Facilities: Neural Network, Statistical, and Expert Opinion Models. Paper SPE 52677 presented at the SPE/EPA Exploration and Production Environmental Conference, Austin, TX, USA, 28 February-3 March.

[36] Skaugrud, Ida, Dahl, Oyvind, Olsen, Espen. 2012. A Qualitative Study of Organizational Factors Influencing Compliance with Procedures. Probabilistic Safety Assessment and Management Conference, Helsinki, Finland, 25-29 June 2012.

[37] Skjerve, Ann Britt, Bye, Andreas. 2012. Towards Guidance for Assessment of "Training" as a Performance-Shaping Factor in Human Reliability Analysis. Probabilistic Safety Assessment and Management Conference, Helsinki, Finland, 25-29 June 2012.

[38] Sneddon, A., Mearns, K., Flin, R. 2006. Safety and Situation Awareness: "Keeping the Bubble" in Offshore Drilling Crews. Paper SPE 98629 presented at the International Conference on Health, Safety, and Environment in Oil and Gas Exploration and Production, Abu Dhabi, U.A.E., April 2-4.

[39] Stafford, Sarah L. 2003. Assessing the Effectiveness of State Regulation and Enforcement of Hazardous Waste. *Journal of Regulatory Economics*, 23 (1): 27-41.

[40] Winter, J., Owen, K., Read, B., Ritchie, R. 2010. How Effective Leadership Practices Deliver Safety Performance And Operational Excellence. Paper SPE 129035 presented at the SPE Oil and Gas India Conference and Exhibition, Mumbai, India, 20-22 January.

Change Impact Analysis as required by safety standards, what to do?

Authors: Thor Myklebust[1,a], Tor Stålhane[b], Geir Kjetil Hanssen[a] and Børge Haugset[a]

[a] SINTEF ICT,
[b] IDI NTNU

Abstract: Change Impact Analysis related to safety of products and systems is used by companies in many industries and is required by several standards. The International Electrotechnical Commission (IEC) has issued several standards with requirements and guidelines for the establishment of analysis like FMECA (IEC 60812), FTA (IEC 61025), Design review (IEC 61160), HAZOP (IEC 61882), Markov (IEC 61165) and RBD (IEC 61078) but no standard for Change Impact Analysis. Based on the aforementioned standards, a literature study and experience from several projects, this paper proposes a Change Impact Analysis Report adapted to the specific characteristics of the Railway and Process industry domains. The purpose of this paper is to serve as a tool to aid manufacturers in performing a Change Impact Analysis at the appropriate level which will be approved by assessors and certification bodies. This is important since the Change Impact Analysis report is one of the main inputs to the assessor/certification body.

The paper starts by presenting and clarifying relevant terms and definitions, as these differ from standard to standard. The main part of the paper structures and describes the relevant topics for a Change Impact analysis report.

Using the described approach will save time and cost and reduce the risk of having to re-issue the Change Impact analysis, thus ending up with a product having hidden defects. Using the mindset from SafeScrum - a method that introduces elements from agile into safety-related software development, will result in further savings.

This work is part of a series of Railway and IEC 61508 certification projects and the SUSS[2] Research projects.

Keywords: Change Impact Analysis, EN 5012X, IEC 61508, Certification

1 Introduction

Change Impact Analysis (CIA) related to safety of products and systems is used by companies in many industries and is required by several standards. The CIA report (CIAR) is one of the main inputs to the assessor. A standardized CIAR will simplify the work both for the manufacturer and the assessor and will also improve the certification process.

The International Electrotechnical Commission (IEC) has issued several standards with requirements and guidelines for the establishment of analysis like FMECA (Failure mode effect and criticality analysis) [1], FTA (Fault Tree Analysis) [2], Design review [3], HAZOP (Hazard and operability studies) [4], Markov[3]

---

[1] Thor.myklebust@sintef.no
[2] Norwegian: Smidig utvikling av Sikkerhetskritisk Software. English: Agile Development of safety Critical Software
[3] Copy from IEC 61165: The Markov techniques make use of a state transition diagram which is a representation of the reliability, availability, maintainability or safety behaviours of a system, from which system performance measures can be calculated. It models the system's behaviour with respect to time.

[5] and RBD[4] (Reliability Block Diagram) [6] but no standard for CIA. Based on the above mentioned standards, a literature study, experience from several projects and study of agile methods like Scrum, this paper proposes a CIAR adapted to the characteristics of the Railway and Process industry domains. The purpose of this paper is to serve as a tool to aid manufacturers in achieving a CIA at the appropriate level that will be approved by assessors and certification bodies. We also provide guidelines on how to perform a CIA.

The guidance for an impact analysis plan and report provided in the present paper is intended to be complementary to the standards and plans. The guidance for an impact analysis report will ensure that the manufacturers document the CIAR in such a way that an assessor will accept the report. It will also ensure that the modifications performed are sufficiently thought-through. Several examples exists where this has not been the case, and an inadequate CIAR has resulted in products that either have not been approved or the time before the product was on the market has been delayed with months and even years.

Furthermore, we relate our guidelines for CIA to ongoing work (see www.sintef.no/SafeScrum that is updated on a regular basis) on improving both the development and assessment efficiency for safety critical software by applying principles and techniques from agile software development methods. In particular, we look at Scrum, which has had a large uptake in industry over the past decade. We have proposed the SafeScrum approach which combines the benefits of agile development with the requirements of the IEC61508 standard. In this paper we show how a process like SafeScrum can be used to make CIA more efficient and coupled to the development process, making it more practical and with reduced costs.

The rest of the paper explains our research approach and the research questions we address (section 2). We present an overview of the key concept of the SafeScrum process as a background for our guidelines for CIA (section 3) and then go on to provide an overview of identified requirements for CIA (section 4), followed by requirements in the analysis and review techniques in relevant IEC standards (section 5) and an insight on CIA plans (section 6) and CIA reports (section 7). Finally we summarize and conclude our work (section 8).

Assumptions: It is assumed that a modification/change request report exists.

Terms and definitions: So far there is not an international consensus on terminology related to CIA, and only a few relevant terms are defined in the standards that we have studied. These terms are presented in Table 1 below. The regulations for the European process industry do not have a directive or regulation related to change, while the railway industry has the CSM (Common Safety Method) regulations [7] and [8]. The risk management and risk assessment processes in the CSM Regulation relate to the processes that are put in place for assessing the safety levels and compliance with the safety requirements of any significant change. The CSM on risk assessment shall be applied by the person in charge of implementing the change under assessment. The EN 5012X series [9] and Railway CSM regulations [7] and [8], do not include any of the definitions mentioned in the Table below.

---

[4] Copy from IEC 61078: A reliability block diagram (RBD) is a pictorial representation of a system's reliability performance. It shows the logical connection of (functioning) components needed for successful operation of the system (hereafter referred to as "system success").

Table 1: List of definitions in different standards and regulations

Term	IEC 61508-4 [10]	ISO 26262-1 [11]	ISO/IEC/IEEE 24765 [12]
Change	Not defined	Not defined	**3.379 Change.** The modification of an existing application comprising additions, changes and deletions.
Modification	Not defined	**1.75 Modification.** Authorized alteration of an item. NOTE 1 Modification is used in ISO 26262 with respect to re-use for lifecycle tailoring. NOTE 2: A change is applied during the lifecycle of an item, while a modification is applied to create a new item from an existing item.	Not defined
Impact analysis	**3.7.5 Impact analysis.** Activity of determining the effect that a change to a function or component in a system will have to other functions or components in that system as well as to other systems.	Not defined	**3.1360 Impact analysis** *1.* Identification of all system and software products that a change request affects and development of an estimate of the resources needed to accomplish the change NOTE This includes determining the scope of the changes to plan and implement work, accurately estimating the resources needed to perform the work, and analyzing the requested changes' cost and benefits.

<u>Limitations</u>: This paper is limited to products and systems such as e.g. a railway signalling system and E/E/E/PES (electrical/electronic/programmable electronic safety-related systems) including railway signalling systems and process industry domains with emphasis on the EN 5012X standards and the IEC 61508 series. Only modification of products delivered by the manufacturer is considered, not modifications of existing products on site.

2   Research method and research questions

We have performed a literature search including a survey of mainly European directives and International and European standards. Relevant directives and information about these directives can be found at www.newapproach.org. The main standardization organizations we searched were ISO, IEEE, IEC and CENELEC. We identified the following regulations, directives and standards: [2-8, 13-19].

The work we have done is related to an ongoing industrially oriented research project (Agile development of safety critical software) were we are developing a new software engineering method for more efficient development and certification of safety critical systems up to safety integrity level 3. This new approach, named SafeScrum [20], is based on the agile software development method Scrum [21] which is extensively used in the software industry. As part of this research effort we have defined the following research questions which are being addressed in this paper:

RQ1: How can we develop a CIAR that ensures that the product is approved by the assessor/certification body?

RQ2: How can agile development improve CIA and the process towards recertification?

3   Agile development of safety critical software

Agile software development is a way of organizing the development process, emphasizing direct and frequent communication, frequent deliveries of working software increments, short iterations, active customer engagement throughout the whole development life cycle and change responsiveness rather than change avoidance. This can be seen as a contrast to waterfall-like models such as the V-model, which emphasize thorough and detailed planning, an upfront design, and consecutive plan conformance. Several agile methods are in use, whereof Scrum [21] is one of the most commonly used. Figure 1 explains the basic concepts of an agile development model.

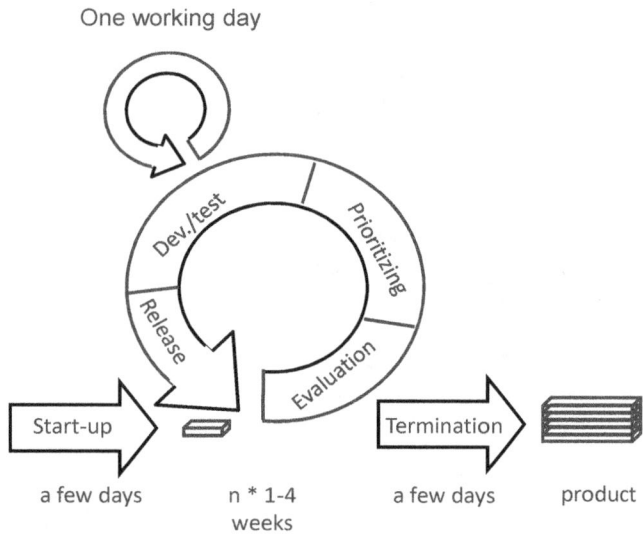

Figure 1: Scrum model

The main constructs of this model are (based on Scrum):
- Initial planning phase is short and results in a prioritized list of requirements for the system called *the product backlog*. Developers also provide *estimates* per item, in essence how much effort it takes to fulfil the requirement.
- Development is organized as a series of *sprints* (iterations) that lasts a few weeks.

- Each sprint starts with a *sprint planning meeting* where the items with the highest priority from the product backlog are moved over to the *sprint backlog* – adding up to the amount of resources available for the period. These requirements will be implemented in the following sprint.
- Each working day starts with a *scrum,* which is a short meeting where each member of the development team (1) explains what she/he did the previous work day, (2) any impediments or problems that need to be solved and (3) planned work for the work day.
- Each sprint *releases* an *increment,* which is a running or demonstrable part of the final system.
- The increment is *demonstrated* for the customer(s), which will decide which backlog items that have been resolved and which that need further work. Based on the results from the demonstration the next sprint is planned. The product backlog is revised by the customer and is potentially changed / reprioritized. This initiates the sprint-planning meeting for the next sprint.
- When all product backlog items are resolved, all available resources are spent and / or time does not permit further development, the final product is released. Final tests can be run to ensure completeness.

In order to realize some of the proven benefits of agile development, such as better quality, more efficient development and closer involvement of customers, we have proposed a method called SafeScrum [20]. This variant of Scrum is motivated by the need to make it possible to use methods that are flexible with respect to planning, documentation and specification while still being acceptable to IEC 61508 and e.g. EN 50128 (Railway), as well as making Scrum a useful approach for developing safety critical systems. The rest of this section explains the components and concepts of this combined approach.

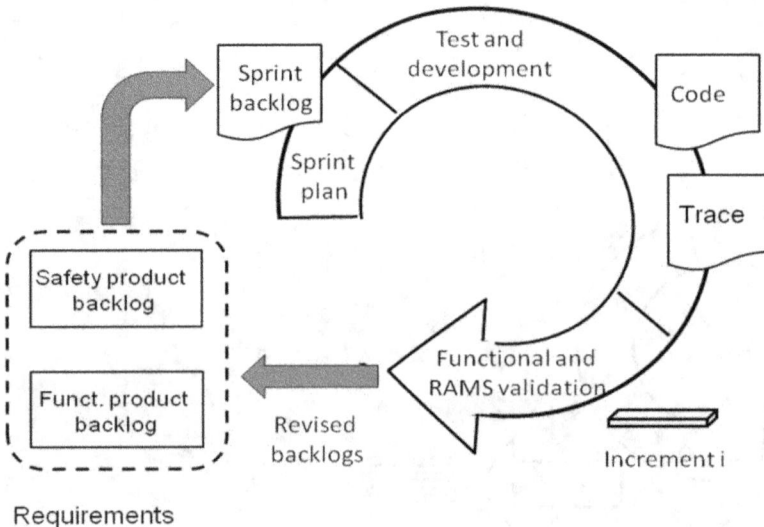

Figure 2: SafeScrum model

Our model has three main parts. The first part consists of the IEC 61508 steps needed for developing the environment description and the SSRS phases 1-4 (concept, overall scope definitions, hazard and risk analysis and overall safety requirements). These initial steps result in the initial requirements of the system that is to be developed and is the key input to the second part of the model, which is the Scrum process. The requirements are documented as *product backlogs.* A product backlog is a list of all of the functional and safety related system requirements, prioritized by the customer or a similar role having the users and/or owners view and interests in mind. It might be practical to also include the developers in the prioritization process. We have observed that the safety requirements are quite stable, while the functional

5

requirements can change considerably over time. Development with a high probability of changes to requirements will favour an agile approach.

Test-driven development (TDD) is a development practice that embraces the principle of never adding or changing code without first having added or changed the runable test case that verifies its success criteria. The method is popular within the agile community and perceived benefits include having to think about design prior to coding, a safety harness for making changes, and less time spent debugging. In essence TDD has, through studies, been shown to increase quality at the possible expense of productivity [22, 23]. We believe this focus on quality could present a benefit in using TDD for safety-critical software development, and that the increased trust in the code could also benefit the assessment.

4 Agile Change Impact Analysis

We propose a CIA-approach that integrates with the SafeScrum process in two phases. Phase 1 analysis is done upfront of the change implementation process, like the present practice but with more detailed guidelines as described in chapters 7-8. In phase 2, analysis is performed as an integrated part of the SafeScrum process and is thus done iteratively throughout the change process itself. For each item that is selected from the product backlog to enter the sprint backlog (to be implemented in the upcoming sprint) the change impact is evaluated per item by checking whether the change will impact the safety level of the system. This happens at the same time as the requirements (often described as a User Story, i.e. "As a *<role>*, I want *<goal/desire>* so that *<benefit>*") are broken down into a set of smaller tasks that are easier to develop. All these evaluations are logged to keep track of the process and will eventually be added to the CIAR. In cases where a change is found to affect the safety integrity level of the system, a strategy needs to be developed on how to tackle it. This strategy may lead to new items to be added to the backlog to change one or more items in order to avoid a conflict. In the figure below, the different CIA are presented in the SafeScrum model. In this paper only the first CIA is described. Phase 2 will be described in a later paper.

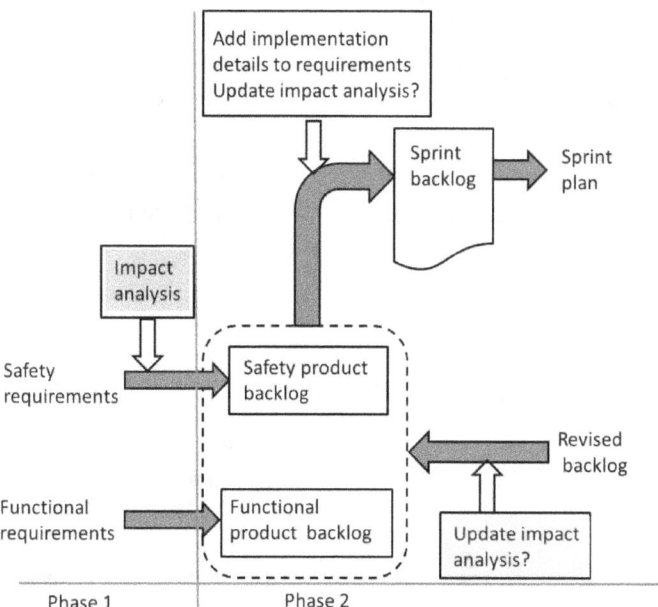

Figure 3: SafeScrum and Change Impact Analysis

5 Important requirements for the analysis and review techniques in relevant IEC standards

The requirements for impact analysis during change are the same in most standards on the development of safety critical systems. The impact analyses requirements in IEC 61508 and in several other standards does, however, not provide any guidance on how this analysis shall be carried out and documented. In one way, this is good, since it is now up to the developers and the assessor to agree on what is needed on a case by case basis, hence opening for methods like SafeScrum to be used. On the other hand, it leaves companies that are new to the trade totally in the dark.

The standards for FTA [2] and FMEA [13] are normally used by the companies developing E/E/E/PES and SIL4 signalling systems. When evaluating relevant chapters for a CIA Report, the following four topics where studied in these six standards [2-6, 13] Experts/teams, Plan, Process and Documentation. Experts/teams and Planning are important for all projects, and our SafeScrum approach emphasize an effective Process and an optimal amount of documentation.

Table 2: Overview of comments to different analysis standards

Topics	Experts/team	Plan	Process	Documentation
HazOp	Selection of a relevant team is an important part of the standard, see e.g. ch. 6.3 in the standard	This is an important part of the standard. What the plan should include is also listed.	A process for the HAZOP is well written and figures are included.	Documentation is presented in ch.6.6 and states that this is one of the strengths by using HAZOP.
FTA	Not emphasized.	Some information is included.	Some information is included.	Requirements for the FTA report are presented in ch. 9.
FMEA	Selection of relevant experts is important part of the standard.	This is an important part of the standard. What the plan should include is also listed.	Some information is included	Requirements for the FMEA report are presented in ch. 5.4.
Design review	Selections of relevant experts (specialists) are an important part of the standard. Top management shall be included.	This is an important part of the standard.	A process for the Design review is well written and a figure is included.	Documentation is mentioned.
RBD	Not emphasized.	Not emphasized.	Not emphasized.	Not emphasized.
Markov	Not emphasized.	Not emphasized.	Not emphasized.	Requirements for the Markov analysis report are presented in ch. 10.

Experts/team: From the table above we see that the HazOp, FMEA and Design review supports our view and experience that selection of a relevant team is important.

7

Plan: See ch.6 below.

Process: Two of the standards, HazOp and Design review, present a description of the process.

Documentation: No relevant information is presented regarding the right amount of documentation.

6 Change Impact Analysis plan

6.1 The need for a plan for change impact analysis

The European Committee for Electrotechnical Standardization (CENELEC), the International Organization for Standardization (ISO) and the Institute of Electrical and Electronics Engineers (IEEE) have issued a series of standards with requirements and guidelines for the establishment of safety plans (EN 50126-1, Railway), quality plans (ISO 10005) and project plans (ISO 10006). IEEE has also issued a standard for software safety plans (IEEE Std 1228-1994). As mentioned earlier, none of these organisations have issued guidelines or requirements for a CIA plan. The closest we get to this is the work done by OLF[5] with their planning matrix in OLF 070 [24]. This section is an attempt to fill this need.

6.2 On change and risk

During a project's lifetime we often need to change either part of existing code or part of an existing requirement. In many cases, we have to do this irrespective of cost and risk. In other cases there are two questions: (1) what are the costs and risks and (2) what are the benefits and regrets. There are several ways to identify cost, risk, benefits and regrets. We will not discuss these methods here but instead assume that the company in question has appropriate methods for assessment of these four values and the ability to make informed decisions.

The most important item that must be in place for a CIA are traceability information and relevant documentation. Without these CIA will be extremely difficult and costly.

6.3 Needed activities

Changes may relate to behaviour (what the system does), quality (how the system does it) and safety (the risk incurred by putting the system into operation). The CIA process needs to go through the following steps:

1. Use the traceability information to identify artefacts that need to be changed. This will include – but is not necessarily limited to
   a. Code and code documentation
   b. User documentation
   c. Tests and test plans
   d. Earlier relevant analysis – e.g. HazOp, FMEA and CIA for related requirements and system parts.
2. We need to assess the risks related to the planned changes. The risk assessment should also include mitigations for all important risks. In addition, it is important to take into consideration that doing nothing will also carry risks.

---

[5] OLF: Norwegian OljeIndustriens Landsforening, English: Norwegian Oil and Gas Association
(www.norskoljeoggass.no/en)

3. Who shall participate? It is important that we have covered all relevant aspects – e.g. code, testing, user needs and domain knowledge.
4. Decide and plan necessary V&V-activities, e.g. document inspections, unit tests, system tests, user tests. The necessary tests will depend on the artefacts changed and the related risks.
5. Run a retrospective when all the activities have been performed. Important issues to consider:
   a. What worked well
   b. What needs improvement

In addition it might be practical to check whether relevant standards have been changed since the last CIA.

6.4 Special considerations for Scrum development

The diagram in section 4 shows how impact analysis can be used for the backlog content in an organization using Scrum. Note that we might need a new CIA when a requirement is detailed for implementation after it has been taken out of the backlog, or when a requirement is returned to the backlog after it has been rejected by the customer or we have found errors in the implementation.

We recommend that all requirements in the functional product backlog – may be also in the safety product backlog – are formulated as user stories. This will help to make sure that we have information on (1) who: user role, (2) what: goal – what we want to achieve and (3) why: rational for user story requirement. The why-part is important when we want to assess the change's effect on customer satisfaction or on system's safety.

The results from the change CIA can lead to updates of the SRS and we may thus need to change both safety requirements and functional requirements in addition to already written backlog elements.

7 Change Impact Analysis report

Below is a suggestion for a report including chapters and corresponding information presented.

Table 3: Table including an overview of relevant chapters

Chapter	CIA report chapter information
Distribution list	This chapter should be in line with ISO 9001:21008 [11] "*4.2.3 Control of documents...f) to ensure that documents of external origin determined by the organization to be necessary for the planning and operation of the quality management system are identified and their distribution controlled*". This also applies for the safety management.
Names of authors and signatories	No further explanation needed
Revision history including version number	Summarize the change in a few sentences. Version number and date has to be included. This is also a practical information for assessors and employees
Table of content	No further explanation needed
Introduction	Purpose, scope, relevant standards and definitions.

9

Chapter	CIA report chapter information
Name of participants including information related to competence and experience	Selection of relevant and sufficient number of experts is an important part of an Impact analysis. Even SW may e.g. have influence on EMC performance [25]. This information is often included as part of other chapters in the CIA, e.g. the chapter containing the names of the participants, analysis dates, meeting days, etc.
Any deviations from normal operations and conditions that occur as a result of this change	Failure behaviour related to the change has to be checked. This can be performed as part of e.g. a HazOp. The condition list or e.g. SRAC (safety related application condition) list should be checked.
Re-entry point of lifecycle	It is required by both IEC 61508 and EN 50128 that all changes shall initiate a return to an appropriate phase of the lifecycle. This is normally not a problem for the SafeScrum approach, as the sprints are part of Phase 10 in IEC 61508 and phase 6 in EN 50126 (Railway).
Required verification and validation	Describe the verifications and validation steps required. This can normally be based on the former verification and validation plan. *When applying SafeScrum and the sprints are completed, a final RAMS validation will be done. Since most of the system has been incrementally validated during the sprints, we expect the final RAMS validation to be less extensive than when using other development paradigms. This will also help us to reduce the time and cost needed for certification. A further decrease in assessment cost is expected if test-driven development is used. An extended decrease in assessment cost is expected when test-driven development is used.*
Assessor, certification and authorization aspects	New certification body or assessor? More countries involved that may affect the authorization? Special interpretations of the standards in the new design that should be discussed with the assessor in the beginning of the project?

10

Chapter	CIA report chapter information
Required document changes	All affected documents shall be updated. The documents that have to be updated should be mentioned in the CIA report. The relevant documents that normally have to be updated are normally listed in the "Document plan", "Safety Case" and/or the "CER[6]". The CER Method: This method is based on the IEC TRF (Test Report Format) method, as described in Worldwide System for Conformity Testing and Certification of Electrotechnical Equipment and Components (IECEE) guide [26]. The IEC TRF system is intended to facilitate certification or approval according to IEC standards. The TRF and CER method seeks to help industry avoid unnecessary obstacles to trade and to encourage different countries to harmonise their national standards and certification activities
Conclusion/summary	The conclusion of a CIAR has to summarize the content and purpose of the analysis. The conclusions should be precise and straight to the point. The conclusion should also briefly state the implications of these analyses. Why should the assessor believe your result? Show evidence that your result is valid or why it will be valid—that it actually helps to solve the problem you shall solve.
Document references	No further explanation needed

8 Summary

We have studied all relevant standards and identified the need for help in planning, performing and reporting on CIA. Even though this is a common problem, especially for those who are new to a standards regime, the problem is of special interest for companies who want to use Scrum in the development of safety-critical software.

Based on the observations above, we have written a guideline for planning a CIA and reporting of the analyses results. We have also considered and given guidelines for dealing with problems that are relevant when using Scrum for development of safety-critical software. These guidelines are of special importance since they help the developers to incorporate the change impact planning and analysis into the backlog administration. They can thus be included in the Scrum process in a simple, seamless way.

9 References

[1] "IEC 60812: Analysis techniques for system reliability. Procedure for failure mode and effects analysis (FMEA), Edition 1," ed, 2006.
[2] "IEC 61025 Fault tree analysis (FTA), ed. 2," ed, 2006.
[3] "IEC 61160 Design review. Ed. 2," ed, 2005.
[4] "IEC 61882 Hazard and operability studies (HAZOP studies) – Application guide. Ed 1," ed, 2001.
[5] "IEC 61165 Application of Markov techniques. Ed. 2," ed, 2006.

---

[6] CER: Conformity Evidence Report

[6] "IEC 61078 Analysis techniques for dependability – Reliability block diagram and Boolean methods (RBD). Ed. 2," ed, 2006.

[7] "CSM 352/2009. Commission Regulation 352/2009 on the adoption of ac common safety method on risk evaluation and assessment as referred to in Article 6(3) of Directive 2004/49/EC of the European Parliament and of the Council," ed.

[8] "CSM 402/2013. Commission Implementing Regulation (EU) No 402/2013 of 30 April 2013 on the common safety method for risk evaluation and assessment and repealing Regulation (EC) No 352/2009," ed.

[9] "EN 5012X series. Railway applications," ed.

[10] "IEC 61508: Part 4. Functional safety of electrical/electronic/programmable electronic safety-related systems – Definitions and abbreviations. Ed 2," ed, 2010.

[11] "ISO 26262-1. Road vehicles – Functional safety – Part 1: Vocabulary. Ed. 1," ed, 2011.

[12] "ISO/IEC/IEEE 24765. Systems and software engineering – Vocabulary. First edition ", ed, 2010.

[13] "IEC 60812:2006, Analysis techniques for system reliability. Procedure for failure mode and effects analysis (FMEA), Edition 1," ed, 2006.

[14] "Directive 89/336/EEC Electromagnetic compatibility (EMC)," ed.

[15] "Directive 2004/108/EC Electromagnetic compatibility (EMC)," ed.

[16] "Directive 2006/42/EC Machine directive (MD)," ed.

[17] "Directive 1999/5/EC radio and telecommunications terminal equipment (RTTE)."

[18] "Directive 2006/95/EC Low voltage directive (LVD)," ed.

[19] "Directive 89/106/EC Construction products (CPD)."

[20] T. Stålhane, T. Myklebust, and G. K. Hanssen, "The application of Scrum IEC 61508 certifiable software," presented at the ESREL, Helsinki, Finland, 2012.

[21] K. Schwaber, Beedle, M., *Agile Software Development with Scrum*. New Jersey: Prentice Hall, 2001.

[22] A. Causevic, D. Sundmark, and S. Punnekkat, "Factors Limiting Industrial Adoption of Test Driven Development: A Systematic Review," presented at the IEEE Fourth International Conference on Software Testing, Verification and Validation (ICST), Berlin, Germany, 2011.

[23] H. Munir, M. Moayyed, and K. Petersen, "Considering rigor and relevance when evaluating test driven development: A systematic review," *Information and Software Technology,* vol. ONLINE, 2014.

[24] "Olf 070. APPLICATION OF IEC 61508 AND IEC 61511 IN THE NORWEGIAN PETROLEUM INDUSTRY. Ed.2," ed, 2004.

[25] T. Williams, *EMC for Product Designers*, 4 ed.: Newness, 2007.

[26] "IECEE CB-SCHEME OD- CB2020-Ed.1.7. TRF – Development, maintenance and use," ed, 2010.

12

Bucket wheel excavators:
past to present experiences in safety operation

Marek Młyńczak
Wrocław University of Technology, Wrocław, Poland

Abstract: In Polish open pit mining there were and exist now 124 machines of 46 different types including: excavators and spreaders. After World War II it was observed 235 breakdowns and 38 of which were classified as major catastrophes related to 95 machines. Undesired events had both design and operational causes. Total number of multiple failures caused by five structural units reaches 205. Another 30 failures had operational and environmental causes. Each of 235 catastrophes was followed by official penetrating inquiry looking for basic causes. Results of investigations were introduced in design process, modernizations and regulations regarding safety operation and maintenance. Cost analysis shows that modernization or rebuilding of failed machine could be even twice cheaper than building a new one. Objective of the paper is to present closed chain of taking knowledge and data from current operation and applying it in the design and modernization process. In the paper there are historical data shown and analysis about catastrophes. Examples of included conclusions and recommendations from catastrophes of open pit mining machines in design and operation show progress in that branch of industry.

Keywords: open pit mine machine, failure, catastrophe, safety

1. INTRODUCTION AND BACKGROUND OF OPEN PIT MINING. ELECTRICAL ENERGY PRODUCTION AND MARKET.

Braun coal deposits are exploited using one of the largest in the world machines of high efficiency, high durability and high initial cost. Faults of these machines may affect as well energy production as supply of heat source. Machines have various structures, various driving and working systems and work usually in unstable conditions. Historical overview of accidents in open pit mining industry shows many catastrophes and less dangerous events influencing machines availability and safety.

Electrical energy is produced in the world in several ways and the distribution of energy sources is shown in the Fig. 1. It is also seen from historical data that domination of different energy sources changes according to technology, geography and political regulations [2].

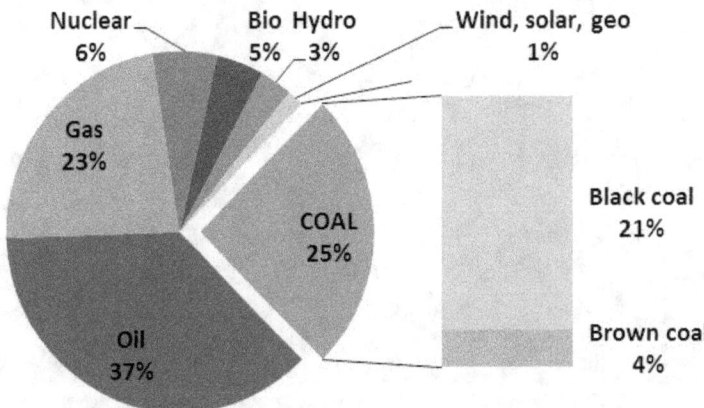

Figure 1. Approximate distribution of energy sources in the world market

It is also expected 56% growth of energy consumption within the next 30 years but still conventional, fossil fuels will be main energy sources. 80% of fuels are oil, gas and coal. Polish energetic market strongly depends on coal and brown coal has growing ratio comparing to traditional black coal. Electrical energy produced of brown coal makes 34% of total electricity while energy made of black coal is 44%. Brown coal mining in Poland exceeds 60 *106Mg what makes 5% of world mining, while Germany mines reached 17% of world mining (24% of electrical energy) as in the year 2011 [3].

2. MACHINES AND MINING TECHNOLOGY

Brown coal mining in Poland is based nowadays on four open pit mines distributed over mid-west part of Poland. Since the beginning of mine operation they produced: Bełchatów (881*106 Mg), Turów (861*106 Mg), Konin (552*106 Mg) and Adamów (187*106 Mg) of brown coal (Fig. 2) [3].

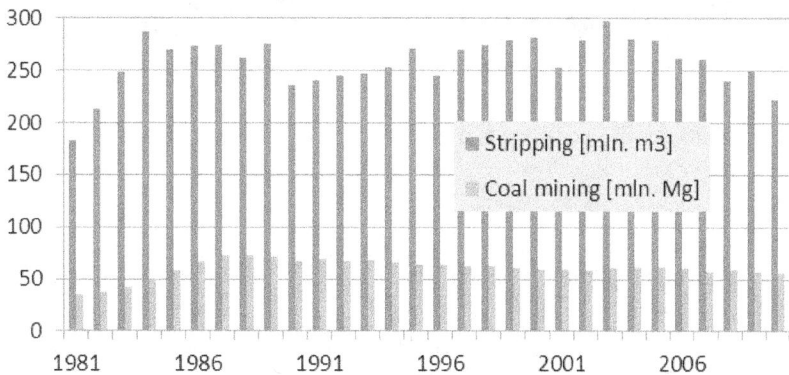

Figure 2. Stripping and brown coal excavation in Polish mines

All mines are located close to power stations and coal is transported directly from open cast to storage yard at power station by the mean of conveyor systems. Important factor in brown coal mining takes geological factors where the measure of mining efficiency is a ratio of overlay (stripping) measured usually in m3 to amount of coal (in Mg) which in Polish mines is kept approximately on the stable level of 3,6 m3 of stripping per each 1 Mg of brown coal. That factor shows the efforts to reach coal and if considering type of stripping (sand, rock, clay, etc.) it is rough assessment of total magnitude of external loads and stresses applied to excavators. Efforts of machine structure depend mainly on its design, excavating forces, natural conditions and way of operation. All these factors are not precise and not constant over operation process what make difficulties in assessment of safety and machine durability. Huge dimensions and mass cause additional loads due to: mass inertia, stiffness of the structure and clearance in all mechanisms. Aging and wearing processes constantly lower strength of the structure and enlarge clearances. Variable cohesion of excavated ground and coal introduces additional uncertainty to reliability prognosis.

There is a great variety of machines operated in open pit mines. Classification criteria may take into account for instance: functions, design and driving structures, dimensions, efficiency, etc. [1,4]. Main criteria of excavating machines classification are:

- type of excavating tool: bucket wheel, bucket chain,
- type of excavating wheel: bucket, semi bucket, bucket less,
- type of boom: withdrawable, without withdrawable,
- type of undercarriage: rail, caterpillar, walking.

Technology of brown coal exploitation in Poland is based on coal excavation in open pit mines because coal deposits are situated up to 310 m deep under the face of the earth. Excavators, most often bucket wheel and chain excavators, mine for stripping and coal creating terraces. Main movements of the machine are: rotating of wheel boom, lifting and lowering it and running along the cut slope. Principle of excavating process is shown in Figure 3. It is seen bucket wheel and slope with excavating wedges on three levels.

Figure 3. Mining process using bucket wheel excavators [3]

Lay-out of excavator movements, while mining, is also presented schematically in Fig. 4.

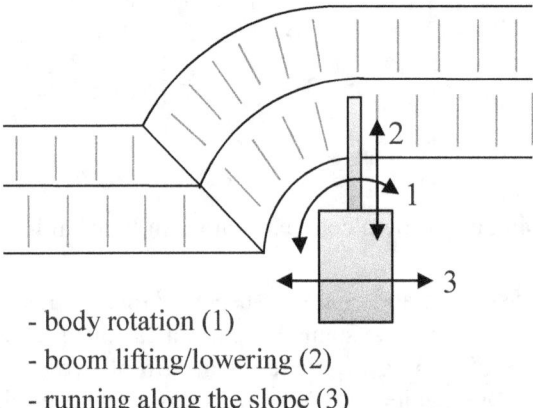

- body rotation (1)
- boom lifting/lowering (2)
- running along the slope (3)

Figure 4. Working motions of bucket wheel excavator

This short description of mining gives an idea of various factors and elements that could be causes of disorder of intended operation process and lead to the catastrophe.

3. OPEN PIT MINES CATASTROPHES OVERVIEW

Operating process of mining machines consists of usage and maintenance processes. Change of state from usage to maintenance may take place due to predefined preventive maintenance or undesired event like mishap, machine component failure or events with disaster consequences called usually catastrophes. It was observed 87 different events classified as minor (59) and major (38) catastrophes. Some of failures repeated more than once and happened even in the same machine. Total number of undesired events has reached 235 and affected 95 machines. Only 23% of all machines were not involved in large breakdown [1].

There are distinguished five main machines' subsystems discovered as initial cause of the catastrophe (supporting structure, wheel boom, switching funnels, supporting bearings and counterweight lifting system). Only these five units are responsible for number of multiple failures. Main hazards in the excavation process are: overload caused by operator, rocks inclusions, slope slide, fire or technical failure due to ageing or wearing. These hazards are divided in two main groups containing hazards arose from the machine failure, called design imperfection causes and from operation process. Historical data show almost equal part of design and operational causes (Tab. 1) [1].

The most frequent catastrophic events dealt with impact load (120 events) and failure of the system rotating machine body (67 events). More precise analysis shows just for these events the following failure modes:

- impact load:
    - break of the bucket wheel boom structure,
    - deformation of the bucket wheel boom structure,
    - break of the beam supporting bucket wheel gearbox,
    - break of gearbox cogs,
    - crack of the wheel boom structure and superstructure masts,
    - crack of the gearbox and bucket wheel shaft,
- system of the machine body rotating (solid-web girders: lower at undercarriage and upper supporting superstructure):
    - crack of undercarriage supporting portal,
    - crack of welds and structure of solid-web girders.

Table 1: Causes of catastrophes classification

Catastrophe cause	Location	No. of events	Percentage
Design imperfection	Supporting structure	10	49%
	Switching funnel	8	
	Rotating system	7	
	Impact load	12	
	Lifting system	6	
Operational error	Various operational	14	51%
	Spreading process	5	
	Slope instability	18	
	Fire	7	

4. CATASTROPHE EXAMPLES (WRONG READJUSTMENT TO WORKING CONDITION AND UNEXPECTED WORKING CONDITION – SLOPE SLIDE)

The following section shows some examples of large-scale catastrophes resulting in destruction of considerably part of the machine (Fig. 5-9).

Figure 5. Destruction of SchRs-1200 excavator caused by crack of the wheel boom structure falling-down of wheel excavating boom (drawings before and after catastrophe, picture on-site, dimensions in m) [1]

5. Recommendations as lessons learned

Though machines of open pit mines do not subject to EU Machine Directive, it was introduced recommendations concerning phase of design and operation. It has been formulated safety regulations which order to perform safety analysis and work out field tests of loads and structure strength. It was introduced long-term program of state monitoring, measuring vibrations and data acquisition and processing. New models of degradation and remaining life time was proposed and verified on collected data. Special attention was paid to human error. To avoid errors due to working conditions it was introduced project aiming in building in the machines new ergonomic, vibro-acoustic isolated operator cabins. Catastrophes caused by ageing are no longer observed but still problems of sudden, environmental hazards and, in slighter degree, human errors exist.

6. CONCLUSION

Heavy environmental conditions, unexpected load and high demands due to efficiency of all open pit mine machine introduce in operation management high uncertainty. Undesired events like described above catastrophes lower availability of the output system and rise total output costs. Numerous catastrophes after World War II were caused; despite of operation conditions, mainly from wearing and aging. As a result, couple of excavators was rebuilding according to new regulations and safety requirements. New machines are also equipped with diagnostic systems which allow for data about load and vibration acquisition and processing.

References

[1] S. Babiarz, D. Dudek, „*Kronika awarii i katastrof maszyn podstawowych w polskim górnictwie odkrywkowym*". Oficyna Wydawnicza Politechniki Wrocławskiej. Wrocław, 2007.
[2] H.L. Gabryś, "*Electric power industry of the year 2012 in Poland in the light of energy balance for 2011 and not only*". Energetyka. 2012.
[3] Z. Kasztelewicz, "*Bulletin-board system: Węgiel brunatny*". Special edition 1/78, 2012 http://www.ppwb.org.pl. (access: July, 2013).
[4] K. Simkiewicz „*Podstawy górnictwa odkrywkowego i otworowego*". http://zasoby.open.agh.edu.pl. (access: July, 2013)

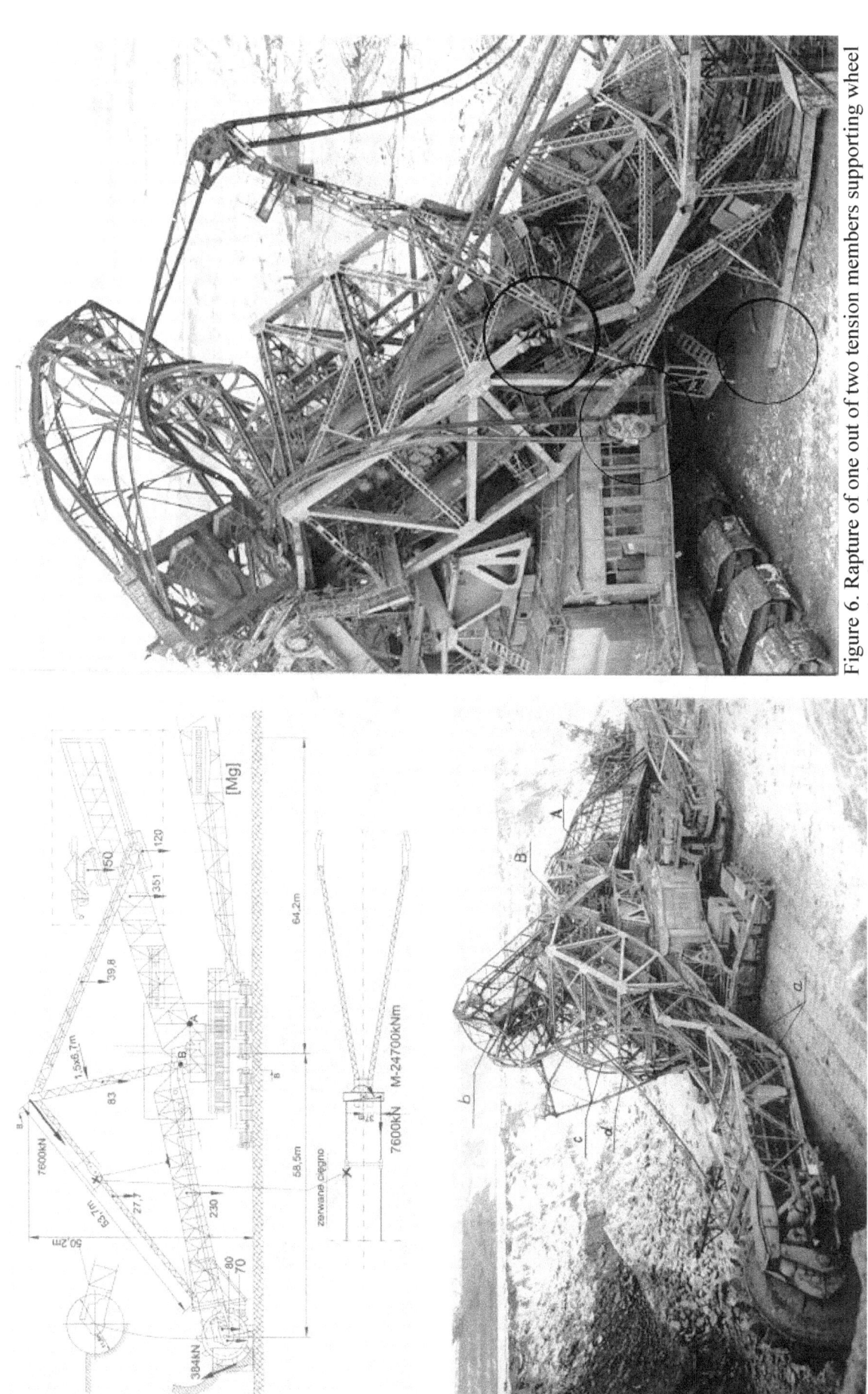

Figure 6. Rapture of one out of two tension members supporting wheel excavating boom of SchRs-1200 [1]

Figure 7. Tilt of chain excavator due to slope sliding [1]

Figure 8. Braking of excavating boom caused by slope creep [1]

Figure 9. Caving of caterpillar vehicle [1]

Verification of Risk Assessment and Treatment model and Software tool in Chemical Establishments in Slovak Republic

Katarina Holla[a] **and Jozef Ristvej**[b]

[a,b] University of Žilina, Žilina, Slovakia

Abstract: The major industrial accidents prevention is one of the principle conditions for ensuring the security and safety of the employees and citizens living close to the industrial establishments. The most hazardous industrial establishments in the EU are called the "SEVESO establishments" and their number is about 10,000 in the framework of all EU countries. Until June 2015 the individual member states are to transpose the new directive SEVESO III to their legal environment. In the implementation framework there is space for the individual member states to identify the problem areas in the existing legal environment and to implement new approaches especially for the risk assessment and treatment. Regarding to the space that was created during the SEVESO III transposition, it is possible to suggest unified procedures and methodologies which could be used by the enterprises for these types of analyses. This paper summarizes research results and recommendations (of University of Žilina in Žilina) in area of industrial accidents prevention in Slovak republic based on ongoing verification of created quantitative risk assessment and treatment model and software tool in two chemical establishments. The advantage is especially utilizing the bow-tie diagrams which link the fault trees and event tress and in this way create a possibility to utilize the generic trees for carrying out an analysis. Also a whole range of other methods and techniques can be utilized in the individual steps of this systematic model. This approach should be considered after its validation to be used in all SEVESO establishments in Slovak republic in the future and therefore it will be possible to compare results of analysis between SEVESO establishments in Slovak republic.

Keywords: ARAMIS, Risk Assessment and Treatment, industrial accident

1. INTRODUCTION

The major industrial accidents prevention is a specific area in the framework of the process of planning and solving the crisis phenomena not only in the Slovak Republic but also on the international level. The European Union decided to solve the area of the major industrial accidents in 1982 through a legal tool called the "SEVESO Directive" which has been amended twice so far. Currently the SEVESO II Directive is valid and until 2015 the new directive SEVESO III has to be transposed to the legal environment of the EU member states. These legal tools determine the rules for handling with the hazardous substances and keeping the rules how to handle them. The Slovak Republic as an EU member has about 80 SEVESO companies in its territory and attempts to fulfil all requirements of the valid directives and to ensure sustainability of the interests protected. There are approximately 10,000 SEVESO companies in the European Union.

2.   BASIC ACTIONS

The transposition of the new SEVESO III Directive creates a space for improving the problem areas which have been identified by the statistical research [3,4]. The most important actions for the proposed changes in the framework of the SEVESO III are as follows:

o   harmonising the Appendix 1 with the EC regulation No. 1272/2008 on classification, labelling and packaging of substances and mixtures (further only CLP regulations), amending and repealing Directives 67/548/EEC and 1999/45/EC to which the SEVESO II Directive refers. The CLP rules will become effective on 1st June 2015,

o   strengthening the effectiveness of regulations concerning the access of general public to information about security and safety, participation in the decision-making process and enforcement of justice as well as an improved method of collecting, managing, sharing information and its accessibility,

o   more strict rules for carrying out inspections of the premises with the goal to ensure effective realisation and enforcement of the stated rules are being stated. [2]

Further changes comprise technical adaptations and simplifications which will reduce the unnecessary administrative burden. In the framework of the transition period there is a space for the member states not only to adapt to the identified changes but also to alter the existing and routine procedures in some areas especially during processing security and safety documentation and its most important parts. Undoubtedly, one of these parts is also the risk assessment and management, an area which creates the basic assumption for reducing the possibility of the rise of a major industrial accident by implementing preventive measures.

The need of solving the area of the risk assessment and management is supported also by the investigations on the European but also national level. [3,4] The selected conclusions which affect the adoption and adaptation of the legal environment in the Slovak Republic are based on a statistical research realised during 2012 – 2013 in the framework of the MOPORI project under the title „Statistical Research of the SEVESO companies". The statistical research started in 2011 by meetings with competent bodies and creating questions for the questionnaire. Subsequently we accomplished the list of persons in the framework of the companies who were sent the questionnaire – this required involvement of several project team members in this action (81 companies were addressed). The information obtained was often inaccessible and incomplete, but even outdated. The project team as well as the firm RISK CONSULT s.r.o. (which actively participated in this action) had to remind repeatedly the addressed to companies to send the questionnaires.

81 SEVESO companies from 26 lines of business in the Slovak Republic were addressed in the framework of the project solution (see the figure 1). The following facts affected the representative character of the research:

- the questionnaire was filled in and returned by companies from 16 lines of business – a success rate of 62%,
- out of 81 SEVESO companies 44 firms sent the filled in questionnaire - a success rate of 54 %,
- 16 lines of business filled in and returned the questionnaires. Out of the 70 addressed companies 42 questionnaires were returned - a success rate of 63 %. [3]

The SEVESO companies can be investigated on the selected representation rate of 50 % and the achieved results and conclusions can be referred to 14 lines of business. [3]

Figure 1 Distribution of establishments falling under the scope of MIAP act in Slovak Republic (Source: http://www1.enviroportal.sk/indikatory/detail.php?kategoria=263&id_indikator=4425 , 2012)

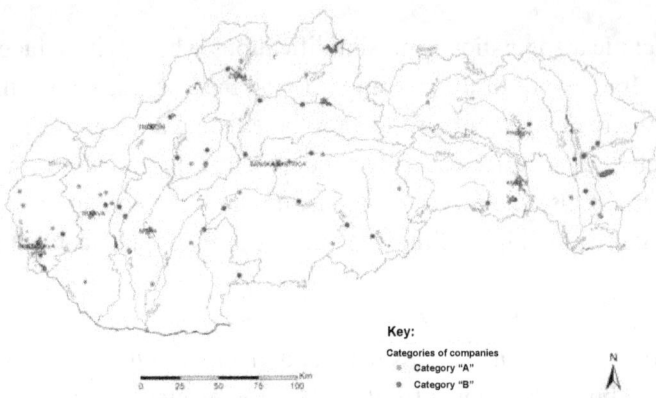

Although the statistical research contained 29 questions, several of them had a principal influence on the model. The utilisation of the methods and techniques for the risk assessment and management is one of the bases and usable results for creating a complex model which results from the investigation.

The most common methods utilised by the experts in the SEVESO companies were ETA – Event Tree Analysis (39 companies), FTA – Failure Tree Analysis (38 companies) and Security and Safety Audit (26 companies). [3] The figure 2 shows a thorough overview about distribution of the methods for analysing the major industrial accidents according to the experts´ knowledge in the area of the major industrial accidents prevention.

Figure 2 Numbers of SEVESO establishments adopting various methods of MIA risk analysis [3]

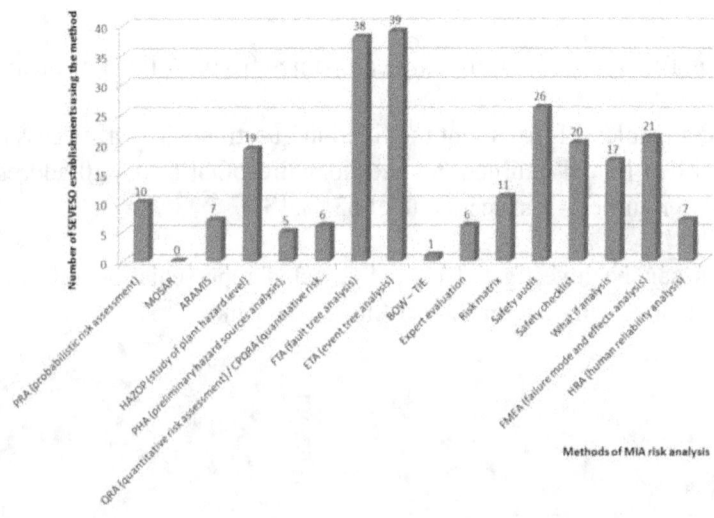

In the framework of the EU regulations as well as the law about major industrial accidents prevention there is no determined software tool used for creating scenarios, modelling the consequences and impacts, however, there are concrete inputs for calculating the summarised expression of the impact seriousness. Currently there are several software products which can be used. They differ by their prices and variedness of inputs as well as different user environment.

The question concerning the usage of the software for assessing the major industrial accidents was answered by all 44 SEVESO companies. The answers show the companies utilise the following software means for assessing the major industrial accidents: Excel, Aloha, TerEx, RiskSpectrum and Saphire. The companies do not use any other software for assessing the major industrial accidents listed in the questionnaire. The companies could also introduce other software they use for assessing the major industrial accidents. The following ones were introduced: ROZEX 2001, ALOFT-FT 3.10, IAEA-TEDDOC-7, Breze-Hass profesional, ATON. The dominant part of the companies (38 companies, 36%) uses EXCEL and Aloha (37 companies, 35%). The programme ATON seems to be introduced by a mistake. The overall distribution of the companies according to the software used for assessing the major industrial accidents is shown in the figure 3.

Figure 3 Number of SEVESO establishments preferring software products for risk assessment [3]

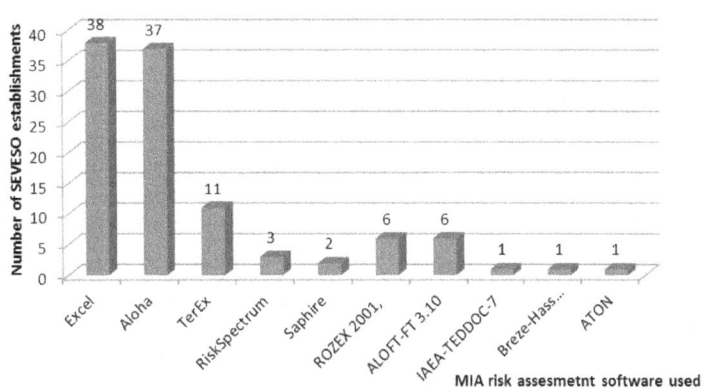

Based on several outputs from the aforementioned statistical investigations, consultations with experts, analyses of the current state in this area and other accessible information we created the **Complex Model for Risk Assessment and Treatment of the Industrial processes** which is explained in the next parts.

3. COMPLEX MODEL FOR RISK ASSESSMENT AND TREATMENT OF THE INDUSTRIAL PROCESSES

Based on a comprehensive analysis of the current state of the approaches to assessing and managing the risks, methods, techniques and tools used in individual steps, the existing legal standards, the research realised in the project framework and bases and specifics characteristic for a particular country, we created a model for assessing and managing the risks of industrial processes. Subsequently individual methods, techniques and tools from the ARAMIS methodology, the created software and other methods and approaches were implemented. Therefore this model can be characterised as a comprehensive one and it can be implemented for a particular technological process.

In the further text we will explain a simplified approach to implementing the model which complies with the routine procedures in this area.

Figure 4 Simplified approach to risk assessment [1]

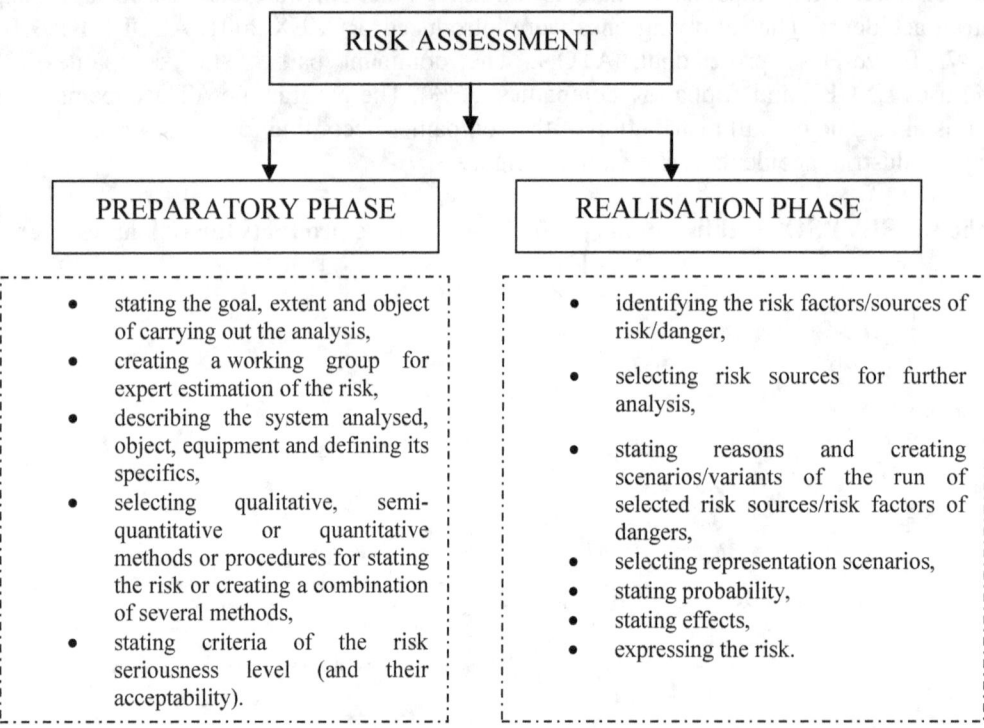

The individual phases together with recommended utilisation of the methods and techniques in the individual steps will be explained in the text to come.

3.1 PREPARATORY PHASE

Stating the goal, extent and object for realising the analysis. The operator of the risk technology is obliged to ensure the assessment of the risks in connection with carrying out all steps to prevent the rise of a major industrial accident in compliance with the law about major industrial accidents prevention.

Creating a working group for an expert estimation of the risk. The given analysis is carried out by professionally capable persons who are dealing with the area of assessing and managing the risks and who utilise the knowledge of specialised workers who understand the technology being assessed (e.g. foremen from production, mechanics, maintenance workers, security and safety technicians, etc.).

Describing the system, object and premises analysed and defining their specifics. The information which is of a general and specific character enters the analysis.[11]

General requirements on the input information from the ARAMIS methodology:

- the documentation of the system/object being assessed,
- describing the running technological processes,
- describing the devices and pipelines,
- the list of hazardous substances used in the assessed object/subject/technology and their connection with the identified equipment,
- summarising the cards of security and safety data and information for the analysis.

Specific information necessary for every identified equipment/device:

- the name of the device,
- the volume and dimensions,
- the service pressure and temperature,
- the substances handled,
- the substance state,
- the quantity of the substance in the identified equipment (in kg or kg/s),
- the boiling point of the hazardous substance.

Selecting the qualitative, semi-quantitative or quantitative methods or procedures for stating the risk or creating a combination of several methods. Based on the previous analyses and investigations the methods HAZOP and individual parts of the ARAMIS methodology were chosen. Based on re-evaluation some procedures of this methodology were adapted for the current conditions.

Stating criteria for the seriousness of the risk level (and their acceptability). In this step we have to determine the social acceptability of the risk of a major industrial accident from the point of view of assessing the possibility of threatening life of one or several persons and it is defined by the acceptable probability or the frequency of occurring a major industrial accident and this is assessed according to the following relation:

o If life of one person is endangered
$F_{pr} = 10^{-5}$ for the existing companies
$F_{pr} = 10^{-6}$ for new companies or premises

o If lives of several persons are endangered
$F_{pr} = 10^{-3}.N^{-2}$ for the existing companies and premises
$F_{pr} = 10^{-4}.N^{-2}$ for new companies and premises. [12]

In the end there is the assessment of the risk acceptability according to values introduced in the law.

3.2   REALISATION PHASE

Identifying the risk factors/ the risk source/the danger. This phase consists of identifying the following risk factors:

- the list of selected hazardous substances (SHS) (comparison with reference values),
- identifying dangerous devices.

First of all it is necessary to choose those hazardous substances (based on the cards with the security and safety data from the preparatory phase) which can be designated as selected/specified. The SHS are then compared with the existing list of the selected hazardous substances listed by the SEVESO II or SEVESO III Directive. If the selected hazardous substance (according to the corresponding legal regulation) exceeds the reference value introduced in the tables of the law about major industrial accident prevention and shows a dangerous property, the company is categorised into the class A and B.

The list of SHS, which are bound to a particular device, should be the output of this step. The devices are introduced in the table – List of Dangerous Devices – in the ARAMIS methodology and the individual devices are designated from EQ1 to EQ16.

Stating the causes and creating the scenarios/variants of the selected risk sources/ risk factors and dangers. The identified SHS and devices are subsequently attached to critical events:

- Decomposition (CE1),
- Explosion (CE2),
- Materials set in motion – entrainment by air (CE3),
- Materials set in motion – entrainment by liquid (CE4),
- Start of fire (LPI) (CE5),
- Crack of casing in vapour phase (CE6),
- Crack of casing in liquid phase (CE7),
- Leak from liquid pipe (CE8),
- Leak from gas pipe (CE9),
- Catastrophic crack (CE 10),
- Vessel collapse (CE11),
- Roof collapse (CE12). [17]

Subsequently, based on the identified critical events in connection with the SHS and devices, we create the bow-tie diagrams which are a principal contribution in the area of the major industrial accidents prevention. In the framework of the project solved at the University of Žilina in Žilina we created software means in the user environment Excel. [3] The basic tree structure was taken over from the ARAMIS and the calculation rules were adapted to the currently used approaches in the Slovak Republic. The selection for implementing barriers to the individual trees was left open; the investigators themselves enter their name as well as the success level. On Figure 5 there are presented created bow – ties within software tool.

Figure 5　　　List of bow-ties created within Software tool and adapted to Slovak republic conditions

	Generic fault tree (FT)	Zložitosť	Critical event	Nr. CE	Event tree (ET) - Substance state (STAT)	Zložitosť
1	FT Chemical decomposition	4	Decomposition	CE1	STAT1 Solid	1
2	FT Decomposition tied to a punctual ignition source	3	Decomposition	CE1	STAT1 Solid	0
3	FT Thermal decomposition	2	Decomposition	CE1	STAT1 Solid	0
4	FT Explosion of an explosive material	5	Explosion	CE2	STAT1 Solid	1
5	FT Explosion (violent reaction)	5	Explosion	CE2	STAT1 Solid	0
6	FT Materials set in motion (entrainment by air)	1	Materials set in motion (entrainment by air)	CE3	STAT1 Solid	1
7	FT Materials set in motion (entrainment by a liquid)	1	Materials set in motion (entrainment by a liquid)	CE4	STAT1 Solid	1
8	FT Start of fire (Loss of Physical Integrity)	2	Start of fire (LPI)	CE5	STAT1 Solid	1
9	FT Start of fire (Loss of Physical Integrity)	0	Start of fire (LPI)	CE5	STAT2 Liquid	1
10	FT Start of fire (Loss of Physical Integrity)	0	Start of fire (LPI)	CE5	STAT3 Two-phase	1
11	FT Start of fire (Loss of Physical Integrity)	0	Start of fire (LPI)	CE5	STAT4 Gas / Vapour	1
12	FT Large breach on shell or leak from pipe	6	Breach on the shell in vapour phase	CE6	STAT1 Solid	0
13	FT Large breach on shell or leak from pipe	6	Breach on the shell in vapour phase	CE6	STAT3 Two-phase	1
14	FT Large breach on shell or leak from pipe	6	Breach on the shell in vapour phase	CE6	STAT4 Gas / Vapour	1
15	FT Medium breach on shell or leak from pipe	3	Breach on the shell in vapour phase	CE6	STAT1 Solid	0
16	FT Medium breach on shell or leak from pipe	3	Breach on the shell in vapour phase	CE6	STAT3 Two-phase	0
17	FT Medium breach on shell or leak from pipe	3	Breach on the shell in vapour phase	CE6	STAT4 Gas / Vapour	0
18	FT Small breach on shell or leak from pipe	3	Breach on the shell in vapour phase	CE6	STAT1 Solid	0
19	FT Small breach on shell or leak from pipe	3	Breach on the shell in vapour phase	CE6	STAT3 Two-phase	0
20	FT Small breach on shell or leak from pipe	3	Breach on the shell in vapour phase	CE6	STAT4 Gas / Vapour	0
21	FT Large breach on shell or leak from pipe	0	Breach on the shell in liquid phase	CE7	STAT2 Liquid	1
22	FT Large breach on shell or leak from pipe	0	Breach on the shell in liquid phase	CE7	STAT3 Two-phase	1
23	FT Medium breach on shell or leak from pipe	0	Breach on the shell in liquid phase	CE7	STAT2 Liquid	0
24	FT Medium breach on shell or leak from pipe	0	Breach on the shell in liquid phase	CE7	STAT3 Two-phase	0
25	FT Small breach on shell or leak from pipe	0	Breach on the shell in liquid phase	CE7	STAT2 Liquid	0
26	FT Small breach on shell or leak from pipe	0	Breach on the shell in liquid phase	CE7	STAT3 Two-phase	0
27	FT Large breach on shell or leak from pipe	0	Leak from liquid pipe	CE8	STAT2 Liquid	0
28	FT Large breach on shell or leak from pipe	0	Leak from liquid pipe	CE8	STAT3 Two-phase	0
29	FT Medium breach on shell or leak from pipe	0	Leak from liquid pipe	CE8	STAT2 Liquid	0
30	FT Medium breach on shell or leak from pipe	0	Leak from liquid pipe	CE8	STAT3 Two-phase	0
31	FT Small breach on shell or leak from pipe	0	Leak from liquid pipe	CE8	STAT2 Liquid	0
32	FT Small breach on shell or leak from pipe	0	Leak from liquid pipe	CE8	STAT3 Two-phase	0
33	FT Large breach on shell or leak from pipe	0	Leak from gas pipe	CE9	STAT1 Solid	0
34	FT Large breach on shell or leak from pipe	0	Leak from gas pipe	CE9	STAT3 Two-phase	0
35	FT Large breach on shell or leak from pipe	0	Leak from gas pipe	CE9	STAT4 Gas / Vapour	0
36	FT Medium breach on shell or leak from pipe	0	Leak from gas pipe	CE9	STAT1 Solid	0
37	FT Medium breach on shell or leak from pipe	0	Leak from gas pipe	CE9	STAT3 Two-phase	0
38	FT Medium breach on shell or leak from pipe	0	Leak from gas pipe	CE9	STAT4 Gas / Vapour	0
39	FT Small breach on shell or leak from pipe	0	Leak from gas pipe	CE9	STAT1 Solid	0
40	FT Small breach on shell or leak from pipe	0	Leak from gas pipe	CE9	STAT3 Two-phase	0
41	FT Small breach on shell or leak from pipe	0	Leak from gas pipe	CE9	STAT4 Gas / Vapour	0
42	FT Catastrophic rupture	4	Catastrophic rupture	CE10	STAT1 Solid	1
43	FT Catastrophic rupture	4	Catastrophic rupture	CE10	STAT2 Liquid	1
44	FT Catastrophic rupture	4	Catastrophic rupture	CE10	STAT3 Two-phase	1
45	FT Catastrophic rupture	4	Catastrophic rupture	CE10	STAT4 Gas / Vapour	1
46	FT Vessel collapse	1	Vessel collapse	CE11	STAT2 Liquid	0
47	FT Collapse of the roof	1	Collapse of the roof	CE12	STAT2 Liquid	1

Probabilistic Safety Assessment and Management PSAM 12, June 2014, Honolulu, Hawaii

Stating the probability. Based on the created bow-ties for the stated critical event it is necessary to determine the probability/frequency of the causalities which are on the FTA side. The partial frequencies/probabilities of causes are written directly into the created software and relations according to the Boolean algebra re-calculate the frequencies which enter other knots. Also the probability/frequency and barriers in every branch are mutually in an interaction.

According to ARAMIS the frequency of the critical event moves in the range of F0 to F4 – the F0 is attached to the event with the highest frequency and F4 with the lowest one.

Stating the consequences and impacts. In the end, the impacts of an accident are determined (based on the created bow-ties). Before we define them, it is necessary to simulate the development of the crisis phenomenon by appropriate software for us to find out the extent of the hit area. Based on this simulation we subsequently determine the impacts on life, property and environment through the C1 – C4 indicators.

The impact classification is based on the assessment of the effects on human targets and effects on the environment.

Expressing the risk (stating the risk and determining the risk perception acceptability). In the end it is necessary to identify if the stated risk value is acceptable or not. In the Slovak Republic we stated these values as the individual and social risks.

4. IMPLEMENTATION OF THE COMPLEX MODEL IN PRACTICE

For the time being, the verification of the model created is under way in two Slovak enterprises. The first is *Mondi SCP, a.s* and the second one *EVONIC Fermas s.r.o.*

The negotiations with the companies lasted several months and the real implementation has started in 2014. The principal aim of this implementation is to assess the possibilities of utilising the created model in the Slovak conditions as an alternative of the currently used PSA method.

In *Mondi SCP Ružomberok* the identification of SHS is currently under way (after studying the necessary documents), the SHS are attached to the devices and subsequently we state the critical events. The whole analysis is realised on the workplace RK2 – the reaction boiler 2. There are several types of the identified hazardous substances the operators come into contact with (e.g. green/black lye, natural gas,...). Now we are studying the cards of the security and safety data and adapt the information to the currently used regulations.

There is an intensive communication with the company regarding to the possibility to publish the results of the model implementation which have not been confirmed yet and therefore it is not possible to publish the sensitive information. We will continue according to the steps introduced in the chapter 3.

Another group is dealing with implementing the created model in the company *EVONIC Fermas,*s.r.o, which works with ammonia water. First of all it was necessary to carry out tests with the ammonia water in the labs of the University of Žilina in Žilina.

The results show that the speed of releasing the NH3 depends on the temperature of the surroundings/sample and the solution concentration, especially during the initial evaporation phase. Due to the endothermic character of evaporation the lotion temperature reduces quickly under the surrounding temperature – this is proved by the approximation of the curves of individual measurements. During the UNI 2 measurements which are carried out under the ambient temperature of 20°C, the sample temperature was only 11°C.

This information enters further analyses. The same is underway in the company *EVONIC Fermas, s.r.o* in the framework of the preparatory phase and beginning of the realisation stage.

5. CONCLUSION

The major industrial accidents prevention is one the basic pillars for ensuring the citizens´ security and safety – both in the position of an employee or concerned general public. Due to the dynamics of the technological processes and the quantity of hazardous substances currently used in production it is necessary to look for such tools and methods which will reduce the danger of the rise of industrial accidents to the lowest possible level.[13]

The basic contribution of this paper is that it states the approach for the risk assessment and management which we recommend to utilise in the SEVESO companies. The designed model can be utilised as a whole but also some of its parts included especially in the systematic approach ARAMIS which complies with the currently used approaches for assessing and managing risk worldwide as well as in the Slovak Republic. Its advantage is especially the utilisation of the bow-tie diagrams which connect the failure trees and the event trees and in this way create a possibility to make use of the generic trees during the analysis. In the individual steps of the created systematic procedure we can utilise also a whole range of methods and techniques presented in this paper.

The issues of a unified approach to assessing and managing risks have their positive but also negative aspects. One of the advantages is especially the possibility to compare the companies from the point of view of the results from the analyses created for the responsible inspection bodies of the state administration. However, on the other hand the companies created their own procedures when the law about the major industrial accidents prevention became effective and purchased their own software which they use for assessing and managing risks and therefore their decision to change their routine procedure has to be justified by the applicability of the procedure and by identifying the advantages against the system being used. The implementation in the companies *Mondi SCP, a.s* and *EVONIC Fermas s.r.o.* was the first step for introducing this model. The assumed completion of the model implementation in both companies is June 2014 and can be possibly presented at the conference. The authors do not assume any detection of different results than in the case of a more utilised PSA method but offer a structural alternative for assessing and managing risks which is at least equally suitable for the chemical factories.

Acknowledgements

„This work was supported by the Slovak Research and Development Agency under the contract No. APVV-0043-10"

and

"This project has received funding from the European Union's Seventh Framework Programme for research, technological development and demonstration under grant agreement no 313308." and "This work was co-funded by the Slovak Research and Development Agency under the contract No. DO7RP-0025-12."

References

[1]   Hollá, K., Kampová, K., Šimák, L., Šimonová, M., Míka, V. 2013. *Major Industrial Accident Prevention*.Žilina. Žilinská univerzita, 2013. 147 s. ISBN 978-80-554-0786-9.

[2]   Proposal for a Directive of the European Parliament and of the Council on control of major-accident hazards involving dangerous substance. European Commission COM(2010)781 final - 2010/0377 (COD).

[3]   Hollá, K. et al. 2013. *Statistical Survey of SEVESO Establishments in Slovak Republic: project APVV-0043-10 Complex model for risk assessment and treatment in industrial processes*, Žilina : Faculty of Special Engineering – University of Žilina, 2013. - 22 s.

[4]   SALVI, O a kol.: F – SEVESO, 2008. *Study of the effectiveness of the Seveso II directive*. Brusel: EU – Vri, 2008.

[5]   MARS & SPIRS. Joint research centre [online]. [cit. 4.5.2012]. Dostupné na: http://mahb.jrc.it/index.php?id=39 .

[6]   Zánická Hollá, K.: *Major industrial accident prevention in the Slovak Republic and the project MOPORI*. In: 11th international probabilistic safety assessment and management conference and the annual European safety and reliability conference. Helsinky, Finland, 25-29 June 2012, ISBN 978-1-62276-436-5.

[7]   Zánická Hollá, K. et al. 2010: *Risk Assessment of Industrial Processes. Bratislava: Iura Edition,* 2010, ISBN 978-80-8078-344-0.

[8]   MARS & SPIRS. Joint research centre [online]. [cit. 4.5.2012]. Dostupné na: http://mahb.jrc.it/index.php?id=39 .

[9]   Informačný systém PZPH [online]. [cit. 4.5.2012]. Dostupné na: http://www.enviroportal.sk/environmentalne-temy/starostlivost-o-zp/pzph-prevencia-zavaznych-priemyselnych-havarii/informacny-system-pzph

[10]   The Framework Programme Accidntal Risk Assessment Methodology For Industries in the Context of the Seveso II Directive [online]. 2004. [cit. 25.6.2012]. Dostupné na: http://mahb.jrc.it/fileadmin/ARAMIS/downloads/ARAMIS_FINAL_USER_GUIDE.pdf

[11]   Paleček, M. a kol.: *Procedures and Methodologies Of Analyses and Risk Assessments for Purpose of Law No 353/1999 Coll., on Prevention of Serious Accidents*. Praha: VÚBP.

[12]   Kandráč, J.: *Metodická príručka pre expertný odhad pravdepodobnosti výskytu priemyselných havárií v podnikoch podliehajúcich režimu zákona o závažných priemyselných haváriách*, 2012 [online]. [cit. 15.2.2013]. Available on: http://www.minzp.sk/files/skody-a-havarie/priemyselne-havarie/metodicke-postupy-a-prirucky/prirucka_vyskyt.pdf .

[13]   Zánická Hollá K., Moricová V.: *Human factor position in rise and demonstration of accidents*. In: Communications: scientific letters of the University of Žilina. - ISSN 1335-4205. - Vol. 13, No. 2, 2011, s. 49-52.

[14]   Kandráč, J.: Personal Interview, Risk consult ltd. 2013.

[15]   Galková, M.: Personal Interview, Slovak environmental agency, 2013.

[16]   Čajková, H., Danečková, T., Rothová, R., Trcka T..: Personal Interview, Ministry of Environment of Slovak Republic 2013.

[17]   Delvosalle Ch. Et al.: ARAMIS project: A comprehensive methodology for the identification of reference accident scenarios in process industries. 2006. [online]. [cit. 11.2.2014]. Available on: http://www.sciencedirect.com/science/article/pii/S0304389405003742

A preliminary accident investigation on a Norwegian fish farm applying two different accident models

Siri Mariane Holen[a*], Ingrid Bouwer Utne[a], and Ingunn Marie Holmen[b]

[a] Department of Marine Technology, NTNU, 7491 Trondheim, Norway
[b] SINTEF Fisheries and Aquaculture, 7465 Trondheim, Norway

Abstract: The aquaculture industry is one of the most dangerous professions with respect to occupational hazards in Norway. Hazardous operations are carried out daily on fish farms and safe operations are crucial. This paper aims to apply two methods of accident analysis for an accident at a fish farm. Accident analysis is necessary for understanding why accidents happen, helps us understanding the system in which the accident happened, and can provide for improvements for a safer system. The two methods, namely STEP and CAST are based on different assumptions of accident causation, and highlights different mechanisms that contributed to the accident happening. STEP provides a systematic guidance to ask the right questions to get a full view on what happened during the accident sequence, and portrays the accident in an easy accessible flowchart. CAST is a more comprehensive method that models all levels in the sociotechnical system to evaluate if there is inadequate control in any of the feedback loops of the different levels. Using CAST for accident investigation is more resource demanding, but will also give more information on safety problems which can be used to improve the risk management system.

Keywords: Aquaculture, Fish Farming, Accident models, STEP, CAST

1. INTRODUCTION

Norway has grown to be the world's largest producer of Atlantic salmon [1], and the industry is of great economic importance in Norway. Especially in the coastal districts of Norway the aquaculture industry is important and in addition to traditional fisheries, aquaculture has become a cornerstone of many local communities. Sea based fish farming has great potential in producing healthy food for consumers worldwide, and the Norwegian government had in 2013 an explicit goal of becoming the world's leading seafood nation [2]. Fish farms in Norway have traditionally been situated in sheltered fjords. Due to challenges faced by the industry and prospects of greater value-creation, the industry is increasing the focus on moving the farms into less protected areas.

The prospects of exposed fish farming will impose several challenges and new hazards. Aspects, such as harsher environmental conditions, availability of the facilities and distance will influence safe operation in exposed fish farming. This also implies a more systematic approach to risk management. The industry is already one of the most dangerous professions with respect to occupational hazards in Norway. The work on a sea-based fish farm is characterized by manual labor assisted by heavy machineries like cranes and work boats. The total work force on a fish farm in one shift, on a normal day varies from 2-5 operators, and when special operations are performed extra personnel can increase the number to 10-15 operators. Many of the demanding operations on fish farms are carried out on work boats moored to the net pens. The forces from wind, currents and waves are making these work platforms unstable relative to each other, and especially tasks involving the use of cranes are difficult. Hazardous operations are carried out daily and ensuring safe operations are crucial. Safety is an issue that needs careful review to prepare the industry for the need for improved health, safety and environment (HSE) performance when moving production to areas with harsher weather conditions, and less accessible sites.

Most research on safety and risks in the aquaculture industry have focused on either food safety, structural safety, and preventing escape of salmon; the latter due to the implications regarding wild salmon [3-5]. Occupational safety has gotten less focus, and a Canadian study on occupational accidents

---

* siri.m.holen@ntnu.no

concludes that despite the fact that the industry is one of the most injury exposed industries there has only been limited focus on occupational safety management in the aquaculture sector [6]. The aquaculture industry is number two on the statistics when it comes to occupational accidents and fatalities per man-labor year, just after fisheries [7, 8]. In the years 1988-2013 over 1400 injuries have been reported, with the assumption that there is a substantial level of under reporting. From 1980-2013 40 deaths have been reported from the aquaculture industry [9]. Hence, there is positively a potential for improvement regarding safety in the aquaculture industry.

Accident models are used in accident investigations to help understanding causal factors. Accident models can also be seen as a type of risk analysis, when defining risk as the systematic use of methods to identify hazards and to estimate the risk to individuals or populations, property or the environment [10]. Different accident models are based on different assumptions on the functions and structures of the sociotechnical systems, which again can lead us to the conclusion that the outcome of using different models is different results in terms of causes to an accident. As the results from an accident investigation should lead to input for improvements of the system, the choice of accident models for an accident investigation is important for the work on accident prevention. Accident models affect the way people think about safety and how they identify and analyze risk factors, and accident models can be used both in reactive and proactive safety management [11].

The main objective of this paper is to apply two different accident models, i.e., STEP and STAMP/CAST, to one accident on a fish-farm and compare the usability of the models. The aquaculture industry has not been subject to extensive occupational or organizational safety research yet. Hence, it is not evident which accident model will provide the most valuable insights into the causes of an accident on a fish farm as basis for determining relevant mitigating measures. The work in this paper analyzes two different approaches to lay the grounds for further systematic safety work and research in the Norwegian aquaculture industry.

The structure of the paper is as follow; Section 2 presents the main categories of accident models. In Section 3 two different methods for accident investigation based on different accident models are presented and applied for the accident case. In Section 4 the usability of the accident models is discussed, and seen in relation to what we can learn from using the different accident models.

2. THREE TYPES OF ACCIDENT MODELS

Accident models can be categorized in many different ways, for example, Hollnagel suggests three main categories of accident models; the sequential models, the epidemiologic models and systemic models [12]. Each accident model has its own characteristics as to the types of 'causal factors' that it highlights [13].

2.1. Sequential accident models

Sequential accident models portray accidents in terms of a chain of discrete events which occur in a particular temporal order where an initial unexpected event (root cause) triggers the following events [12, 14, 15]. The initial model was developed by Heinrich in the 1940s and named the Domino theory. This model uses the analogy of domino pieces to demonstrate that different factors/events will contribute to an accident, but if you remove one of these factors, such as an unsafe act, you can also prevent the accident from happening. An assumption for the sequential accident models is that accidents happen in a linear and deterministic manner. Some types of risk analysis can be seen as sequential accident models, like fault tree analysis (FTA), event tree analysis (ETA) and failure mode and effect analysis (FMEA) [14]. Other accident models belonging to this category is the Loss Causation Model, Rasmussen and Svedungs model and the Sequentially Timed Events Plotting (STEP) which will be introduced in detail later in this paper [15].

2.2. Epidemiologic accident models

Epidemiologic accident models portrays accidents analogues to the spreading of a disease, i.e. as the outcome of a combination of factors, some manifest and some latent, that happen to exist together in space and time [14]. The Swiss cheese model is an example of an epidemiologic accident event developed by James Reason, where slices of Swiss cheese are used as an analogy to barriers. The Swiss cheese has holes of different sizes at different places, and these holes are referred to as latent failures or latent conditions in the barriers [16]. There can be several layers of barriers (defence in depth) and an accident happens if there are latent failures in all barriers. The theory of man-made disasters by Barry Turner is another epidemiologic accident model consistent with the ideas that accidents happen because of latent failures and conditions in an organization or system. The model suggests that accidents are the result of a long series of events where errors, misconception of hazards and lack of information flow in the end will accumulate into an accident [17].

2.3. Systemic accident models

Systemic accident models are based on systems theory. Systems theory assumes that some systems cannot be separated into subsystems without losing information on the systems interaction and relationship between technology and social aspects [18]. Accidents must thus be seen in relation to the system as such, and the thought is that if you single out parts of the system to model how and why an accident happened information will be lost. A basic property of the system is also that it consists of hierarchical levels, and that each level will put constraints on the lower level. A system is held in balance through these constraints and accidents happen when control over these constraints is lost. As Leveson [19] argues, it is in the interaction of system components that safety of a system can be determined, and that accidents occur when component failures, external disturbances, and/or dysfunctional interactions among system components are not adequately handled by the control system. As opposed to sequential and epidemiologic models the focus in a systemic accident model is not on the actors or events. This means that it is to a lesser extent possible to assign blame when using a systemic model, because instead the focus is on context and conditions. Systemic accident models are Rasmussen's hierarchical sociotechnical framework, Hollnagels FRAM (Funtional Resonance Accident Model) based on cognitive systems engineering and Leveson's STAMP (Systems-Theoretic Accident Model and Processes) on which CAST (Causal Analysis based on STAMP) is based, introduced in more detail in the next chapter [12, 14, 20, 21].

3. APPLYING ACCIDENT MODELS

3.1. The accident case

The accident modeled in this paper involves the capsizing of a workboat at the coast of Norway, fall of 2013. The boat sunk, but was later lifted and moored at quay. The workboat has been out of service since the lifting, as the material damages after the accident were severe. Two operators were involved in the accident. Both operators were exposed to imminent danger as they had to jump from the sinking workboat and swim approximately 250 meters to shore in cold water at night. The air temperature was -3 °C and the sea was calm.

A short description of the accident is as follows: two operators manning the workboat started their shift at 0300 a.m. Their task was to load the workboat with fodder from quay, then leave for the fish farm to start the feeding. When the workboat was 800-1000 meters from the quay, heeling was observed by one of the operators, and shortly thereafter water on deck. The operators turned the workboat to return to the quay, but before they reached land the workboat capsized and eventually sank. One of the operators jumped off the boat when it capsized, while the other operator fell and got hit in the face, nevertheless he managed to swim away from the sinking workboat. The operators swam to shore and were rescued. The material damages after the accident was a sunken workboat and lost fodder sacks. Both operators were chilled and worn out after the long swim in the cold water, and one had a small cut in the lip. The operators tell that they have experienced some psychological after-effects from the capsizing.

The main source of information when analyzing the accident has been the accident investigation documents, provided by the company management. These documents include notes from some interviews with key personnel and a preliminary technical analysis ordered from a consultancy company to analyze the technical causality of the sinking. The labor authorities also initiated a separate investigation and some of the documents related to this investigation were used for information about the accident[†]. In addition, an independent research institute was hired to assist in the company's internal investigation, which is documented in a report describing the event and the most likely causalities. None of these documents states to have been based on a particular accident model.

3.2. Sequentially Timed Event Plot (STEP)

[22] gives a thorough description on how to conduct an accident investigation, where guidance on how to collect information through interviews and documents is provided. The book focuses on how to acquire and treat information related to an accident, and how to use this information to reconstruct the events leading to an event. The core of the accident model is the STEP diagram. This is a matrix type diagram where each row is designated one specific actor and each column is one time unit. All actors involved in the accident are identified; these can be both human and technical. The actors relevant for the investigation are those whose actions initiate changes during the accident. The time aspect is also noted as important as it can explain why further actions in the diagram are taken, and it highlights at what times critical actions are made. The STEP building block events are then identified and placed according to actor and time in the matrix. The first and the last event should be identified first. The first event should be the event that initiates the changes that leads to the accident; this is the first event when change is not countered with an adaptive response to keep the expected course of events. The last event should be the event when homeostasis has again been restored and no more harm is occurring. From the first event the consecutive event are then identified by asking "Which actors must do what to produce the next event?" [15] An event is the situation when one actor is performing one action. The events may be physical and observables or they can be mental if the actor is a person. Events can in the model occur at the same time when actuated by different actors, and the method is thus multi-linear. When all events have been identified the events are connected through arrows. The arrows between events can be both converging and diverging showing that several events can lead to one single event and that one single event can trigger multiple other events or actions. The linking of the events also tests the relationship between the identified events as the logical sequence must follow the arrows.

When the STEP matrix diagram is completed, the event sequence can be analyzed for safety flaws. Then each event set and their linking arrows in the flowchart is used to find the effect that one event has on another event. The thought is that it is the flow of the events that will disclose the problems, and that the flow of events that triggers harm must be changed or better controlled [22].

3.2.1 Applying STEP

When analyzing the accident according to the STEP method, the first task is to decide the first and last event in the accident sequence. The first event should be in some proximity in time to the accident. The first event was chosen to be the start of the shift of the two workers, and the end event was the notification of emergency response team (ambulance). The actors in the STEP workflow sheet are identified as the two operators (O1 and O2), and the workboat (WB). An excerpt of the STEP diagram can be seen in Figure 1.

[†] These documents are confidential and the accident has to some extent been generalized to avoid revealing sensitive information. Please contact the authors for more information.

Figure 1: STEP diagram of the accident

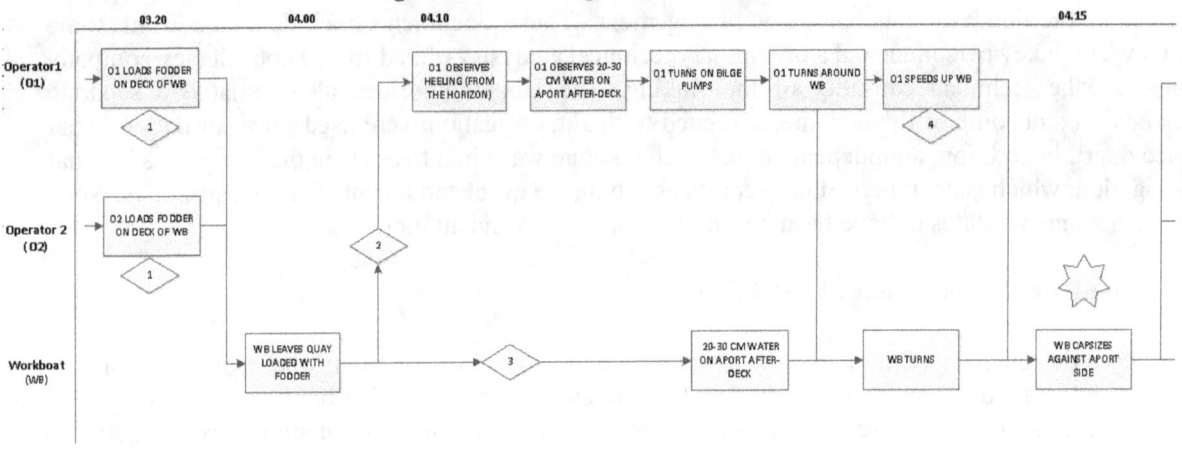

From the flowchart, safety problems are identified by analyzing the event pairs, which is marked in Figure 1 with diamond-squares. The capsizing is marked with a star. Problems identified:

- The loading of the fodder sacks

It is not obvious from the flow chart that the loading of the fodder sacks was a direct cause to the capsizing of the ship. It is, however, part of the changes leading to capsize, and one cannot neglect the possibility of a relation. In hindsight it has been difficult for the involved operators to identify exactly how many fodder sacks that were loaded onto the boat, and where on the deck they were placed. Only four sacks of fodder were found when heaving the sunken boat, while 12 sacks was claimed to be loaded on board. A preliminary analysis performed by a research institute after the accident identified the amount and the height of the fodder sacks due to stacking as a factor likely to have negatively influenced the stability of the boat. The operators were not sufficiently trained to understand which factors that influence a vessel's stability, and they had no available documentation on how to load the workboat in order to maintain the stability.

- Observation of heeling

The operators had no technical aids onboard that notified the intake of water. The first observation of heeling was made by operator 1 as he looked at the horizon; while operator 2 did not notice this change. Operator 1 explains that he had to climb on top of a fodder sack to see the water on the aft deck. Any technical aid monitoring the heeling of the workboat, or sensors and alarms that can notify water levels on the workboat could have prevented the operators from leaving quay with an unstable boat, or at least alarmed the change in stability earlier.

- Water on deck

The water on deck is the first obvious symptom of the following capsizing event in the accident sequence. The accident sequence in the STEP flowchart cannot identify any technical reason for why the boat started to take in water, however, as stated earlier the accident analysis suggests stability problems due to the placement of fodder sacks and intake of water due to negative trim. This again suggests that the workboat had openings in the hull permitting water to enter.

- Speeds up

After turning the boat, operator 1 speeds up to get faster back to the quay. In hindsight, this has been identified as a reason that might have contributed to the fast capsizing of the boat, as the ship then could have been pressed lower into the sea, and thus taking in water faster. This phenomenon should be taught to workboat operators.

- Capsizing

According to the operators, the time from discovering water on deck until the capsizing was a fact, was very short. The workboat was a catamaran which is a vessel design known to have a good initial stability, but which may lose its stability quickly if there is water below deck and the freeboard is lost. This might implicate that there is a need for some technical safety barriers like safety bulkheads to prevent distribution of water inside the hull.

- Procedures for man over board (MOB)/Personal Protective Equipment (PPE)

Both operators wore thermal suits and life vests, which is company standard for working at sea. None of the operators had time to put on the immersion suits available in the workboat. There was a life raft on top of the wheel house, but this was not released when hitting the water, and was still attached to the boat when it was raised after the accident. Obviously, the routines for regular maintenance and function checks of the safety equipment should be reviewed by the company.

These above safety related issues are not in themselves answers to why the accident happened. They do, however, highlight which areas could be further investigated to find more concrete mitigating solutions into prevention of similar accidents.

3.3. Causal Analysis based on STAMP

The systemic accident model to be applied in this paper is STAMP (Systems-Theoretic Accident Model and Process) [21]. Systems are here viewed as interrelated components that are kept in a state of dynamic equilibrium by feedback loops of information and control [19]. The most basic component of the model is constraints, as control is always associated with the imposition of constraints. Control loops and process models are essential parts of systems quality functions, and these are also in the core of the STAMP model where different modes of inadequacies will provide for answers as to how a system failed in the path towards an accident. [21] presents a causality model used in accident or incidents analysis named CAST (Causal Analysis based on STAMP). The CAST framework does not aim to appoint blame, but rather to analyze the sociotechnical system design to identify weakness and to find changes that will eliminate not only symptoms but also causal factors. In STAMP there is no root cause, but inadequate safety control structure. The time aspect should be large as adaptions and changes in the system can be a contributing cause for accident.

The first step of using STAMP for accident analysis is to identify the systems and hazards involved in the loss. [22] defines a hazard as a system state or set of conditions that, together with a particular set of worst case environmental conditions, will lead to an accident or loss. Hazards should be identified at systems level first, and then decomposed into lower level hazards if needed. For all of the hazards identified system-level safety requirements and design constraints should be found. With this information as background material, the system safety control structure should be constructed as it was designed to work. The control structures are made up of components and their responsibilities, in addition to the control actions and feedback loops of a standard control loop (controlled process, sensors, controller and actuators). Starting from the physical process and working up the levels of control, a STAMP analysis examines each level for the flaws in the process at the level that provided inadequate control of safety in the process level below. The process flaws at each level are then examined and explained in terms of a potential mismatch in models between the controller's model of the process and the real process, incorrect design of the control algorithm, lack of coordination among the control activities, deficiencies in the reference channel and deficiencies in the feedback of monitoring channel. For human decision making analysis must involve information about the context, information available and not available. When the system have been analyzed for flaws in the control loops, any coordination and communication deficiencies should be examined. Coordination of tasks is especially important whenever there are two or more components or actors with the same responsibility, and conflicting objectives concerning control coordination which is a major risk for unsafe behavior should be considered. When the above mentioned steps have been conducted for all levels of the sociotechnical system, recommendations for improvements can be suggested on the basis of the analysis. There is no algorithm on how to generate recommendations as political and situational factors always will be involved in these decisions [21]

3.3.1 Applying CAST

The first step of applying CAST for accident investigation is defining the system hazard(s) and constraint(s). [22] defines a hazard as a system state or set of conditions that, together with a particular set of worst case environmental conditions, will lead to an accident or loss. In the accident analyzed it is the safety of the fish farming workboat, and its loss of buoyancy that is the hazard to control. Constraints related to this can be identified as keeping sufficient freeboard and prevent water filling.

The control structure is seen in Figure 2, where the analysis stops above company management. It is possible to continue the analysis to the top level of the sociotechnical system, the government. The squares represents components and the arrows between the components represent control actions and feedback loops, an explanation of constraints and feedback of the loops is found in Table 1. Loop 1 illustrates the interaction between the operators and the workboat. Loop 2 portrays the hand-over of the workboat from the fish farm where the workboat previously had been used (Fish Farm 2) to the fish farm where the accident happened. This hand-over happened the day before the accident. Loop 3 is the interaction between the operators and the closest management level, Operations management, and loop 4 is the interaction between the Operations management and the top management level, Company management. Loop 5 portrays the interaction between the acquired companies through which the work boat came into company possession. This happened some years prior to the accident, but is relevant to the control structure as it can contribute to understanding why there was very little knowledge and documentation on loading capacities and stability in the company when the accident happened.

Table 1: Constraints and feedback of control loops

Control loop	Higher level	Lower level	Constraints/control	Feedback
1	Operators Fish farm 1	Workboat	Loading	Response
2	Operators Fish farm 2 (same level)	Operators Fish farm 1 (same level)	Informal exchange/request of information the workboat changed operators/location	Informal exchange of information when the workboat changed operators/location
3	Operations management	Operators Fish farm 1	Procedures, Training requirements, Planning	Deviation reporting, Experience feedback, Change requests
4	Company management	Operations management	Policy, Resources	Reports
5	Company management	Acquired company	Information request on acquired workboat	Documentation of workboat history etc.

The next step of CAST is to examine the physical process relating to the event. This would comprise of a thorough analysis of the ship and its functions. A preliminary analysis of the workboat have been conducted by a consultancy company, however, as there is no documentation on the workboats capacities and measurements, the boat must be further investigated onshore to understand the physical processes of what happened when the ship sank. Each control loop should then be investigated for flaws in control, as an example control loop 1 will be closer examined, see Table 2. In this loop, two operators have been included at the "Fish farm" level, the shift leader and a new operator working his first shift.

Table 2: Control flaws of control loop 1

	Control flaws
Safety related responsibility	Keep control over workboat Correctly load the workboat with sacks of fodder Shift leader had the responsibility of guiding the new employee
Context	Shift leader had one year of experience working on fish farms Shift leader had only limited knowledge of the workboat Shift leader had more knowledge from a similar work boat with larger capacity New operator had his first day at work, no previous experience from fish farming New operator had some experience from fishing Shift started at 0300 night, dark outside
Unsafe Decisions and Control Actions	Loads fodder sacks on workboat without adequate knowledge on how to load the workboat to maintain stability and freeboard Noticed water on deck too late? Sped up workboat to reach quay increases intake of water?
Process Model Flaws	There is no formal procedure on how to load the workboat, operators on a neighboring fish farm has through experience learned how the workboat must be loaded There is no technical aids to monitor stability of the workboat Shift leader had no formal training that should allow him to predict the stability loss of the work boat

Last, the system should be examined for coordination and communication problems. Some communication issues can, be pinpointed by the available information. The history of the workboat in question seems to be adverse as, included the aforesaid accident, four incidents of sinking have been recorded for the workboat. of these events happened when the boat was owned by the current company, and information on the other sinking incident is a part of the very limited documentation that was handed over during takeover. The current company management describes a generally chaotic situation during takeover where very little documentation followed. Furthermore, at the time when the workboat was built this vessel category had no mandatory requirement by the maritime authorities to test and document stability and loading capacities. This is why there was no documentation on loading or stability capacities of the workboat in the current company. This led to a situation where the company relies on informal communication channels to deliver information on loading capacity from operator to operators. This was also what happened when the workboat was handed over from the previous fish farm the day before the accident.

Figure 2: CAST control structure

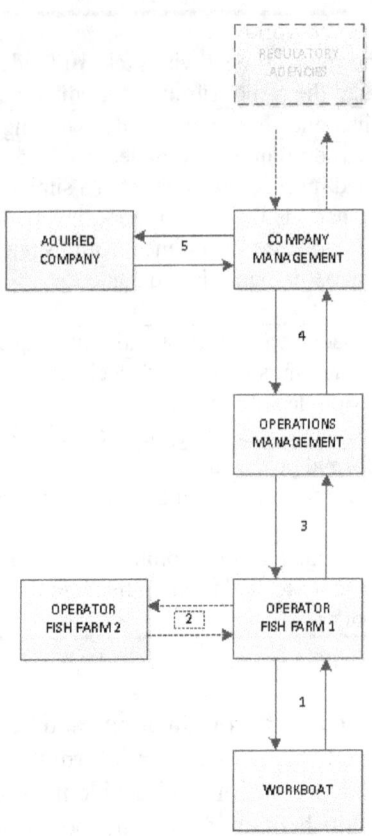

4. DISCUSSION

An advantage of using accident models for investigating accidents is that they provide a systematic approach for collecting and examining information about the accident. It also helps finding relevant questions to pose during an investigation. Using both methods to model this accident raised several questions that were not covered in the original accident investigation. However, where STEP is a method where interview techniques and self-evaluation can be consulted in [22] the CAST framework does not claim to be an accident investigation technique, but only a way to document and analyze the results of an accident investigations process [21]. This might, however, be the only point at which the STEP method takes a more holistic approach than CAST. The main focus of STEP is to portray the close events leading up to the accident; thus, several relevant aspects are lost. For example, the time limit in STEP leads to a focus away from latent conditions in the system that contributed to the accident. There is no integration of these factors in the model, and they must be included separately when analyzing the safety problems through the STEP chart. The latent conditions are, however, not excluded from the analysis, because some of these will naturally come up during an investigation, but there is no guidance provided as to how to find such conditions or how to use them in analysis. Other factors that get less focus in the STEP method are questions aiming to find out why actors acted as they did, higher level decisions, feedback mechanisms, coordination problems and other organizational, social and human factors. These factors can be identified in the STEP model and [23] identifies STEP as an accident model suited for analyzing events influenced by the regulators and the government. However, the inclusion of these factors is more arbitrary than for CAST. Also, when identifying problem areas using STEP, there is a sense of blame in all the identified areas as the problems identified are connected to events, which again are connected to actors. The goal of CAST is to avoid pointing blame, especially to those in the lower part of the hierarchical system and often closest to the actual loss events [21]. CAST enables us to look for reasons why the operators acted as they did, as it gives larger focus to the context in which the decisions are made and to process model flaws on which the operators not necessarily have any influence. An advantage of the STEP model is its simplicity and lucidness. CAST is much more complicated to use, and it does not capture the essence of the model in its graphic form to the same degree as the STEP chart does. The use of these two models support the view that the choice of accident model can to a certain extent determine the results of the investigation. In [24] several accident investigation manuals were analyzed and the author found that since the accident models used in the manuals all took an epidemiologic accident model approach, they all lost certain aspects that would have been covered by a systemic approach. This phenomenon is named "what you look for is what you find" (WYLFIWYF). Also, this will affect any further remedial actions after an investigation as "what you find is what you fix" (WYFIWYF).

The two accident models applied are based on different assumptions for accident mechanisms. Hence, they will provide different feedback for improvement of the system. [13] gives two important theories to a well-functioning safety management system based on feedback; the Van Court Hares hierarchy of order of feedback and Ashbys law of requisite variety. Van Court Hare distinguishes between different orders or levels of feedback, where the lowest level of feedback is no feedback, leading to a "quick fix" of a problem at operator level, with no feedback to higher organizational levels. The level of feedback then goes through foremen, middle management, president and staff and ending with the highest level, the board of directors. These levels consequently leave us with the possibility of changes in the systems according to the decision power of the different organizational levels, where the most fundamental change is possible at the level of board of directors with a change in safety policy and goals [13]. Ashbys law of requisite variety states "For an analyst to gain control over a system, he must be able to take at least as many distinct actions, i.e., as great a variety of countermeasures, as the observed system can exhibit". Kjellén states that one way of increasing variety in possible measures is to" train the organization to conduct more comprehensive accident investigation". These requirements to a safety system could also be applied to the case of accident investigations and accident models; a more in depth and holistic perspective on accident causation could give a more appropriate set of tools in future prevention of accidents.

5. CONCLUSION

This paper analyses an accident in the aquaculture industry using two different accident models, namely STEP and CAST. The accident occurred onboard a workboat, where the consequences included loss of the workboat and a potentially fatal situation for two operators having to swim to shore. STEP provides an easy to understand flow chart over the accident sequence and helps pinpoint six areas where further investigation could contribute to understand why the accidents happen, and possibly could have prevented the accident if they had not occurred. The focus of this method was mainly on the workboat itself and the operators manning the workboat. CAST, which is based on a systemic accident model, takes a wider approach to accident modelling and is also more comprehensive. Areas of risk management improvements are thus identified also at higher levels of the sociotechnical system in the company, e.g. the need for procedures assuring that all relevant documentation on old workboats is gathered, and that crucial information in some cases relies on informal distribution channels in the company. Also, in this paper, only a small part of the comprehensive CAST framework have been explored. Further analysis in the higher levels of the sociotechnical system of the accident could provide for even higher level improvements measures.

All industries go through different stages of appointing accident causation to different factors. Technical measures, human factors and organizational factors have all been the focus of causation in different eras and accident models. The three perspectives should not replace each other, but should be seen as complimentary [11]. Accident investigation in the aquaculture industry should reflect this view, and through investigating accidents from several perspectives we can learn more about the accident mechanisms in the industry, but also more about the industry itself, beyond the direct causes of the accident.

Acknowledgement

This work was funded by the Norwegian Research Council through the SUSTAINFARMEX project.

1. FAO, *The State of World Fisheries and Aquaculture - 2012*. 2012. p. 209.
2. FKD, *Verdens fremste sjømatnasjon* Fiskeriogkystdepartementet, Editor. 2013.
3. Jensen, Ø., et al., *Escapes of fishes from Norwegian sea-cage aquaculture: causes, consequences and prevention.* Aquaculture Environment Interactions, 2010. **1**: p. 71-83.
4. Bourret, V., et al., *Temporal change in genetic integrity suggests loss of local adaptation in a wild Atlantic salmon (Salmo salar) population following introgression by farmed escapees.* Heredity, 2011. **106**(3): p. 500 - 510.
5. Glover, K., et al., *Atlantic salmon populations invaded by farmed escapees: quantifying genetic introgression with a Bayesian approach and SNPs.* BMC Genetics, 2013. **14**(1): p. 74.
6. Moreau, D.T.R. and B. Neis, *Occupational health and safety hazards in Atlantic Canadian aquaculture: Laying the groundwork for prevention.* Marine Policy, 2009. **33**(2): p. 401-411.
7. Størkersen, K.V., *Fish first: Sharp end decision-making at Norwegian fish farms.* Safety Science, 2012. **50**(10): p. 2028-2034.
8. McGuinness, E., et al., *Fatalities in the Norwegian fishing fleet 1990–2011.* Safety Science, 2013. **57**(0): p. 335-351.
9. Aasjord, H., *Upubliserte data over dødsulykker i havbruk.* 2013.
10. Harms-Ringdahl, L., *Relationships between accident investigations, risk analysis, and safety management.* Journal of Hazardous Materials, 2004. **111**(1–3): p. 13-19.
11. Hovden, J., E. Albrechtsen, and I.A. Herrera, *Is there a need for new theories, models and approaches to occupational accident prevention?* Safety Science, 2010. **48**(8): p. 950-956.
12. Hollnagel, E., *Barriers and accident prevention.* 2004: Ashgate.
13. Kjellén, U., *Prevention of Accidents Through Experience Feedback.* 2000: Taylor & Francis.
14. Qureshi, Z.H. *A Review of Accident Modelling Approaches for Complex Socio-Technical Systems.* in *Twelfth Australian Conference on Safety-Related Programmable Systems (SCS 2007).* 2007. Adelaide, Australia.
15. Rausand, M., *Risk assessment: theory, methods, and applications.* 2011: John Wiley & Sons, Inc., Hoboken, New Jersey.
16. Reason, J., *Managing the Risks of Organisational Accidents.* 1997: Ashgate.
17. Rosness, R., et al., *Organisational Accidents and Resilient Organisations: Six Perspectives. Revision 2.* 2010.
18. Blanchard, B.S. and W.J. Fabrycky, *Systems engineering and analysis.* Third Edition ed. 1998: Prentice-Hall, Inc.
19. Leveson, N., *A new accident model for engineering safer systems.* Safety Science, 2004. **42**(4): p. 237-270.
20. Rasmussen, J., *Risk Management in a Dynamic Society: A Modelling Problem.* 1997.
21. Leveson, N.G., *Engineering a safer world*, in *Systems Thinking Applied to Safety.* 2011, The MIT Press.

22.   Hendrick, K. and J. Ludwig Benner, *Investigation Accidents with STEP*. 1987, New York: Marcel Dekker, Inc.

23.   Sklet, S., *Comparison of some selected methods for accident investigation.* Journal of Hazardous Materials, 2004. **111**(1–3): p. 29-37.

24.   Lundberg, J., C. Rollenhagen, and E. Hollnagel, *What-You-Look-For-Is-What-You-Find – The consequences of underlying accident models in eight accident investigation manuals.* Safety Science, 2009. **47**(10): p. 1297-1311.